AF761787

THE OSTRACOD FAMILY ENTOCYTHERIDAE

DABNEY G. HART
and
C. W. HART, JR.
Department of Limnology
Academy of Natural Sciences of Philadelphia

MONOGRAPH 18
THE ACADEMY OF NATURAL SCIENCES OF PHILADELPHIA
19th AND THE PARKWAY
PHILADELPHIA, PENNSYLVANIA 19103

MONOGRAPH 18
THE ACADEMY OF NATURAL SCIENCES
OF PHILADELPHIA
1974

Printed by the George M. Carpenter Fund for the
Encouragement of Original Scientific Research

Published October 25, 1974
Fulton Press Inc., Lancaster, Pennsylvania 17603

For Horton H. Hobbs, Jr.
our favorite crawdaddy

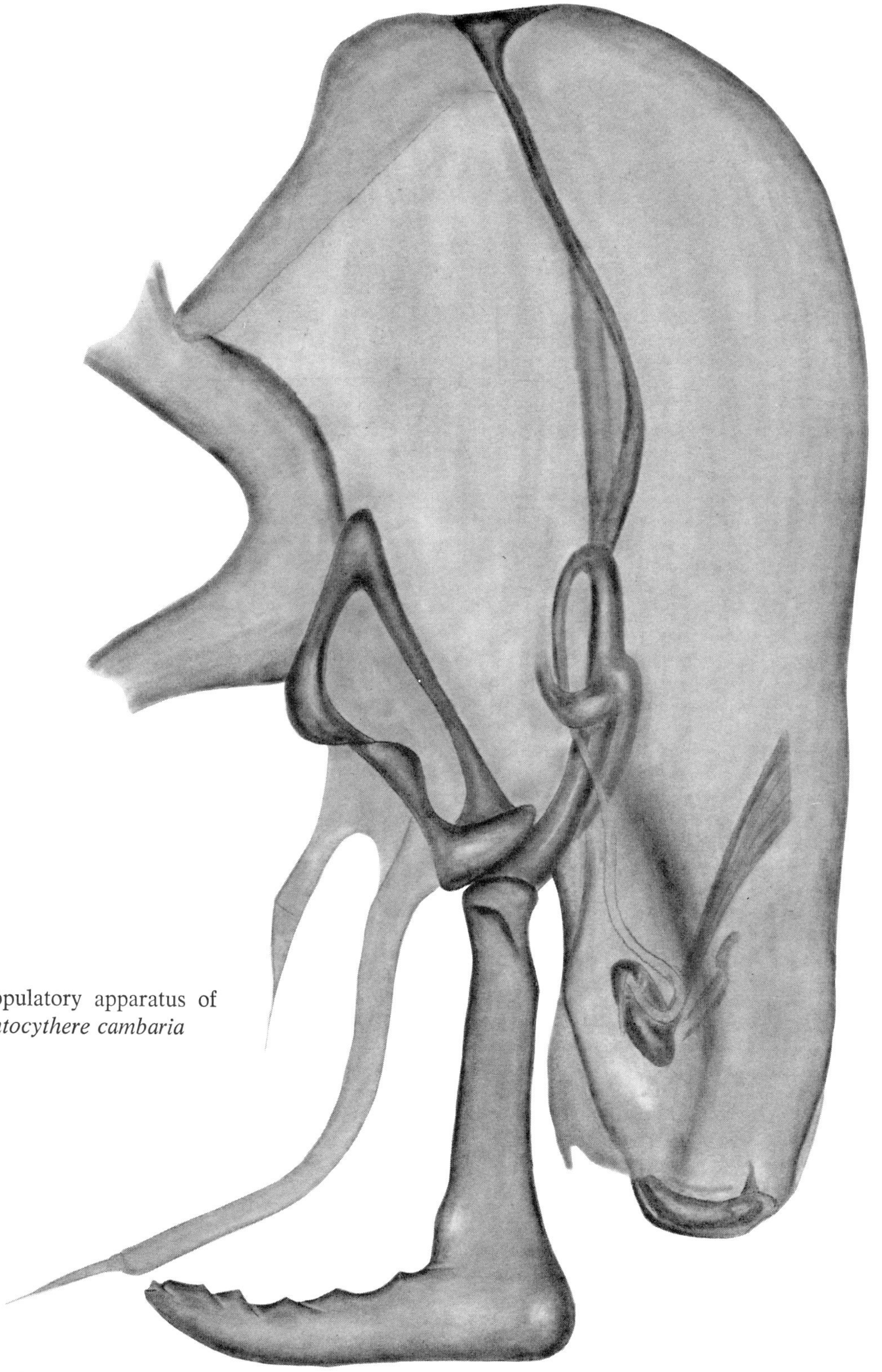

Copulatory apparatus of
Entocythere cambaria

COVER DRAWING:
Copulatory apparatus of *Geocythere geophila*

PREFACE

The purpose of this volume is to summarize, as concisely as possible, our knowledge of the ostracod family Entocytheridae from 1903, when Marshall described the first entocytherid, through those papers published in 1973. Included in this compilation are appropriate synonymies, diagnoses, keys, illustrations of diagnostic characters (chiefly male genitalia), notes on relationships, locality records, hosts, and distribution maps.

This work was made possible largely by grants GB-1463, GB-4197, GB-6943, and GB-6943x from the National Science Foundation to the authors. In addition, substantial help has come from the Limnology Department of the Academy of Natural Sciences and from the Smithsonian Institution Foreign Currency Program. For all of these we are most grateful.

We are particularly indebted to Horton H. Hobbs, Jr., Senior Scientist, Smithsonian Institution, for his help, advice, and encouragement over the years that we have worked with him on the entocytherids. We also thank his son, Horton H. Hobbs, III, of Christopher Newport College; Joseph Fitzpatrick of the University of South Alabama; Perry C. Holt and Daniel J. Peters of Virginia Polytechnic Institute; Margaret Walton of Danville, Virginia; Thomas Cheng of Lehigh University; L. Denis Delorme of Burlington, Ontario, Canada; Robert P. Higgins of the Smithsonian Institution; E. F. Riek of C.S.I.R.O., Canberra, Australia; J. Frances Allen of the Environmental Protection Agency (formerly of the National Science Foundation); Philip Sandberg of the University of Illinois; and Lipke B. Holthuis of the Rijksmuseum van Natuurlijke Historie, Leiden, for help in various phases of this project.

We also admit to having taken some diagnostic and key material *in toto* (or in some cases, almost so) from certain papers of Dr. Hobbs — in particular his exhaustive paper on the entocytherid ostracods of Mexico and Cuba.

By far the greater part of the ostracod material used in our work on the entocytherids was made available to us by the Smithsonian Institution, which allowed us access to and use of its extensive collection of crayfishes — from which we separated the commensal ostracods. The Smithsonian's patience and cooperation in many ways is greatly appreciated.

C. W. Hart, Jr.

TABLE OF CONTENTS

BIOLOGY

Ostracods of the family Entocytheridae (Podocopa) are commensals (free-loaders) on other crustacea — as far as is known passing all or most of their developmental stages upon the bodies of crayfishes, isopods, amphipods, and crabs.

Whereas there are several references to entocytherid ostracods being found living free in crayfish burrows (for review, see Young, 1971), absence of entocytherids in plankton or benthic samples suggests that they are essentially tied to a commensal relationship with one host or another.

Baker (1969) pointed out that whereas it does not appear that *Ankylocythere sinuosa* requires the presence of a crayfish to survive, when it is associated with a crayfish it feeds upon algae extracted from the water or adhering to the crayfish. He further stated that it has not been demonstrated whether it can reproduce and complete its life cycle away from the crayfish host. According to Young (1971) this ostracod is dependent upon the host for reproduction and development (see below).

Entocytherid ostracods have been shown to be present on crayfishes (North America, Mexico, Cuba, Hawaii, New Zealand, Australia, New Guinea), cave-dwelling isopods (Mexico, France, Italy, Yugoslavia), epigean free living isopods (Australia), wood-boring isopods (coastal waters of India), amphipods (coastal waters of France and Italy), and a freshwater crab (Mexico).

LIFE HISTORY

The only studies devoted to the development of entocytherids which include observations of the life histories of members of the subfamilies Entocytherinae, Sphaeromicolinae, and Microsyssitrinae are those of Marshall (1903), Paris (1920), Rioja (1940b), Stamper (1957), Hart, Nair, and Hart (1967), Roelofs (1968), and Young (1971).

Subfamily Entocytherinae

The observations of Marshall on the development of *Entocythere cambaria* Marshall (1903) were limited to only a few remarks. The newly hatched young "have the two pairs of antennae and the mandibles fully developed. Well back on the body are a pair of appendages, and near them two long setae which finally disappear . . . . In all of the earliest stages of *Entocythere* . . . . I found the first three pairs of appendages similar, except in size, to the appendages of the mature individual. The pair of appendages on the body, the prolegs, disappear but I am unable to say whether they disappear entirely or become changed into the first pair of permanent legs. Their position on the body is not so far forward as that of the first pair of legs."

Rioja (1940a) found that in *Ankylocythere heterodonta* four stadia precede the attainment of the stage he considered to have reached the adult form. The first is characterized by having antennules of five articles, antennae of three articles with two apical claws, mandibles, maxillae, a single pair of larval feet consisting of two podomeres, the distal one of which is multiarticulate, and the shell bearing a pair of large, conspicuous setae posteroventrally. The second stadium differs from the first chiefly in the acquisition of a second pair of larval feet. The third stadium is characterized by the acquisition of a sixth article in the antennule, the loss of the larval feet, and the development of a single pair of legs resembling those of the adult but having only three podomeres. The fourth stadium differs from the third in possessing two pairs of legs resembling the subadult foot of the third stadium. A fifth stadium, representing the penultimate instar was not described as such by Rioja, but the female was designated "tipo biunguis" (1940b). This instar may be characterized by the loss of the paired large setae on the posteroventral portion of the shell, the appearance of the third pair of legs, and, in the male, of two pairs of lappets (rudiments of the copulatory complex) projecting posteroventrally slightly posterior to the posteriormost pair of legs. These lappets were apparently misinterpreted by Rioja (1943b: 577) as constituting the genital apparatus of the biunguis female. The ultimate instar differs from the fifth in the subdivision of the penultimate podomere of the antenna and the addition of a third apical claw; the legs consist of four articles, and the clasping apparatus of the male has become completely differentiated.

Young (1971), in a fine study of *Ankylocythere sinuosa* from a farm pond in Texas, presented evidence that entocytherids are commensals, since they survive for long periods in the absence of a host but are dependent upon it for reproduction and development. He showed that the incubation for that species at 20-22°C was eight to ten days, and that transfer between hosts was readily accomplished but that body contact between crayfishes was not necessary for the transfer. In addition, he showed that population densities are at a minimum in winter, increase in spring and early summer, and reach a maximum in late summer — the lowest density of ostracods coinciding with the period of greatest number of crayfish individuals. He demonstrated that *A. sinuosa* reproduces at all seasons (in Texas), and that a major proportion of all populations is immature.

Stamper (1957) noted that 90% of the specimens of *Orconectes r. rusticus* collected from rivers and streams in central Ohio and central Kentucky carried eggs of "*E. illinoisensis*" attached to the hooked setae and/or to the coxal filaments. He noted that the eggs were "whitish in appearance, ovoid in shape, and average 200 microns by 188 microns in thickness." He described the eggs as they developed, and noted that hatching occurred in 23 days at 28°C.

Subfamily Sphaeromicolinae

Paris recognized seven instars preceding the adult stadium in *Sphaeoromicola topsenti* Paris (1920). The first three, referred to as "metanaupliens," are characterized by well-developed antennules, antennae, mandibles, and maxillae; its maxillipeds (the future first pair of feet) consist of styliform, segmented appendages ending in a simple hook. The two following stadia differ from the first only in an increase in size. The fourth through the sixth stadia are designated juveniles, and the fourth differs from the third in that the metanaupliar feet are converted to three-jointed ones which do not differ otherwise from those of the adult. In the fifth stadium, an additional pair of three-jointed feet is added, and the first feet become four-jointed. The sixth stadium is characterized by the addition of a third pair of three-jointed feet; the first two pairs are now four-jointed. The seventh stadium, of which only the female is mentioned and which Paris termed "nubile," may be recognized by all three pairs of feet consisting of four articles, and the external genital organs are evident. He found one nubile female in which the third pair of legs had only three articles. Hobbs rather strongly suspects that the external genitalia attributed by Paris to the nubile female are actually those of the penultimate instar of the male.

Roelofs (1968) found in her investigation of the development of *Sphaeromicola dudichi* Klie (1938) (=*Hartiella dudichi*) that the adults are attached to their amphipod host, *Chelura terebrans* Philippi, by a filament produced by the antennal glands. The females deposit 1 to 16 eggs in packets on the gills or oöstegites of the amphipod, each egg being affixed by two filaments. An almost synchronous development occurs in the eggs comprising a packet, and they hatch within a maximum of 15 days at 18°C. The first instar possesses five-segmented antennules and three-segmented antennae, the latter bearing a well-developed flagellum, two terminal, and one subterminal claw; the mandibles and maxillae are well developed; the future first legs consist of a pair of styliform biarticulate appendages with hooked extremities; and the posterior portion of the shell bears several pairs of long hairs. Roelofs did not observe a stage comparable to the second stadium of *Ankylocythere heterodonta,* but did briefly describe and figure one corresponding to the third stage, this characterized by possessing two pairs of tri-articulate legs. A third preadult stadium was also found in which three pairs of legs were present, the anterior two pairs consisting of four segments and the posterior with only three.

Subfamily Microsyssitrinae

Hart, Nair, and Hart (1967) found six instars preceding the ultimate molt in the development of their *Microsyssitria indica,* the first through the fifth corresponding to those stages in *A. heterodonta.* The sixth apparently is intruded into the development and involves maximum increase in size and is the stadium in which the copulatory complex of the male becomes differentiated.

Despite these contributions, little else is known of the life cycle of any entocytherid. No one has expressed an estimate as to the life span of a member of the family; furthermore, no one has determined how many or how frequently eggs are laid, or whether or not a female produces eggs throughout her adult life. Although an occasional entocytherine triunguis female (see below) has been observed in copula, as a rule, only those females in the penultimate instar (biunguis females) are paired with males. Yolk deposition in the oöcytes apparently follows the ultimate molt in the entocytherines, and in no individual have more than four or five oöcytes been observed to be markedly enlarged at one time, and no more than three seem to approach maximum size approximately simultaneously.

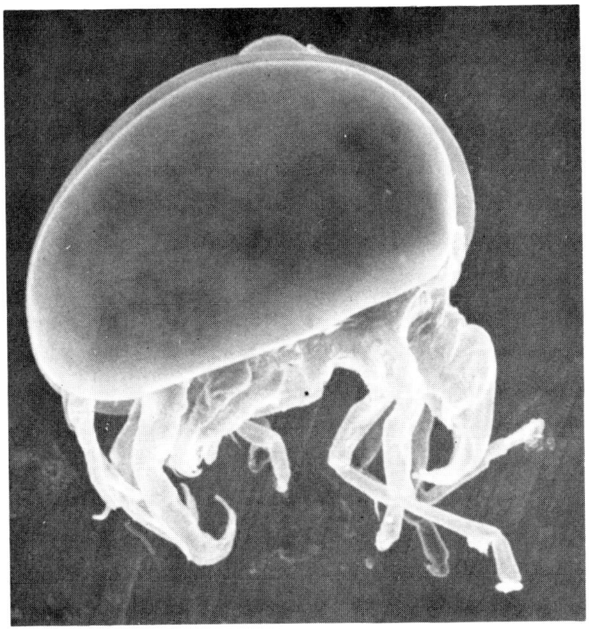

Figure 1. — Scanning electron photomicrograph of ostracod of genus *Ankylocythere,* showing smooth unadorned shell and appendages typical of entocytherid ostracods. Copulatory apparatus in active copulatory configuration.

MORPHOLOGY

Shell

The thin, almost completely transparent shells of entocytherid ostracods are of little taxonomic value, and because of this illustrations of shells are usually limited to mere outline drawings. Aside from the anterior and posterior spines on a few North American species (*Uncinocythere caudata, U. thektura, U. cassiensis,* and *Dactylocythere amphiakis,* for example), size, and scattered setae, the shells bear little in the way of ornamentation — and the shells of European, Australasian, and North American species are usually indistinguishable from one another. Figure 1 is a scanning electron photomicrograph of a member of the genus *Ankylocythere,* showing the typical smooth unadorned shell. Figure 2 is of a single submarginal seta, similar ones of which are found in varying numbers on the shells of virtually all entocytherids.

Figure 2. — Scanning electron photomicrograph of seta near ventral margin of shell.

Muscle Scars

Muscle scars are usually so ill-defined among entocytherids as to render them useless in the systematics of the group. However, under proper conditions they are occasionally seen. Figure 3 depicts their arrangement in *Rhadinocythere serrata.* No comparative study of the pattern of muscle attachment has been undertaken for other members of the family.

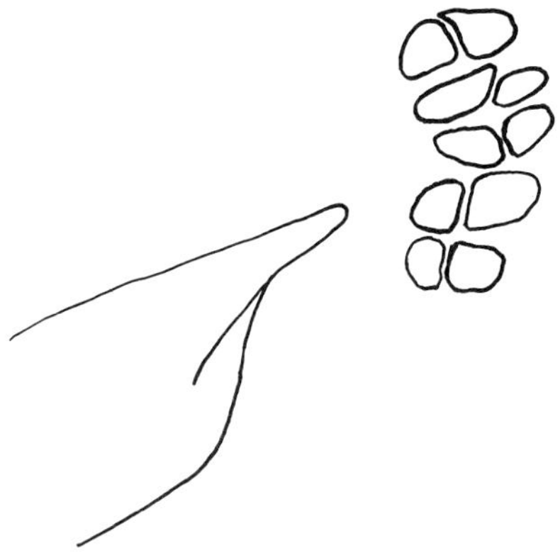

Figure 3. — Diagram of muscle scar pattern on left valve of *Rhadinocythere serrata,* showing relation to proximal portion of mandible (see Fig. 4).

Hinge

Harding (1964) figured transverse sections of the entocytherid hinge, and credited Geoffrey Fryer with drawing his attention to the fact that when sections of the hinge are examined under a microscope in the ordinary way no difference in structure can be seen in the appearance of the valves and the soft cuticle joining them which is of about the same thickness; but if they are examined between crossed nicols the polarized light causes the valves to show up brightly owing to the presence of calcite while the soft cuticle between them remains dark. He points out that a similar type of hinge appears to be present in the freshwater cyprid *Chlamydotheca.*

Appendages

In the absence of obvious diagnostic shell characteristics, studies of the entocytherids have been concentrated on the appendages, which have proved to have many valuable characters.

The antennule (Figs. 4, 5, 7) consists of from five to seven podomeres (depending upon the species), tapers gradually distally, possesses varying numbers of fine setae projecting from the distal margins of certain podomeres, and terminates in a distal brush of several fine flagelliform setae.

The antenna (Figs. 4, 5, 8) possesses several important diagnostic features, with variations in the distal claws (Fig. 8) showing certain consistent features attributable to some species. Hart, Nair, and Hart (1967: 2) pointed out that whereas it has usually been considered that entocytherid antennae consist of five podomeres, it seems likely that the "fifth podomere" is actually only a complex area in the distal portion of the fourth podomere where the terminal claws originate. In adult males the terminal claws may or may not have marginal setae, and may be either styliform or truncate.

Typical of the females of all entocytherids is copulation in the penultimate instar. These "copulating females" (Hoff, 1943: 276) or "hembras del tipo biunguis" (Rioja, 1943b: 568) have antennae consisting of four (three according to the interpretation of Hart et al., 1967: 2) podomeres and two terminal claws. Following copulation and the ultimate molt, the penultimate (ultimate, Hart et al., loc. cit.) podomere of the antenna becomes divided so that there are apparently five (four, Hart et al., loc. cit.) podomeres, and a third claw is added at the distal extremity. These females were termed "gravid females" by Hoff, in reference to the fact that most contain large oöcytes; and "hembras del tipo triunguis" by Rioja, alluding to the three apical claws of the antenna. In members of the genus *Entocythere,* but in no others, one or two small pectinate appendices are added at the base of the apical claws of the triunguis females.

Most members of the genus *Entocythere* bear a mandible (Figs. 4, 5, 6, 9) of which the distal tooth is not pectinate; however, *E. ruibali* has a pectinate distal tooth, as does *E. tyttha.* Pectinate distal teeth are common among other genera.

Respiratory filaments of the mandible and the respiratory lobe of the maxilla (Figs. 4, 5, 6) are often difficult to distinguish, either because of similar refractive indices with mounting media or because other appendages and organs block them from view.

Entocytherid ostracods possess three pair of walking legs (Figs. 4, 5), peculiarly adapted terminally to hold onto the host animals. Typically these legs are made up of four podomeres and a distal claw-like device which is apposable against a projection

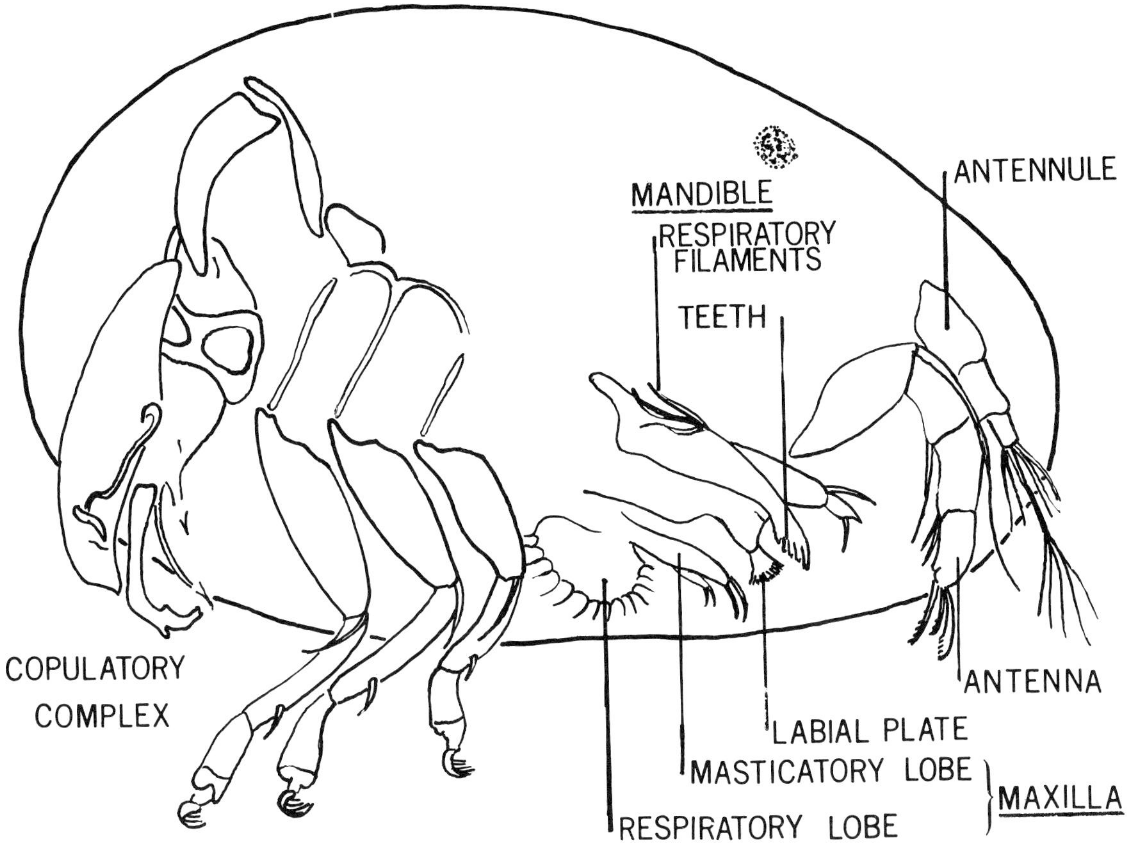

Figure 4. — Generalized entocytherid ostracod with right valve removed. After Hobbs, 1971: 5.

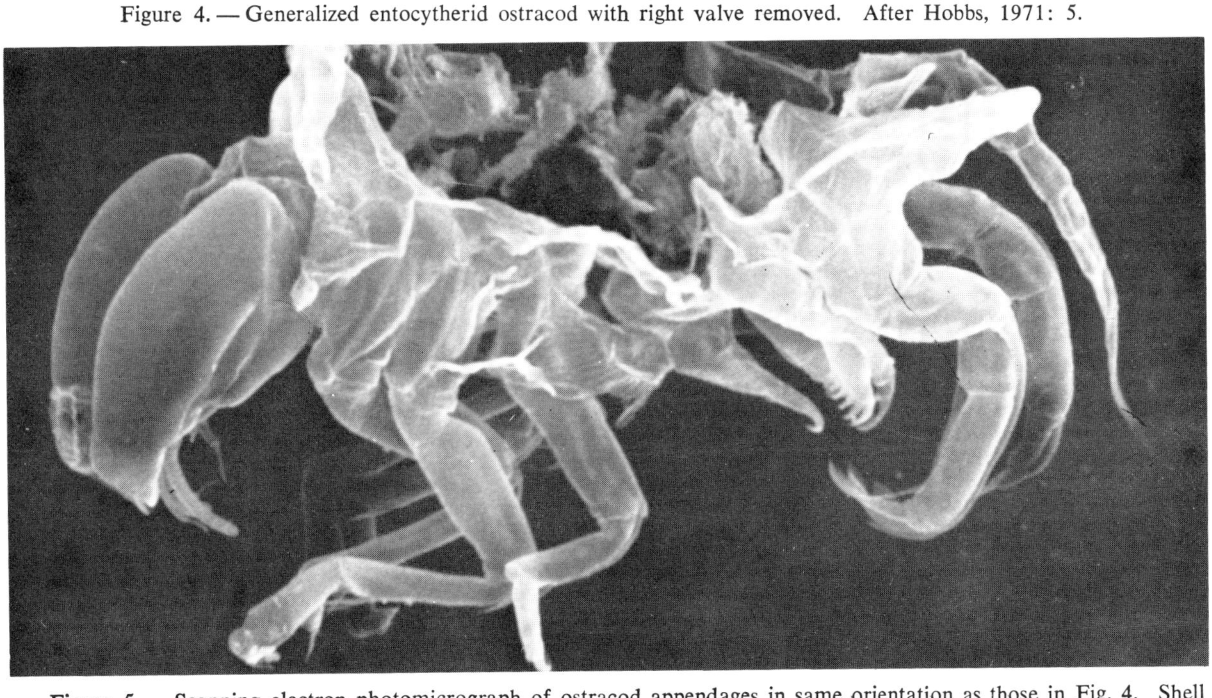

Figure 5. — Scanning electron photomicrograph of ostracod appendages in same orientation as those in Fig. 4. Shell is completely removed.

from the ultimate podomere. It seems likely that the relative lengths of the podomeres to one another may be an important diagnostic feature at the generic level, but this idea has not been pursued.

Over the years a rather extensive terminology dealing with the copulatory apparatus (=copulatory complex) of the male entocytherids has been developed, and this terminology, together with the structural details of the apparatus, is here reviewed:

While essentially similar copulatory apparatuses are present in all entocytherids, for the sake of clarity the large and heavily sclerotized apparatus of the entocytherid genus *Geocythere* is used for Figures 11-14.

The copulatory apparatus of male entocytherids is a conspicuous, paired organ situated in the posterior shell region, and is of great importance as a diagnostic feature in the group (Figs. 1, 4, 5, 10-14). While it is conventional to illustrate only one half of this apparatus, it should be understood that each part exists in duplicate. Figure 10 shows a composite copulatory apparatus, including all parts presently recognized. (Those parts indicated in italics were proposed by Hart and Hart, 1968b).

For the sake of a uniform terminology, the anterior and posterior directions indicated in Figure 10 are used to designate portions of the peniferum and its associated fingers and clasping apparatus,

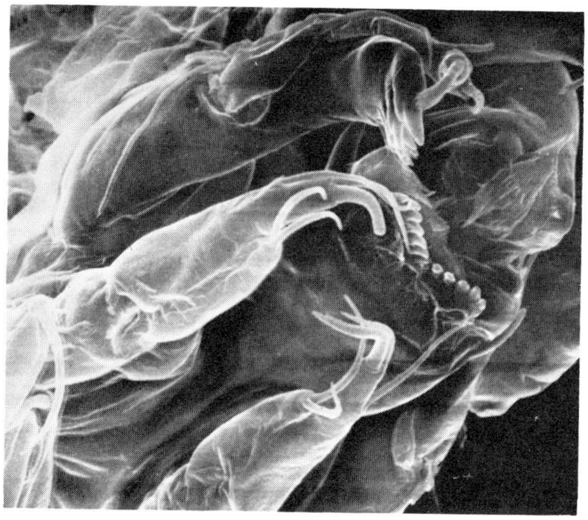

Figure 6. — Scanning electron photomicrograph of ventral aspect of maxillae and mandibles.

Figure 8. — Scanning electron photomicrograph of antennal claws.

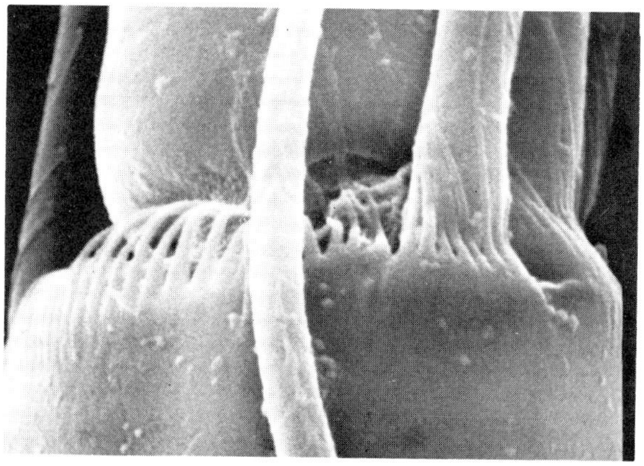

Figure 7. — Scanning electron photomicrograph of junction of two antennular podomeres, showing large seta originating on distal-edge of podomere together with small fringing setae distributed circumferentially.

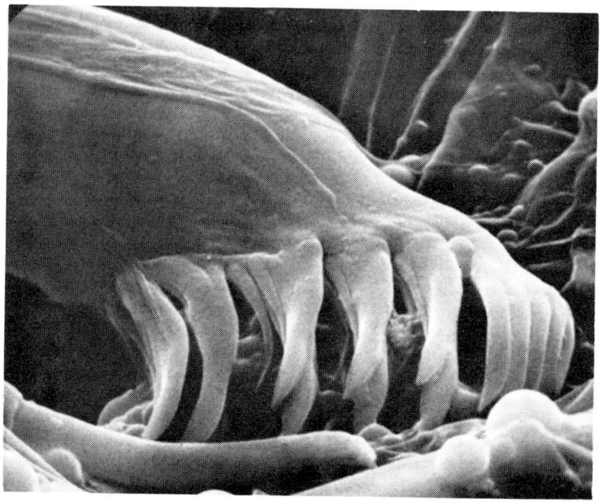

Figure 9. — Scanning electron photomicrograph of mandibular teeth (see Fig. 4).

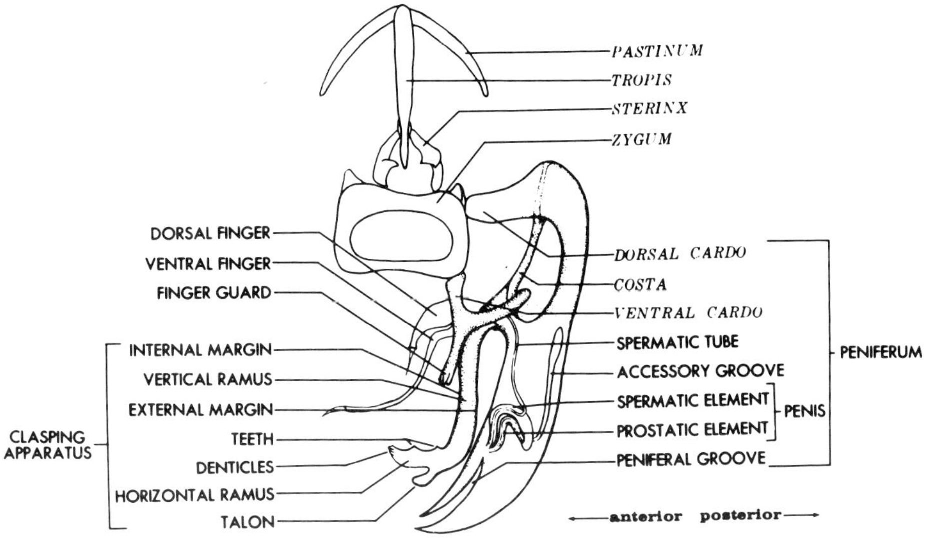

Figure 10. — Composite entocytherid copulatory apparatus, showing all parts currently recognized. Parts labeled in Roman type were recognized before 1969; those in Italic type were proposed by Hart and Hart (1969b). The view is posterior, with the right peniferum at the mid-point of its 180° movement to or from the copulatory position. The left peniferum is not shown. The anterior-posterior directions refer only to reference positions with regard to the peniferum. See Fig. 13.

Figure 11. — Posterior portion of entocytherid ostracod showing copulatory apparatus held up inside shell in non-copulatory position. Note vertical position of pastinum.

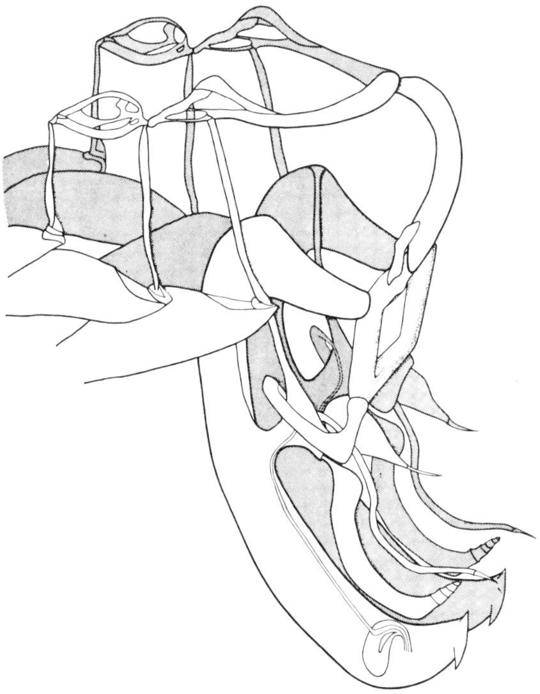

Figure 12. — Three-dimensional drawing of copulatory apparatus and adjacent legs, with copulatory apparatus lowered so that distal portion extends beyond shell. Note horizontal position of pastinum.

although those directions will be "true" directions only when the apparatus is in its copulatory position (see discussion below, and also Figure 14). It should be remembered that these directions apply only to the *peniferum,* fingers, and clasping apparatus; not to the *zygum, sterinx, tropis,* and *pastinum.*

Terminology: The *peniferum* is that portion of the copulatory apparatus which bears the penis (usually in its distal portion) and its proximal portion forms a hinge on which the entire apparatus may be turned through an arc of 180° about the *zygum* (see below and also Figs. 10 and 13). Although the overall plan of the peniferum is similar in all species, variations in the proportions and modifications in details are numerous and bizarre — affording features that permit the delineation of species, genera, and subfamilies. The proximal portion of the peniferum serves as a place of attachment of a complex system of muscles, but until more is known of the homologies of these muscles we feel it advisable not to illustrate them, as they serve only to complicate an already intricate picture.

Marshall (1903: 133-135, 140-141) figured and briefly discussed the copulatory apparatus of *Entocythere cambaria,* but, following Zenker's earlier work (1854: 47), apparently misinterpreted the junction of the *zygum* (see below) — believing it to be a half of the so-called "basal piece." While Marshall understood that "all parts (of the copulatory apparatus) are paired," it seems probable from his illustrations that he believed the zygum to be paired also, while, as far as we can determine, it is a single structure. What we now call the *peniferum* was considered by Marshall to be the other half of the "basal piece."

The peniferum as a discrete structure was discussed in some detail by Hoff (1944: 328-329), who apparently introduced the terms *penis, horizontal ramus, vertical ramus, internal border, external border,* and *teeth* at that time. However, Hoff followed Marshall's terminology, calling the more conspicuous structure the "base," and referring to "distal" and "proximal portions of the base" when discussing what we now call the *peniferum* — a shorter term introduced by Hobbs and Walton (1961: 379) to encompass the distal portion of the base, but which, by usage, has already come to include both Hoff's distal and proximal portions of the base.

So that homologies can be more clearly discussed, Hart and Hart (1969b) introduced the terms *dorsal*

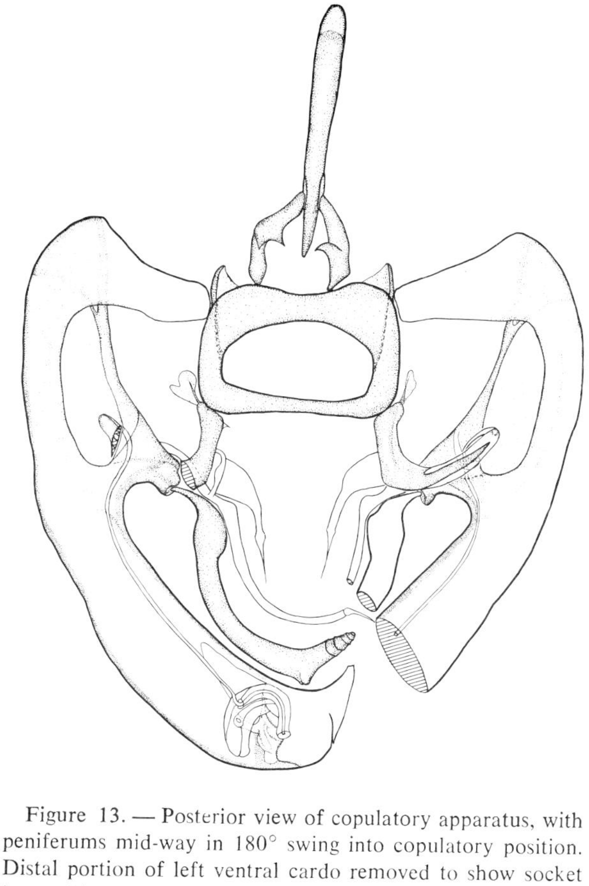

Figure 13. — Posterior view of copulatory apparatus, with peniferums mid-way in 180° swing into copulatory position. Distal portion of left ventral cardo removed to show socket into which clasping apparatus fits; distal portion of right peniferum removed for clarity.

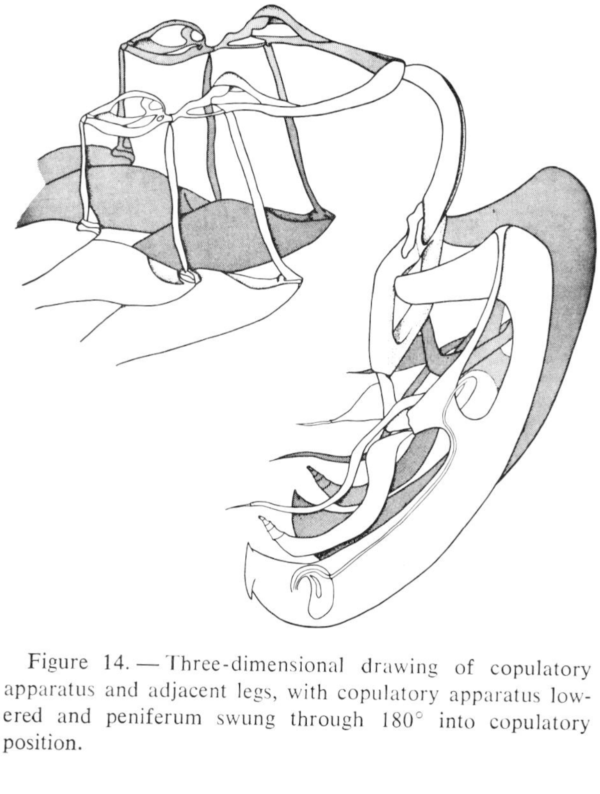

Figure 14. — Three-dimensional drawing of copulatory apparatus and adjacent legs, with copulatory apparatus lowered and peniferum swung through 180° into copulatory position.

cardo and *ventral cardo* (Latin *cardo*=hinge) for those parts of the peniferum (Hoff's "proximal portion of the base") which serve as hinges by which it articulates with the *zygum*. The term *costa* (Latin *costa*=rib) may be used for any thickened portion of the peniferum which appears to perform a strengthening function (Fig. 10).

In North American species (subfamily Entocytherinae), the penis (Figs. 10, 13, 16), which is a distinct sclerified organ, may be variously situated on the peniferum, either in a sub-horizontal position and only slightly curved (as in *Uncinocythere ambophora*) or being curved to a greater degree — sometimes almost approaching a full circle (as in *Geocythere gyralea*) — and it appears to be attached in its proximal portion to a muscle extending dorsally toward the top of the peniferum. It is probable that this muscle is used to pivot the penis at the time of copulation. Depending on the species, the penis is situated anywhere along the distal one-third of the peniferum. The penis of Australasian and European entocytherids (subfamilies Microsyssitrinae, Notocytherinae, and Hartiellinae) appears to be invariably vertically disposed — exhibiting differences in length relative to the entire peniferum and varying from straight to slightly sinuous.

Because it is not associated with the sclerified structure, mentioned above, encasing the terminal portion of the spermatic tube, we do not believe the penis of the Microsyssitrinae, Notocytherinae, or Hartiellinae to be homologous with the penis of the Entocytherinae. We believe that the penis of North American entocytherids (Entocytherinae) is homologous with the movable anterior peniferal claw found in the Indian and certain Australian species, and that in North American species this claw has become integrally associated with the terminal portion of the spermatic tube (which lies adjacent to it in the Indian and Australian species) and has migrated proximally to varying levels on the peniferum.

The term *penis* was first used by Hoff (1944: 329) for the sclerified structure encasing the terminal portion of the spermatic tube, and later Hobbs III (1965: 161) noted that in the North American genus *Plectocythere* the penis was actually divisible into two separate elements — the dorsal one of which he designated the *spermatic element*; the ventral one, the *prostatic element* (Fig. 10).

Later Hobbs and Hart (1966) pointed out that the penis in the genera *Ascetocythere, Cymocythere,* and *Phymocythere,* is also two-parted.

The accessory groove, first illustrated by Hobbs (1955: 327) in the genus *Dactylocythere,* may

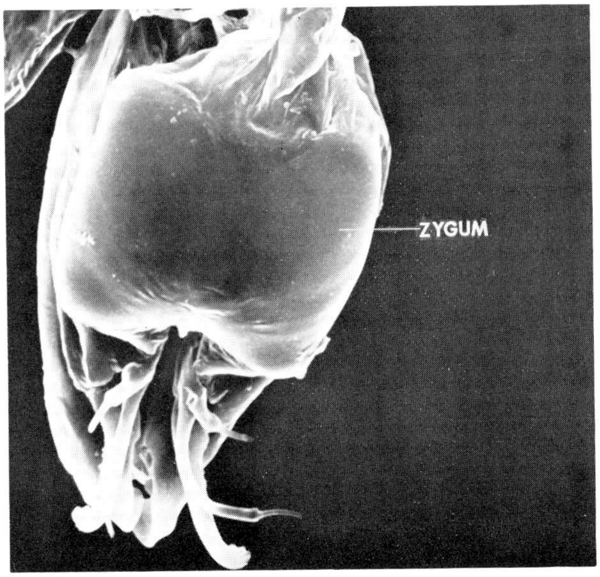

Figure 15. — Anterior scanning electron photomicrograph view of copulatory apparatus.

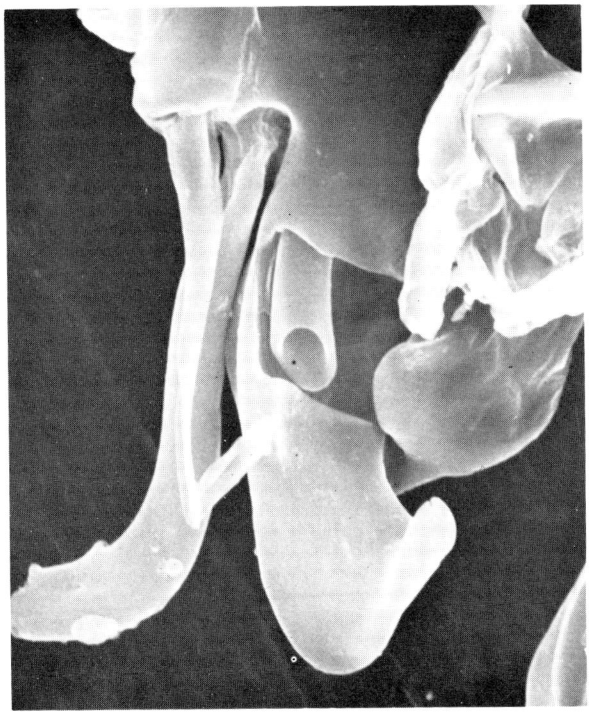

Figure 16. — Posterior scanning electron photomicrograph view of distal portion of peniferum of genus *Donnaldsoncythere,* showing orifice of penis.

serve some reproductive function, but at the present this is unclear.

Hobbs (1966: 8) proposed the term *peniferal groove* for a gap between the arms of the terminal bifurcation present in some species of the genus *Dactylocythere.* This bifurcation also appears to be present in some species of the genera *Uncinocythere* and *Ankylocythere* (Fig. 10).

While the terminal portion of the peniferum in all known North American entocytherids apparently bears no movable structures (with the exception of the sub-terminal penis), certain Australian entocytherids (*Chelocythere* and *Lichnocythere*) appear to bear at least one movable claw-like structure, and the only known Indian entocytherid, *Microsyssitria indica,* bears two apposed terminally located claws.

The term finger guard was proposed by Hobbs (1955: 329) for an extension of the *ventral cardo* alongside the dorsal and ventral fingers (Fig. 10). This structure is present in the North American genera *Dactylocythere, Sagittocythere, Phymocythere, Harpagocythere, Cymocythere,* and *Ascetocythere* — and a possible homolog is found in the Australian genus *Riekocythere.*

The spermatic tube of the entocytherids forms a proximal loop at approximately the level of the *ventral cardo,* and appears to be somehow interconnected with the distal portion of that *cardo.* A bifurcation of this tube, at varying locations in the loop, leads proximally in the direction of the *zygum,* but we have been unable to trace this for any considerable distance beyond the place where it leaves the loop. Distally the spermatic tube terminates in the sclerified penis mentioned above (Fig. 16). Marshall (1903: 134) asserted that one of the fingers — possibly the ventral one — contained the sperm duct, and Rioja (1940: 605) illustrated the loop of the spermatic duct as terminating in the dorsal finger. This interpretation probably arose from Marshall's relying too heavily on the earlier work of Zenker, who described the copulatory apparatus of distantly related ostracods in which the copulatory duct (=spermatic duct) is not an integral part of the peniferum, but a separate entity the homologies of which we do not understand at present. Subsequently Hoff (1944: 329) showed that there was no duct in either of the two fingers, and postulated that the fingers are probably tactile organs and that they may also assist in directing or holding certain structures in position during copulation.

The term *clasping apparatus* (or clasping piece), used by Marshall (1903: 133) was probably taken from Zenker's *Apparat zum Festhalten* (1854: 47). This refers to the heavily sclerotized, apparently movable sclerite articulating with the midportion of

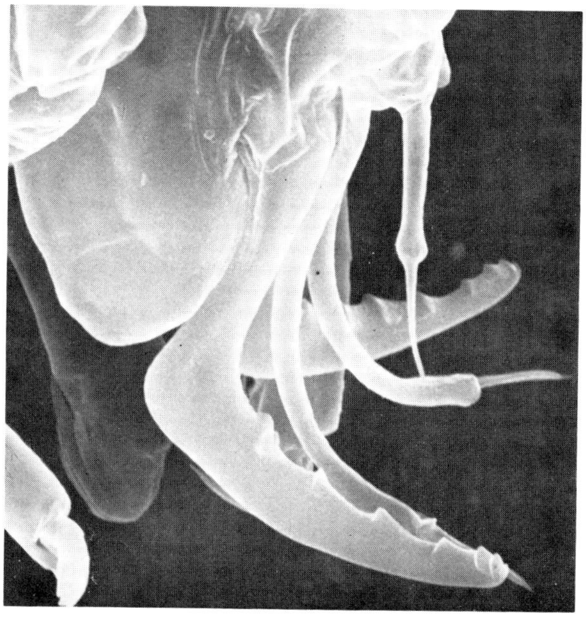

Figure 17. — Lateral scanning electron photomicrograph of distal portion of copulatory apparatus, showing distal portion of peniferum, clasping apparatus, dorsal and ventral fingers.

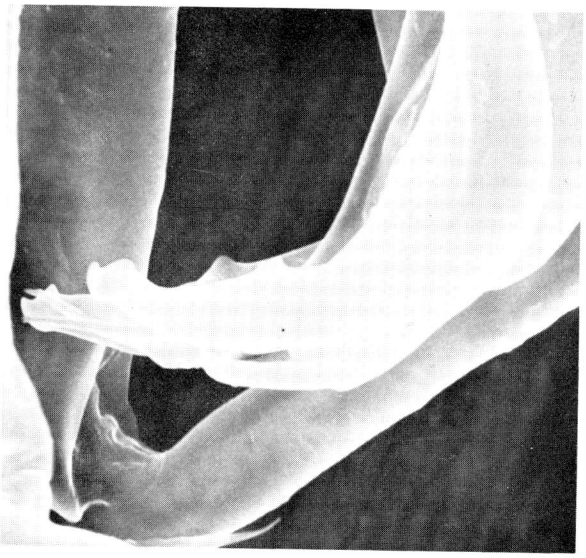

Figure 18. — Lateral scanning electron photomicrograph of clasping apparatus horizontal ramus, showing talon, tooth, and terminal denticles.

the peniferum in a socket near the *ventral cardo* and in the vicinity of the loop of the spermatic tube (Figs. 10-18). It, too, probably serves as a tactile organ, and because of its characteristic shapes — often used as a taxonomic characteristic — probably constitutes one part of a key-lock mechanism. It may also be used in helping to abduct the valves of the female so that other elements of the copulatory apparatus can be inserted between them. The apparatus may be either straight or, more often, L- or C-shaped — and the *horizontal ramus* may bear *teeth* on its internal margin, *denticles* on its tip, and/or one or more *talons* on its external margin (Figs. 17, 18). This usage of "teeth," "denticles," and "talons" to designate the excresences on the parts of the horizontal ramus, has, in spite of its apparent redundancy, proved to be useful in discussing the horizontal ramus.

The term zygum (Latin *zygum*=yoke) was proposed by Hart and Hart (1967) as a name for what Zenker (1854) apparently called the "Basis," Marshall (1903: 140) called "half of the basal piece," Rioja (1941: 184) called the "pieza intermediaria," and Hoff (1944: 329) called the "connecting piece." The *zygum* is a single chitinous piece, suspended in the posterior shell region by a system of chitinous rods (see three dimensional drawings, Figures 11-14, for visualization of its position). It is probably homologous with the fused first podomeres of adjacent walking legs — a supposition supported by its relationship to the *tropis* and the *pastinum* (see below), and by the presence of setae on the *zygum* of the hemicytherid ostracod discussed by Hart and Hart (1969b). These setae, single in the dorsal portion and double in the ventral, may be homologous with the single seta in the posteroventral, portion of the first podomere of the first walking leg — although admittedly considerable torsion would have had to take place to make this true.

While the zygum figured in most of the entocytherid illustrations herein is the comparatively simple, heavily sclerotized form found in the genus *Geocythere,* between genera it is tremendously variable. Similar heavy *zygums* are found in the genus *Donnaldsoncythere,* but the part seems to be so variable that it could, were it easier to see, probably be useful as a specific character.

While the *zygum* of the genus *Geocythere* (Figs. 11-14) is heavy and shows no median bars, that of *Entocythere heterodonta* (=*Ankylocythere heterondonta*) figured by Rioja (1941: 183, fig. 1) and called the "pieza intermediaria," shows traces of median bars.

The *tropis* (Latin *tropis*=keel), a heavy chitinous rod connecting the *zygum* to the *pastinum,* is probably formed by the fusion of the posterior vertical internal chitinous skeletal rods that support the legs. When viewed laterally the tropis may appear as a single rod in some species (Figs. 10-14). When viewed from a caudal aspect it usually appears to be a single rod, but in ostracods other than entocytherids (see Skogsberg 1928: 155, fig. 1, for instance) it sometimes appears as two unfused rods.

The paired *sterinx* (Greek *sterinx* = support) form connecting or buttressing pieces lateral to the tropis, and probably serve to stabilize the zygum in a horizontal position, while at the same time allowing some freedom of anterior-posterior angular motion.

The pastinum (Latin pastinum = fork) appears to be homologous with the horizontal chitinous skeletal rods connecting the vertical rods supporting the legs. When viewed from a caudal aspect it can be seen to be fork-shaped — anteriorly connected with the lateral skeletal apodemes and posteriorly fused to form a structure from which the *tropis* (and hence the entire copulatory complex) hangs. This is most easily understood by referring to Figures 12 and 14.

Function: The copulatory organs of the entocytherids are complex, but possibly more easily understood than those of some other ostracods. Indeed some of the parts appear to be simpler than homologous parts in other ostracods, and we believe that a better understanding of the function and homologies can be gained by beginning our comparisons with entocytherids and working backward (or forward, depending on your outlook) into other groups. Admittedly this is probably due to our ignorance of other ostracod genitalia, but because the knowledge of entocytherid copulatory apparatus homologies is highly systematized, it is a logical point of departure.

In simple terms, the copulatory apparatus of entocytherids apparently serves to raise, lower, and position its integral organs of sperm transfer so that they are able to introduce sperm into the seminal receptacle of the female.

When the ostracod is not concerned with copulation, the apparatus is held high within the shell and usually somewhat between the bases of the last pair of legs (Fig. 11). To accomplish this position the *pastinum* may be raised to almost a vertical position, thus holding the tropis and the associated apparatus in that position.

Then, before copulation can take place, the *tropis* probably swings slightly caudally, allowing the proximal portion of the *peniferums* to clear the legs — at which time the two *peniferums* swing 180° laterally to each side on the *zygum* (Figs. 12-14), thus reversing their non-copulatory position and making it possible for the organs of sperm transfer to face anteriorly and be inserted between the valves of the female (Fig. 19).

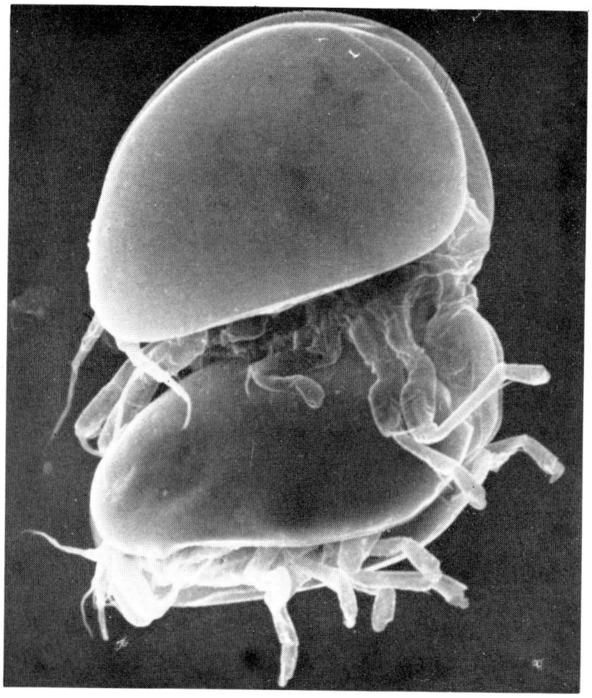

Figure 19. — Scanning electron photomicrograph of entocytherid ostracods in amplexus.

Hart and Hart (1969b) discussed in some detail their ideas on homologies existing between the male copulatory apparatus of entocytherids and hemicytherid ostracods.

Whereas in the crayfishes it is the male that is dimorphic, in the entocytherids the female exists in two "forms": (1) the copulatory or biunguis female, which, to our knowledge, is the form that is almost always the partner in amplexing pairs (and perhaps should be considered the last "larval instar") may be recognized by the 4-segmented second antenna bearing only two terminal claws; (2) the gravid or triunguis female, which frequently contains eggs, may be recognized by the apparently 5-segmented antenna bearing three terminal claws.

Unfortunately, few female entocytherids can be identified unless they are found in amplexus or can definitely be associated with the male. Few generic determinations of females are possible; however, the triunguis females of species of the genus *Dactylocythere* may be recognized by the presence of an amiculum and occasionally by the posteroventral projection on the shell, those of *Donnaldsoncythere* by the small corneous tubercles in the posterodorsal portion of the body, of *Ascetocythere* by the small rounded prominence partially surrounded by an amorphous mass of refractile material in a corresponding position, and of *Entocythere* by the pectinate at the base of the terminal claws of the second antenna.

HOSTS AND ENTOCYTHERID ASSOCIATES

Within the range of the subfamily Entocytherinae, frequently one host animal is infested with two or more ostracods, and there is evidence that in some instances one species tends to be concentrated on one area of the exoskeleton of the host while another is the predominant form elsewhere on the body (Hobbs, Holt, and Walton, 1967: 7).

However, most collections of host animals have been made in such a manner as to preclude any definite statement about which ostracods were really associated with which crayfish. Crayfish have mostly been collected from any given locality by merely putting all available specimens into one container. There, the ostracods have become detached, fallen to the bottom of the container, and subsequently had to have all crayfish species in the bottle assigned to them as their hosts.

Data on host specificity are more reliable, however, when they are concerned with burrowing and troglobitic crayfishes. When such crayfishes are collected, they are usually found in small numbers and it is unlikely that more than one species is placed in any one container.

Walton and Hobbs (1971), in an investigation of the composition of the ostracod populations on individual crayfishes, demonstrated that 28.9 per cent of the ostracods infesting *Cambarus carolinus* were adult members of *Dactylocythere chalaza,* 16.0 per cent were adults of *Ascetocythere asceta,* and the remaining 55 per cent consisted of juveniles of the two. On the crayfish *C. b. bartonii* the respective percentages were 44.2 of adult *Donnaldsoncythere ileata,* 6.7 per cent of *Phymocythere phyma,* and 49 per cent juveniles.

In instances in which two entocytherids share the same host (*Dt. chalaza* and *A. asceta* on *C. carolinus*, and *Dn. ileata* and *P. phyma* on *C. b. bartonii*), the adults of both ostracods occur in most setiferous areas of the exoskeleton; however, the two species tend to be concentrated in different body regions: *Dt. chalaza* and *Dn. ileata* on the abdomen and *As. asceta* and *phyma* in the gnathal area. Even in infestations involving one member of the pair, a preference, though not so strongly marked, for the same region of the host was demonstrated by *Dt. chalaza* and *P. phyma*. In contrast, the juvenile ostracods were more abundant in the combined gnathal-antennal areas than in the more posterior regions of the host. In both infestations, that species concentrated in the gnathal area was markedly less numerous than its ostracod associate.

That larger crayfishes are able to support larger ostracod populations seemed evident, but equally obvious was the fact that there was no correlation between the size of the host and the number of ostracods it harbors. Furthermore, female crayfishes tended to support larger populations than did males. The size of the population appeared to be influenced by the molting cycle of the crayfish. Juvenile crayfishes, being smaller and molting more frequently than adults, supported smaller populations.

Adult females, presumably molting only once a year, bore during the intermolt stages larger populations than did the adult males, which molted twice a year. Both adult males and females had reduced populations during premolt, molt, and postmolt stages.

Although eggs of the ostracods were deposited on most setiferous areas of the exoskeleton of the host, they seemed to be concentrated in the gnathal area and on the ventrolateral pleural sclerite between the bases of the fourth and fifth pereiopods.

On amphipod and isopod hosts, ostracods are usually found associated with the pleopods in the ventral abdominal region, and in one instance over 200 ostracods were removed from 10 isopod specimens (the ostracod *Microsyssitria indica* from the wood-boring isopod *Sphaeroma terebrans*).

Detailed data on the systematics and ecology of crayfish hosts belonging to the families Astacidae and Cambaridae may be found in Hobbs (1974).

TAXONOMIC TREATMENT

HISTORICAL SUMMARY

The family Entocytheridae Hoff 1942 (originally proposed as the subfamily Entocytherinae in the family Cytheridae) now contains five subfamilies (Entocytherinae, Hartiellinae, Microsyssitrinae, Notocytherinae, and Sphaeromicolinae), 31 genera, and 178 species.

The first species, *Entocythere cambaria,* was described by Marshall from Madison, Wisconsin, in 1903, and from then until 1926, when Sars described *Cytherites insignipes* (= *Entocythere insignipes*) from Canada, no further contributions to a knowledge of the genus were made.

In 1931, Klie described *E. donnaldsonensis* (= *Donnaldsoncythere donnaldsonensis*) from Donaldson's Cave (*sic* Donnelson's Cave), Lawrence Co., Indiana. Although he did not associate this ostracod with a crayfish host, it is probable that it was associated with the crayfishes *Cambarus* (*Faxonius*) *pellucidus* (Tellkampf) (= *Orconectes pellucidus inermis* Cope) and *C.* (*Bartonius*) *bartoni tenebrosus* Hay (= *C. tenebrosus* Hay), discussed in another paper pertaining to the fauna of the same cave (Fage, 1931: 373). Allen (1933) reported *E. cambaria* from Durham, Hibbard, and streams near Charlotte, North Carolina.

In 1940 Rioja initiated his study of the entocytherid fauna of Mexico by describing, in detail, *E. heterodonta* (= *Ankylocythere heterodonta*). Dobbin (1941) described *E. columbia* "from crayfish branchiae . . ." from several localities in the State of Washington. Rioja (1941a) continued his morphological study of *E. heterodonta* by presenting a detailed account of its skeletal structure, and in another paper (1941b) reviewed the available data on *E. cambaria, E. donnaldsonensis,* and *E. heterodonta.* In the same year, Hobbs (1941) erroneously reported *E. cambaria* from La Cueva Chica, San Luis Potosi, Mexico.

Hoff (1942a), in describing two new members of the genus *Entocythere* from Illinois — *E. illinoisensis* and *E. copiosa* (= *Ankylocythere copiosa*) proposed two subgenera, *Entocythere* and *Cytherites,* and placed the genera *Entocythere* and *Sphaeromicola* together in a new subfamily, Entocytherinae. In his report on the ostracods of Illinois, Hoff (1942b) mentioned *E. illinoisensis* and *E. copiosa* and referred to his earlier work.

Rioja (1942a) described two new entocytherids, (= *Ankylocythere claytonhoffi*) and *E. heterodonta sinuosa* (= *Ankylocythere sinuosa*), from La Cueva Chica, San Luis Potosi, Mexico. Later in the same year, Rioja (1942b) proposed a third subgenus, *Donnaldsoncythere,* and accorded his subspecies, *sinuosa,* specific rank. Rioja (1943a) described *E. mexicana, E. dobbini dobbini* (= *Uncinocythere dobbinae*), and *E. dobbini bicuspide* (= *Uncinocythere bicuspide*) from two localities in the State of Puebla, Mexico. In a second paper (1943b), he gave a detailed account of the dimorphism in the females, together with a re-evaluation and revised definitions of the subgenera. Hoff (1943) described *E. riojai* (= *Thermastrocythere riojai*) from Illinois and Arkansas and *E. humesi* (= *Donnaldsoncythere humesi*) from New York. Additional locality records were also cited for *E. copiosa* and *E. illinoisensis.* Hoff's most recent contribution (1944b) included a discussion of taxonomic characteristics of the entocytherids and of the inadvisability of recognizing subgenera until the genus was better known. Also included were the descriptions of six new species: *E. serrata* (= *Rhadinocythere serrata*) from Illinois; *E. equicurva* (= *Uncinocythere equicurva*) from Georgia; *E. elliptica* and *E. dorsorotunda* from Florida and Georgia; *E. talulus* (= *Ankylocythere talulus*) from Florida; and *E. hobbsi* (= *Ankylocythere hobbsi*) from Georgia, Florida, and South Carolina.

A discussion of the "*pieza copuladora*" of *E. dobbini,* together with descriptions of two new subspecies, *E. dobbini cuadricuspide* (= *Uncinocythere cuadricuspide*) and *E. dobbini multicuspide* (= *Uncinocythere cuadricuspide*), were given by Rioja in 1945. In 1949 he gave additional locality records for several species of entocytherids and described *E. talirotundi* (= *Ankylocythere heterodonta*) from Lago Lerma and Lagunas de Zempoala, Mexico; *E. bidentata* (= *Ankylocythere bidentata*) from three localities in Vera Cruz; *E. recurvata* (= *Uncinocythere dobbinae*) from Puebla and Vera Cruz; and *E. uncinata* (= *Uncinocythere bicuspide*) from Puebla. In addition, he elevated *E. bicuspide* and *E. cuadricuspide* to specific rank — species that had previously been considered subspecies of *E. dobbini.* He also placed *E. dobbini multicuspide* in synonymy with *E. cuadricuspide.*

Sphaeromicola cirolanae was described by Rioja in 1951 from a cave in San Luis Potosi. Until 1973, when Hobbs described *S. coahuiltecae,* this was the only member of that genus described from the western hemisphere. Hobbs (1971) re-figured *S. cirolanae,* gave a diagnosis of it, and discussed its variations and distribution. In 1973 he added new locality data.

E. occidentalis (= *Uncinocythere occidentalis*) was described by Kozloff and Whitman (1954) from Oregon, and in 1955 Kozloff described *E. caudata* and *E. ericksoni* (= *Uncinocythere caudata* and *U. ericksoni*) from the same state. Rioja's most recent contribution (1955) was his description of *E. ruibali* and *E. heterodonta cubensis* (= *Ankylocythere cubensis*) from the Province of Camagüey, Cuba. Hobbs (1955) described *E. daphnioides* and *E. runki* (= *Dactylocythere daphnioides* and *D. runki*) from the Virginias and North Carolina, and in 1957 described *E. hamata* (= *Ankylocythere hamata*) from the Province of Las Villas, Cuba — also listing new Cuban entocytherid locality records. In 1959, new entocytherid ostracods were described from California [(*E. neglecta Westervelt and Kozloff* (= *Uncinocythere neglecta*)], Florida (*E. geophila* Hart (*Geocythere geophila*) *E. torreya* Hart (= *Hartocythere torreya*), *E. ambophora* Walton and Hobbs, *E. lucifuga* Walton and Hobbs) (= *Uncinocythere ambophora* and *U. lucifuga*), and South Carolina (*E. striophylax* Crawford (= *Dactylocythere striophylax*), *E. suteri* Crawford (= *Dactylocythere suteri*), *E. telmoecea* Crawford (= *Ankylocythere telmoecea*), and *E. tiphophila* Crawford) (= *A. tiphophila*). In 1960, Hobbs and Walton described *E. simondsi* (= *Uncinocythere simondsi*) from Georgia and *E. cyma* and *E. mecoscapha* (= *Cymocythere cyma*) and *E. mecoscapha* (*Cymocythere cyma* and *Dactylocythere mecoscapha*) from Tennessee, and Hart described *E. pennsylvanica* (= *Donnaldsoncythere pennsylvanica*) from Pennsylvania.

With only two exceptions (Howe, *in* Moore and Pitrat 1961: 300 and Howe 1962: 80), Hoff's suggestion (1944b) that the three subgenera — *Entocythere* Hoff 1942a, *Cytherites* Hoff 1942a, and *Donnaldsoncythere* Rioja 1942b — be discarded has been accepted and followed by workers in the field.

Howe (*in* Moore and Pitrat 1961: 300) created the family Entocytheridae by elevating the subfamily Entocytherinae Hoff 1942. The latter, containing the genera *Entocythere* and *Sphaeromicola,* had been placed in the family Cytheridae.

In 1962, Hart revised the family Entocytheridae, recognizing the genera *Ankylocythere, Ascetocythere, Cymocythere, Dactylocythere, Donnaldsoncythere, Entocythere, Geocythere, Rhadinocythere, Sagittocythere,* and *Uncinocythere* in the subfamily *Entocytherinae; Sphaeromicola* in the subfamily *Sphaeromicolinae.* The subgenus *Cytherites* Hoff 1942a, based on the original genus *Cytherites* of Sars (1926) who described *C. insignipes* (= *E. insignipes*) on the basis of three female specimens bearing inadequate locality data and no host reference, was considered *incertae sedis.*

Since 1962, some 75 papers dealing with the entocytherids have been published — nearly equaling the total of 85 papers that had appeared in the preceding 60 years.

Space allotted for this volume precludes detailed summaries here, except to say that the major contributions in that time have been those of Horton H. Hobbs, Jr., Horton H. Hobbs III, Margaret Walton, and the authors.

During that time knowledge of the extent of the geographical distribution of the family has also increased — showing entocytherids to be present in Hawaii (Entocytherinae), New Zealand, Australia, New Guinea (Notocytherinae), and India (Microsyssitrinae), as well as in North and Middle America (Entocytherinae and Sphaeromicolinae) and Europe (Sphaeromicolinae and Hartiellinae). Papers detailing these findings are included in the synonymies under the various subfamilies.

SYNONYMIES AND REFERENCES

To the best of our knowledge, all references to the family Entocytheridae through 1973, are included, and annotations are provided to point out various aspects of the citations or to indicate the kinds of data presented.

Accordingly, the following system of abbreviations has been used in parentheses following page references or in data on the locations of type specimens:

A — Ostracod associates.
B — Biology, ecology (except host relationships).
C — Compilations, indices, bibliographies, classifications, historical resumes, systematics. (Usually little or no new data.)
D — Distribution.
E — Evolution.
H — Host relationships.
K — Key.
M — Descriptions or diagnoses, morphology, measurements.
P — Stratigraphic distribution.
R — Relationships within the family and comparative morphology.
S — Synonymies.
T — Methods and Techniques.
Z — Zoogeography.

Locations of Type Material

AM — Australian Museum (Sydney).
ANSP — Academy of Natural Sciences of Philadelphia.
RMNH — Rijksmuseum van Natuurlijke Histoire (Leiden).
USNM — United States National Museum (Washington).
MCZ — Museum of Comparative Zoology, Harvard (Cambridge, Mass.).

Throughout the sections devoted to distribution records, certain abbreviations occur many times and are used consistently (or nearly so) without following periods. Most common among these are:

ck (or Ck)	= creek(s)
coll	= collection (or collector)
jct	= junction
mi	= mile(s)
R	= river
Rd	= road
rt	= route(s)
trib	= tributary

Interchangeable with "rt" are occasional more specific notations such as US (for United States route), St (for State route), or MO, TN, etc. for specific state routes.

FAMILY ENTOCYTHERIDAE HOFF, 1942

Because of the state of the knowledge, at any one time a discussion of "Entocythere" may well have been valid for what is now recognized as the entire family Entocytheridae, the presently recognized subfamily Entocytherinae, or the genus *Entocythere, sensu strictu.* No attempt has been made here to interpret these discussions or to go beyond synonymizing below the subfamily level.

Entocytherinae Hoff, 1942a: 63-65 (R). — Hoff, 1942b: 159, 166. — Rioja, 1942a: 203. — Rioja, 1942b: 685, 686 (M). — Hoff, 1944b: 330. — Rioja, 1953: 289. — Pennak, 1953: 421 (K). — Howe, 1955: 66, 67 (C). — Howe, *in* Moore and Pitrat, 1961: 300 (C). — Howe, 1962: 80 (C). — Hart, 1962: 125 (S, M). — Hartmann, 1964: 586 (M, C).

Entocytheridae Moore, 1961 (*in* Moore and Pitrat, 1961): 96, 98. — Howe and Sylvester-Bradley (*ibid.*): 255, 257, 300. — Hart, 1962: 125 (S, M). — Hart and Hart, 1966: 564 (C). — Hart and Hart, 1967b: unpaged (B). — Crocker and Barr, 1968: 42 (C). — Hart and Hart, 1968: 315-318 (T). — Hart and Hart, 1969a: 182, 183 (R), fig. 5. — Hobbs, 1971: 18 (S, M).

Diagnosis. — Shell reniform to elliptical, thin chitinous, laterally compressed; valves slightly unequal in size, with or without protuberances, and with marginal or submarginal setae. Eyes, when present, fused and pigmented. Antennules consisting of 6 or

7 podomeres. Antennae with 4 podomeres, armed distally with 2 or 3 claws and basal podomere bearing long setiform flagellum (claw). Mandible well developed with all podomeres of palp fused except distal one. Thoracic legs similar to one another, each terminating in large curved claw. Maxillae variable, consisting of a palp, and only one masticatory process. Copulatory apparatus of male complex; consisting of peniferum, penis, clasping apparatus, 1 or 2 fingers, and with or without finger guard. Five subfamilies: Entocytherinae Hoff, 1942a (restricted); Sphaeromicolinae Hart, 1962; Microsyssitrinae Hart, Nair, and Hart, 1967; and Notocytherinae Hart and Hart, 1967a; and Hartiellinae Danielopol, 1971b.

Distribution. — The five subfamilies of the family Entocytheridae currently recognized are geographically distributed as follows: Entocytherinae (North America), Sphaeromicolinae (North America and Europe), Microsyssitrinae (Asia), and Notocytherinae (Australia, New Zealand, and New Guinea), and Hartiellinae (coastal waters of France and Italy).

Hosts. — Whereas the principal hosts of the entocytherid ostracods appear to be crayfishes (North America, Australia, New Zealand, and New Guinea) they are also found on freshwater crabs (Mexico), cave-dwelling isopods (Mexico, France, northern Italy, and Yugoslavia), marine amphipods (coastal waters of Italy and France), wood-boring marine isopods (two estuarine localities in Kerala State, India), and phreatoicoid isopods (Australia).

Key to the Subfamilies of Entocytheridae

(Modified from Hobbs, Jr., 1971)

1 Dorsal claw of antenna without setae on flexor surface (commensal on crayfishes and phreatoicoid isopods) .... NOTOCYTHERINAE

1′ Dorsal claw of antenna with setae on flexor surface 2

2(1′) Maxilla with respiratory plate and well developed masticatory lobes; penis usually strongly curved (commensal on crayfishes and a species of freshwater crab) .... ENTOCYTHERINAE

2′ Maxilla without respiratory plate or masticatory lobes (occasionally vestigial); penis straight or only slightly curved 3

3(2′) Antennule consisting of five podomeres; distal flexor margin of ultimate podomeres of legs with brush of setae on finger-like projection opposing terminal claw (commensal on wood-boring isopods) .... MICROSYSSITRINAE

3′ Antennule consisting of six or seven podomeres; distal flexor margin of ultimate podomeres of legs lacking setiferous finger-like projection opposing terminal claw 4

4(3′) Clasping apparatus with terminal denticles (commensal on fresh water isopods) .... SPHAEROMICOLINAE

4′ Clasping apparatus without terminal denticles (commensal on marine amphipods) .... HARTIELLINAE

Subfamily Entocytherinae Hoff, 1942

SUBFAMILY ENTOCYTHERINAE HOFF, 1942

Entocytherinae Hoff, 1942a: 63-65. — Hart, 1962: 125-126 (C, M). — Hart and Hart, 1969a: 182, 183 (R). — Hart and Hart, 1969b: 157 (M). — Hobbs, 1971: 19 (C, M, D, H). — Danielopol, 1971a: 190 (D).

Tribe Entocytherini. — Danielopol, 1971a: 190 (D).

Type of the subfamily. — *Entocythere cambaria.*

Diagnosis. — Distal spine of mandibular palp never spatulate; respiratory plate of maxilla well developed; maxilla well developed; maxilla with masticatory lobe ending in setae. Peniferum elongate, bearing curved to angular penis, with only relatively slightly convex posterior margin. Commensal on crayfishes and a crab; found only in fresh water.

Distribution. — Southern Canada; mainland United States plus Oahu, Hawaii; Mexico; Cuba plus Isla de Pinos. Hosts, freshwater crayfishes of the family Astacidae and, in one instance, the pseudothelphusid crab *Pseudothelphusa* (*Tehuana*) *veracruzana* Rodriguez and Smalley.

Key to the Genera of the Subfamily Entocytherinae

(Modified from Hobbs and Hobbs, 1970)

1 Penis with prostatic and spermatic elements widely separated along much of their lengths 2
1′ Penis simple; or if two elements recognizable, contiguous along their entire lengths 4
2(1) Ventral portion of peniferum tapering with tip of penis reaching, or almost reaching, apex .... *Plectocythere*
2′ Ventral portion of peniferum usually rounded or with one or more prominences, seldom tapering; if tapering, tip of penis never approaching apex 3
3(2′) Ventral portion of peniferum rounded, without prominences .... *Phymocythere*
3′ Ventral portion of peniferum with one or more prominences ventrally and/or anteriorly .... *Ascetocythere*
4(1′) Penis directed posteroventrally from base .... *Lordocythere*
4′ Penis directed anteroventrally from base 5
5(4′) Finger guard present 6
5′ Finger guard absent 10
6(5′) Peniferum with accessory groove, except in *Dactylocythere leptophylax* where finger guard is always slender and trifid 7
6′ Peniferum without accessory groove; finger guard never slender and trifid 8
7(6) Posteroventral portion of peniferum terminating in barbed point .... *Sagittocythere*
7′ Posteroventral portion of peniferum variable, but never ending in barbed point .... *Dactylocythere*
8(6′) Ventral portion of peniferum bulbiform, clasping apparatus never extending so far ventrally as does peniferum .... *Cymocythere*
8′ Ventral portion of peniferum slender or strongly flattened; clasping apparatus extending ventrally to or beyond ventral extremity of peniferum 9
9(8′) Ventral portion of peniferum slender, terminating in small recurved projection .... *Harpagocythere*
9′ Ventral portion of peniferum flattened and with concave ventral border .... *Litocythere*
10(5′) Anteroventral portion of peniferum with acute beak-like projection .... *Ornithocythere*
10′ Anteroventral portion of peniferum never with beak-like projection 11
11(10′) External border of horizontal ramus of clasping apparatus with one or more excrescences 12
11′ External border of horizontal ramus of clasping apparatus entire or with a few shallow subapical grooves 15
12(11) Anteroventral portion of peniferum produced ventrally in rounded lobe 13
12′ Anteroventral portion of peniferum never produced ventrally in rounded lobe; if produced, apex acute or truncate 14
13(12) Spermatic loop horizontal; peniferum distal to dorsal margin of spermatic loop at least twice as long as portion dorsal to loop; clasping apparatus with external border bearing single tubercle and terminating in fan-like cluster of serrations .... *Saurocythere*
13′ Spermatic loop vertical; peniferum distal to dorsal margin of spermatic loop much less than twice as long as portion dorsal to loop; clasping apparatus with external border broadly serrate and terminating in annulations .... *Okriocythere*
14(12′) Anteroventral portion of peniferum with conspicuous anterodorsally directed projection 19
14′ Anteroventral portion of peniferum never with conspicuous anterodorsally directed projection .... *Ankylocythere*
15(11′) Internal border of clasping apparatus usually with no more than three teeth; if more than three, with only two apical denticles or vertical ramus strongly convex posteriorly 16
15′ Internal border of clasping apparatus with more than three teeth, apical cluster with more than two denticles, vertical ramus straight 18
16(15) Clasping apparatus not clearly divisible into vertical and horizontal rami, extremities directed at angle of at least 100 degrees .... *Donnaldsoncythere*
16′ Clasping apparatus usually divisible into vertical and horizontal rami, extremities directed at angle of no more than 90 degrees 17

17(16′) Penis large, S-shaped or sinuous, and with curved posteroventral thickening of peniferum giving forcipate appearance to ventral portion of peniferum ...... *Thermastrocythere*
17′ Penis of moderate size, L-shaped, and never so disposed as to contribute forcipate appearance to ventral portion of peniferum .... *Uncinocythere*
18(15′) Ventral portion of peniferum tapering to slender tip .... *Rhadinocythere*
18′ Ventral portion of peniferum never slender nor tapering .... *Entocythere*
19(14) Anteroventral portion of peniferum entire .... *Geocythere*
19′ Anteroventral portion of peniferum bifid .... *Hartocythere*

ANKYLOCYTHERE HART, 1962

Heterodonta Group Rioja, 1943a: 564 (R). — Hoff, 1944b: 330 (R) (Rioja 1943, *not* Rioja 1944), 352 (M, R) (*sic* Rioja 1944), 356 (R) (*sic* Rioja 1944). — Rioja, 1949: 327 (R). — Rioja, 1955: 196 (R). — Hobbs, 1956 (1957): 431 (R, D). — Hart, 1959: 201-202 (M). — Crawford, 1959: 173 (M, R) (*sic* Rioja 1944).

Ankylocythere Hart, 1962: 126 (S), (type species, *Entocythere heterodonta* Rioja, 1940b: 594). — Hart, 1963: 51 (R, D). — Crawford, 1965: 148 (R). — Hobbs III, 1965: 160 (K). — Hobbs, 1966: 67 (R). — Hart and Hart, 1967a: 4 (M). — Hobbs, Holt, and Walton, 1967: 35 (M). — Hart and Hart, 1969a: 182-185 (D, E, R). — Hart and Hart, 1969b: 158 (M, R). — Hobbs and Hobbs, 1970: 2 (K).

Diagnosis. — Terminal tooth of mandible pectinate. Copulatory complex of male without finger guard or accessory groove. Spermatic and prostatic elements of penis contiguous. External border of clasping apparatus with talon or excrescence; internal border with one to four teeth or serrations; apex with two or three denticles. Triunguis females without amiculum; antennae lacking comb-like appendix at base of terminal claws.

Type-species. — *Entocythere heterodonta* (= *Ankylocythere heterodonta*).

Distribution. — Mexico (Campeche and Chiapas) to Wisconsin and eastward to Maryland and Florida. Also present on Cuba, the Isla de Pinos, and Oahu (Hawaii).

Hosts. — Freshwater crayfishes of the family Astacidae.

Figure 20. — Distribution of the genus *Ankylocythere*.

Key to the Species of the Genus Ankylocythere

(Modified from Hobbs, Jr., 1971)

1 Clasping apparatus with low rounded excrescence 2
1′ Clasping apparatus with well developed talon 10
2(1) External border of horizontal ramus of clasping apparatus with single rounded excrescence situated below major tooth on internal border 19
2′ External border of horizontal ramus of clasping apparatus with one or more excrescences situated on external border; if only one, always situated distal to tooth on internal border 3
3(2′) External border of clasping apparatus with two or three excrescences 4
3′ External border of clasping apparatus with single excrescence 8
4(3) Two excrescences on external border of clasping apparatus subequal in size .... *A. talulus*
4′ Two excrescences on external border of clasping apparatus unequal in size 5
5(4′) Clasping apparatus bent to resemble fish-hook and with three to many small teeth (serrations) on internal border .... *A. hamata*
5′ Clasping apparatus never so strongly bent and seldom with more than one tooth on internal border 6
6(5′) Clasping apparatus with vertical ramus longer than horizontal .... *A. hyba*
6′ Clasping apparatus with vertical ramus no longer, often shorter, than horizontal 7
7(6′) Shell length of male greater than 0.34 mm; that of female greater than 0.36 mm .... *A. heterodonta*
7′ Shell length of male less than 0.34 mm; that of female less than 0.36 mm .... *A. cubensis*
8(3) Peniferum truncate or slightly concave distally with acute anteroventral extension 9
8′ Peniferum rounded or tapering distally 7
9(8) Internal border of clasping apparatus with serrations or small teeth at junction of vertical and horizontal rami .... *A. maya*
9′ Internal border of clasping apparatus without serrations or small teeth at junction of vertical and horizontal rami .... *A. toltecae*
10(1′) External border of clasping apparatus with two elongate projections (talons) .... *A. tridentata*
10′ External border of clasping apparatus with only one elongate talon 11
11(10′) Length of talon more than twice least vertical diameter of horizontal ramus 12
11′ Length of talon less than twice least vertical diameter of horizontal ramus 15
12(11) Ventral margin of peniferum bifid 13
12′ Ventral margin of peniferum acute or irregular but never bifid 14
13(12) Talon extending distally at least to level midway between tooth on internal border and proximal terminal denticle .... *A. ancyla*
13′ Talon never extending distally to level midway between tooth on internal border and proximal terminal denticle .... *A. copiosa*
14(12′) Clasping apparatus tapering from level of tooth on internal border of horizontal ramus to terminal denticles; base of talon entirely proximal to level of tooth on internal border .... *A. harmani*
14′ Clasping apparatus expanding between level of tooth on internal border of horizontal ramus and terminal denticles; part of base of talon usually opposite tooth on internal border .... *A. hobbsi*
15(11′) Ventral margin of peniferum bifid 21
15′ Ventral margin of peniferum not bifid, but acute or rounded 16
16(15′) Clasping apparatus with three terminal denticles .... *A. tiphophila*
16′ Clasping apparatus with two terminal denticles 17
17(16′) Gap between upper distal end of talon and horizontal ramus distal to base of talon greater than height of distal portion of ramus .... *A. villalobosi*
17′ Gap between upper distal end of talon and horizontal ramus distal to base of talon less than height of distal portion of ramus 18
18(17′) Vertical ramus of clasping apparatus distinctly longer than horizontal ramus .... *A. bidentata*
18′ Vertical ramus subequal in length, or shorter than, horizontal ramus 7
19(2) Excrescence developed into rounded lobe posterior to notch .... *A. tallapoosa*
19′ No notch present adjacent to excrescence 20

29(19′) Excrescence rounded, discreet, limited to area below major tooth [see 2(1)] .... *A. telmoecea*
20′ Excrescence developed into ridge on external border of horizontal ramus .... *A. ephydra*
21(15) Peniferum about 175 μ in height .... *A. burkeorum*
21′ Peniferum about 75 μ in height .... *A. sinuosa*

Ankylocythere ancyla Crawford, 1965

Plate I, figures 1-6
Plate XLV

Ankylocythere ancyla Crawford, 1965: 148-149, 153 (M, R, A, D), figs. 1-3, 6, 7. — Hobbs, 1966: (K, D, H), fig. 13. — Hobbs, Holt, and Walton, 1967: 7 (B), 33 (F), 34 (K), 35 (M, D, H, A, S), 69 (D, E), 18, 20, 25, 41, 43, 45, 49, 54, 55, 61, 63, 72, 73 (A, H), fig. 7g. — Hobbs and Perkins, 1967: 146 (H, A, D). — Ferguson, 1968: 501 (C, D). — Hart and Hart, 1971: 111.

Diagnosis. — MALE: Shell length 340-390 μ (average, 370 μ); height, 220-240 μ (average, 230 μ). Distal portion of peniferum bifid; anterior protuberance longer than posterior. Clasping apparatus with horizontal ramus longer than vertical ramus and forming approximately 90° angle with one another; external margin of horizontal ramus bearing large subparallel talon approximately one-half length of ramus; internal margin with one tooth at mid-length; distal extremity with two denticles.

FEMALE: Shell averaging 390 μ in length; 250 μ in height. Triunguis females with posteriorly placed short sclerotized tubular genital papilla, directed ventrally.

Type-locality. — City of Greensboro, Guilford Co., North Carolina, U.S.A. (types, USNM: holotype, 113502; allotype, 113503; morphotype, 113504). Host, *Cambarus catagius.*

Distribution (other than type-locality). —

United States

ALABAMA — CONECUH CO. — 3.9 mi NE of Castlebury on rt 31 (hosts, *Procambarus a. acutus* and *P. versutus*). LAUDERDALE CO. — 18¾ mi W of Athens (host, *Orconectes spinosus*). MOBILE CO. — Spring Creek, about 15 mi SW of Mobile on rt 90 (host, *P. lecontei*).

FLORIDA — BAY CO. — Swamp adjoining Econfina R, W of Bennet (host, *Procambarus rogersi* intergrades). ESCAMBIA CO. — 2 mi E of Bluff Springs on Escambia R (hosts, *Cambarus diogenes* subsp., *C.* sp., *P. a. acutus, P. bivittatus,* and *P. shermani*). GULF CO. — 5.8 mi W of Wewahitchka on rt 52 (hosts, *P. pycnogonopodus, P. latipleurum,* and *P. kilbyi*).

GEORGIA — ATKINSON CO. — 3.1 mi S of Douglas on rt 441 (hosts, *Procambarus advena* and *P. seminolae*); 6 mi S of Coffee Co. line on rt 441 (hosts, *P. advena* and *P. seminolae*). BURKE CO. — 3 mi SW of Waynesboro in Walnut Branch on Blount farm (host, *P. a. acutus*); about 7 mi SW of Waynesboro in Rosemary Branch on Blount farm (hosts, *P. troglodytes* and *P. spiculifer*); 9.6 mi N of Millen on rt 25 (host, *P. troglodytes*); Rock Creek, 5.1 mi W of Waynesboro on rt 24 (host, *Cambarus latimanus*). COFFEE CO. — 0.4 mi SE of Jeff Davis Co. line on rt 221 (hosts, *P. seminolae* and *P. pubischelae*); 1.8 mi W of Broxton on rt 268 (host, *P. seminolae*). EMANUEL CO. — Jacks Creek at Lexy (host, *P. enoplosternum*). EVANS CO. — 5.5 mi N of Claxton on rt 301 (host, *P. litosternum*). JEFF DAVIS CO. — 3.6 mi SW of Hazelhurst on rt 221 (hosts, *P. seminolae* and *P. pubischelae*). LAURENS CO. — Rocky Creek, trib to Oconee R, 15.3 mi NE of Cochran on rt 26 (hosts, *P. pubescens* and *P. spiculifer*). LOWNDES CO. — 14.1 mi E of Valdosta (host, *P. pubischelae*). LIBERTY CO. — 7.6 mi SW of Midway on rt 38 (hosts, *P. advena, P. troglodytes,* and *P. barbatus*). LONG CO. — 10.1 mi NW of Ludowici on rt 25 (host, *P. advena*). MADISON CO. — 5.4 mi SE of Jackson Co. line on rt 98 (hosts, *C.* (*Depressicambarus*) sp. and *P. raneyi*). MERRIWETHER CO. — 7.2 mi E of Merriwether Co. line on rt 109 (hosts, *C. latimanus* and *C. sp.*). MONTGOMERY CO. — 1.3 mi S of Treutlen Co. line on rt 221 (hosts, *P. a. acutus* and *P. enoplosternum*). SCREVEN CO. — South Ogeechee Creek, 6.8 mi S of Sylvania on rt 301 (host, *P. epicyrtus*). TREUTLEN CO. — 16 mi SE of East Dublin on rt 29 (hosts, *P. enoplosternum* and *P. a. acutus*); 19.5 mi SE of East Dublin on rt 29 (hosts, *P. truculentus* and *P. enoplosternum*). WARE CO. — 12 mi N of Waycross on rt 1 (host, *P. pubischelae*). WHEELER CO. — 5.4 mi W of Glenwood on rt 280 (hosts,

P. howellae and *Faxonella clypeata*). WORTH CO. — 2.5 mi E of Poulon on rt 83 (hosts, *P. advena* and *P. seminolae*).

NORTH CAROLINA — BLADEN CO. — roadside ditch 7.7 mi N of Clarkton on rt 701 (host, *Procambarus p. pearsei*); 3.7 mi S of Dublin on rt 41 (host, *P. p. pearsei*). CHATHAM CO. — 9.8 mi N of Sanford on rt 501 (hosts, *Cambarus reduncus, C. acuminatus,* and *Fallicambarus uhleri*). COLUMBUS CO. — 0.6 mi E of Hallsboro (hosts, *P. p. pearsei, P. blandingii*). ORANGE CO. — NE edge of Chapel Hill (host, *C. b. bartonii*); 5 mi E of Chapel Hill (host, *C. reduncus*); trib to New Hope Creek, E of Piney Mountain (host, *C. b. bartonii*). RANDOLPH CO. — 2.8 mi NE of Randolph Co. line on rt 49 (hosts, *P. a. acutus* and *C. acuminatus*). WILSON CO. — 4.4 mi NW of Saratoga on rt 264 (hosts, *P. a. acutus* and *C. latimanus*). YADKIN CO. — 5.5 mi S of Elkin on rt 21 (hosts, *C. acuminatus* and *C. b. bartonii*).

SOUTH CAROLINA — DORCHESTER CO. — ditch 1.4 mi NW of Harleyville on rt 178 (hosts, *Procambarus troglodytes* and *Fallicambarus uhleri*). LANCASTER CO. — spring draining into ditch on Potter Rd, 2 mi N of Pleasant Hill (hosts, *C.* sp. and *P.* sp.); pond 2 mi N of Pleasant Hill on Potter Rd (hosts, *C.* sp. and *P.* sp.); branch on Marshall property, 2 mi N of Pleasant Hill on Potter Rd (hosts, *C.* sp. and *P.* sp.). RICHLAND CO. — Crane Creek at bridge on rt 215, Denny Terrace area of Columbia (hosts, *P. blandingii* and *C. latimanus*).

VIRGINIA — AMHERST CO. — Buffalo R at forks of Buffalo (hosts, *Cambarus longulus* and *C. b. bartonii*). HALIFAX CO. — Grassy Creek on rt 58, 4 mi E of South Boston (host, *C. acuminatus*). LOUDON CO. — Little R, about 1.5 mi NE of Aldie on rt 15 (host, *Orconectes limosus*). MECKLENBURG CO. — about 2 mi S of Brodnax on rt 659 (hosts, *C. acuminatus* and *P. a. acutus*); trib to Roanoke R 3.2 mi NE of Ebony (host, *C. acuminatus*); St. John's Creek on rt 49, 7 mi N of Chase City (host, *C. acuminatus*); Pole Branch on rt 58 between rt 1 jct and Boyton (hosts, *P. a. acutus* and *C. acuminatus*). PATRICK CO. — 1.6 mi S of Patrick-Franklin Co. line on rt 40 (hosts, *C. longulus* and *C. acuminatus*). PITTSYLVANIA CO. — Shockoe Creek on rt 832, 7 mi E of Chatham (host, *C. acuminatus*); Whitehorn Creek on rt 29 between Gretna and Chatham (hosts, *C. acuminatus* and *C. longulus*).

Ankylocythere bidentata (Rioja, 1949)

Plate I, figures 7-10
Plate XLVI

Entocythere bidentata Rioja, 1949: 320-321 (M, H, R, D), figs. 9-11. — Hobbs, 1956 (1957): 431 (R). — Crawford, 1959: 173 (R). — Hart, 1962: 122 (C).

Entocythere heterodonta. — Rioja, 1949: 316 (M, in part).

Entocythere sinuosa. — Rioja, 1949: 322 (M, in part).

Ankylocythere bidentata. — Hart, 1962: 126 (C, S, D). — Hobbs, 1966: (K, D, H), fig. 14. — Ferguson, 1968: 501 (C, D). — Hobbs, 1971: 29-31 (M, H, R, A, D), figs. 15, 16.

Diagnosis. — MALE: Shell length 300-330 μ (average, 320 μ); height, 190-200 μ (average, 200 μ). Distal portion of peniferum entire, but with small acute projection anteroventrally. Clasping apparatus with well-developed talon never projecting distally more than distance equivalent to its own width; internal border of horizontal ramus usually with two teeth (one above base of talon, other proximal to it), occasionally only one, and frequently irregularly serrate along and adjacent to junction of horizontal and vertical rami; vertical ramus longer than horizontal ramus, distal extremity of horizontal ramus with two denticles.

Type-locality. — Tomatlan, 15 km SW of Huatusco, Veracruz, Mexico (types, not extant). Host, *Procambarus aztecus*.

Distribution (*other than type-locality*). —

Mexico

CHIAPAS — Cerro Hueco, 8 km WNW of Tuxtla Gutierrez (host, *Procambarus mirandai*).

OAXACA — Monte Bello, Tuxtepec (host, *Procambarus mexicanus*); Leguna Escondida, Ojitlan (host, *P. veracruzanus*).

TABASCO — Villa Hermosa (host, *Procambarus llamasi*).

VERACRUZ — Sinkhole between Huatusco and Jalapa (host, *Procambarus aztecus*); ? Coscomatepec (Rioja, 1949: 321) (host, *P. aztecus* [*Ankylocythere sinuosa* also reported by Rioja (1949: 322) on *P. mexicanus* from same locality, but this needs confirmation]; Mirador de Zacuapan (Rioja, 1949: 321) (host, *P. mexicanus*); Hacienda Jalapilla, 5 km SE of Orizaba (reported by Rioja (1949: 322) as *Entocythere sinuosa*) (host, *P. mexicanus*); Playa Norte de la Laguna de Catemaco, San Andres, Tuxtla (reported by Rioja (1949: 316) as *Entocythere heterodonta*) (host, *P. vazquezae*); Zapoapan de la Cabaña, 11 km SSE of Catemaco) (host, *P. ruthveni zapoapensis*); Cueva de Ojo de Agua Grande, Paraje Nuevo (host, *P. rodriguezi*); 32 km carretera a Juan

Díez Covarrubias (host, *P. r. ruthveni*); Rincon de la Doncella, Ciudad Mendoza (host, *P. mexicanus*); 3 km S of Coscomatepec (host, *P. aztecus*); Zapoapan de Cabaña, Catemaco (host, *P. ruthveni zapoapensis*); El nacimiento de la Laguna Encantada, San Andrés, Tuxtla (host, *P. vazquezae*); Spring at Tlilapan (host, *P. mexicanus*); Hacienda de Tenejapa (reported by Rioja (1949: 316) as *Entocythere heterodonta*) (host, *P. mexicanus*); Ojo de Agua, Tlilapan (host, *P. mexicanus*); Sinkhole between Huatusco and Jalapa (host, *P. aztecus*).

Ankylocythere burkerorum Hobbs III, 1971

Plate I, figures 11-13
Plate XLV

Ankylocythere burkerorum Hobbs III, 1971: 137-140 (M, H, R, A, D), figs. 1a-c.

Diagnosis. — MALE: Shell length 350-360 μ (average, 360 μ); height, 210-220 μ (average, 210 μ). Distal portion of elongate peniferum slightly enlarged, possessing small terminal bifurcation. Penis situated in distal seventh of peniferum. Clasping apparatus with distinct vertical and horizontal rami, vertical ramus approximately twice length of horizontal ramus; horizontal ramus slender, sometimes tapering, bearing single tooth on internal border proximal to mid-length; distal extremity terminating in two denticles; external border bearing talon arising proximal to, or at level of, tooth on internal border and extending subparallel to horizontal ramus for distance approximating $\frac{1}{6}$ to $\frac{1}{5}$ length of horizontal ramus; external and internal borders of vertical ramus entire.

FEMALE: Unknown.

Type-locality. — Crayfish burrows along bank of Tallahatta Creek, near Duffee, Newton County, Mississippi, U.S.A. (types, USNM: holotype, 135617; paratype, 135618). Hosts, *Cambarus d. diogenes* and *Cambarus striatus*.

Distribution (in addition to type-locality). —

United States

MISSISSIPPI — NOXUBEE CO. — crayfish burrows in ditch 1 mi W of jct of rts 14 and 45 on rt 14 (hosts, *Cambarus d. diogenes* and *C. striatus*) (Hobbs III, 1971: 139).

Ankylocythere copiosa (Hoff, 1942)

Plate I, figures 14-16
Plate XLV

Entocythere copiosa Hoff, 1942a: 66 (K), 67-71 (M, H, R, D), 73 (F).

Entocythere (*Cytherites*) *copiosa.* — Rioja, 1942b: 688 (K), 689, 696 (R), fig. 3. — Hoff, 1942b: 166 (D). — Rioja, 1942a: 203 (H, A). — Rioja, 1942b: 688 (K), 689, 696 (R), fig. 3. — Hoff, 1943: 280, 281, 285, 286 (H, D). — Rioja, 1943a: 564 (R). — Rioja, 1943b: 569 (R), 576 (R, M), fig. 18.

Entocythere copiosa. — Hoff, 1944b: 330, 356 (R), 332 (K). — Tressler, 1947: 705 (C). — Rioja, 1949: 328 (R). — Tressler, 1959: 731 (K), fig. 28.193. — Crawford, 1959: 173 (R). — Hart, 1962: 122 (C).

Ankylocythere copiosa. — Hart, 1962: 126-127 (C, D, S). — Crawford, 1965: 149 (R). — Hart and Hart, 1966a: 5 (A). — Hobbs, 1966: 70 (K, D, H), 72 (R), fig. 10.

Diagnosis. — MALE: Peniferum elongate, terminating in bifurcation comprised of acute anteroventrally directed anterior process. Small anteroventrally directed penis situated in distal $\frac{1}{6}$ of peniferum. Clasping apparatus with horizontal ramus approximately $\frac{1}{2}$ length of vertical ramus; single tooth situated at mid-length of internal border of horizontal ramus; anteroventrally directed talon on external border arising proximal to tooth and extending for approximately $\frac{1}{3}$ length of horizontal ramus; horizontal ramus terminating in three denticles.

FEMALE: Length, 380-480 μ; height, 210-240 μ.

Type-locality. — Crawford Co., Illinois, U.S.A. (types, USNM: holotype, 81087; allotype, 81088). Hosts, *Orconectes immunis, O. virilis, O. propinquus,* and *Procambarus a. acutus.*

Distribution (in addition to type-locality). —

United States

ARKANSAS — PULASKI CO. — Ivesville (hosts, *Orconectes meeki* and *Procambarus a. acutus*). GREENE CO. — trib of St. Francis R, 4.9 mi NE of Paragould on rt 1 (hosts, *Cambarellus puer, Procambarus a. acutus, P. viaeviridis, Faxonella clypeata,* and *Fallicambarus hedgpethi*). MADISON CO. — headwaters of White R, 2.8 mi E of Pettigrew on rt 16 (host, *O. nais* ?). PERRY CO. — Little Cove Creek at intersection on rt 7 (host, *O. palmeri longimanus*); 5.5 mi S of Hallis on rt 7 (host, *O. p. longimanus*). SCOTT CO. — 14.1 mi SE of Boles (host, *O. p. longimanus*); 6.0 mi E of Cauthron on rt 28 (host, *O. p. longimanus*). YELL CO. — no locality given (host, *O. p. longimanus*); Arkansas R, 2.8 mi S of Dardanelle on rt 7 (host, *P. a. acutus*);

8.6 mi S of Centerville on rt 7 (host, *O. p. longimanus*).

ILLINOIS — CHAMPAIGN CO. — no further locality data (Hoff, 1942a) (hosts, *Orconectes immunis, O. virilis, O. propinquus, Procambarus a. acutus*); no further data (host, *P. a. acutus*). HARDIN CO. — Sugar Creek, 2 mi NE of Creal Springs (host, *O. indianensis*). HENRY CO. — Hoff, 1942a, no locality data (hosts, *O. immunis, O. virilis, O. propinquus,* and *P. a. acutus*). McLEAN CO. — 2 mi NW of Bellflower (host, *Cambarus diogenes* subsp.). WABASH CO. — no data on USNM coll (host, *C. diogenes* subsp.).

INDIANA — WABASH CO. — North Manchester (host, *Procambarus a. acutus*).

IOWA — SCOTT CO. — Davenport (host, *Cambarus diogenes* subsp.).

KENTUCKY — CHRISTIAN CO. — Sinking Fork of Little R, 7.1 mi W of Hopkinsville on rts 68 and 80 (host, *Orconectes tricuspis*).

LOUISIANA — CALCASIEU PARISH — trib of Black Bayou at W city limits of Vinton on rt 90, 1 mi W of center of town (hosts, *Procambarus hinei* and *P. clarkii*). GRANT PARISH — near Dry Prong (hosts, *P. a. acutus* and *Faxonella clypeata*). IBERVILLE PARISH — near White Castle (Hoff, 1943) (host, *P. clarkii*). JACKSON PARISH — stream 4 mi S of Clay (host, *P. a. acutus*). LINCOLN PARISH — near Ruston (host, *P. a. acutus*); Indian Creek (hosts, *P. clarkii, P. a. acutus,* and *P. natchitochae*); Sugar Creek (hosts, *P. vioscai* and *P. natchitochae*). POINTE COUPEE PARISH — near Livonia (Hoff, 1943) (hosts, *P. clarkii* and *P. hinei*). RAPIDES PARISH — near Alexandria (host, *P. a. acutus*). SABINE PARISH — Little Bayou Sara (host, *P. a. acutus*); Beech Bayou on rt 1215 (hosts, *P. a. acutus, P. dupratzi,* and *P. clarkii*). WEST BATON ROUGE PARISH — near Port Allen (Hoff, 1943) (hosts, *P. clarkii* and *P. a. acutus*); near Westover (Hoff, 1943) (hosts, *P. a. acutus* and *P. clarkii*).

OKLAHOMA — BECKHAM CO. — stream at bridge 8.1 mi NE of Sayre on rt 66 (host, *Procambarus s. simulans*). LATIMER CO. — Bondy Creek, 1 mi S of Wilburton (host, *O. difficilis*); Turkey Creek, 8.7 mi E of Wilburton on rt 270 (host, *O. palmeri longimanus*). LEFLORE CO. — Bushy Creek, 4.2 mi from Cedar Creek Recreation Area on road to rt 270 in Ouachita National Forest (host, *O. p. longimanus*). McCURTIN CO. — Eagle Creek, trib to Mountain Fork R, NW of Smithville (host, *O. p. longimanus*).

TENNESSEE — DAVIDSON CO. — Richland Ck at Bellemeade, W of Nashville (host, *Cambarus tenebrosus*). HARDIN CO. — Tally Pond near Savannah College (host, *Procambarus viaeviridis*). OBION CO. — ck at end of Scott's orchard (host, *P. a. acutus*); Indian Ck near Samburg Rd (hosts, *S. diogenes* subsp. and *O. p. palmeri*); Indian Ck near Old Samburg (host, *O. p. palmeri*); Lake Center (Hoff, 1943) (host, *P. clarkii*); Reelfoot Lake (Hoff, 1943) (hosts, *Orconectes i. immunis* and *P. clarkii*); Walnut Log (Hoff, 1943) (host, *P. clarkii*); S of Walnut Log (Hoff, 1943) (hosts, *C. diogenes* subsp. and *O. immunis*).

WISCONSIN — JEFFERSON CO. — no locality data (host, *O. propinquus*).

Ankylocythere cubensis (Rioja, 1955)

Plate I, figures 17-22
Plate XLVII

Entocythere heterodonta. — Hoff, 1944a: 370 (D, E).

Entocythere heterodonta cubensis Rioja, 1955: 193-194 (M, H, D), fig. 1. — Hobbs, 1956 (1957): 431, 433 (D, M), 435, 436 (R). — Hart, 1962: 122 (C). — Straskraba, 1969: 9 (D, H, S).

Ankylocythere heterodonta cubensis. — Hart, 1962: 127 (C, D, S). — Hobbs, 1966: 67-68 (S). — Ferguson, 1968: 501 (C, D).

Ankylocythere cubensis. — Hobbs, 1966: 71 (K). — Hobbs, 1971: 31-32 (M, H, R, A, D), fig. 17.

Diagnosis. — MALE: Shell length 320-330 μ (average, 320 μ); height, 200-220 μ (average, 210 μ). Clasping apparatus of male with well developed or adnate talon, latter never projecting distally more than distance equivalent to its own width; internal border with single tooth above talon and with or without irregular serrations in area of junction of vertical and horizontal rami; apex with two upturned denticles. Ventral portion of peniferum rounded with very small acute anteroventral projection.

FEMALE: Shell similar to that of male, but slightly larger; length, 340-350 μ (average, 340 μ); height, 210-220 μ (average, 210 μ).

Type-locality. — Fourteen mi W of Camagüey, Camagüey Province Cuba. Host, *Procambarus cubensis cubensis*. Types not extant.

Distribution (*in addition to type-locality*). —

Cuba

CAMAGÜEY — Arroyo pequeño en La Gloria (Hobbs, 1957: 432) (host, *Procambarus c. cubensis*).

LA HABANA — Isla de Pinos — La Ceiba, Los Indios, and Nueva Gerona (Hobbs, 1957: 432) (hosts, *Procambarus atkinsoni* and *P. c. cubensis*).

MATANZAS — Unión de Reyes (Hobbs, 1957: 432) (host, *Procambarus c. cubensis*).

Ankylocythere ephydra Hart and Hart, 1971

Plate II, figures 1-5
Plate XLVI

Ankylocythere ephydra Hart and Hart, 1971: 105-106 (M, H, R, A, D), figs. 1a-e.

Diagnosis — MALE: Shell length 340-400 μ (average, 370 μ); height, 190-230 μ (average, 210 μ); subelliptical, with greatest height $\frac{1}{3}$ shell length from posterior end. Eye pigmented, located $\frac{1}{5}$ shell length from anterior end.

Copulatory complex with peniferum of almost uniform width, ending anteroventrally in acute angle. Anteroventral area sclerotized. Penis horizontal and occupying median $\frac{3}{4}$ of width of peniferum with tip directed anteroventrally.

Clasping apparatus with length of vertical ramus $1\frac{1}{2}$ times length of horizontal ramus and joining it at approximately 90° angle. External border with one distinct emargination slightly distal to midlength of horizontal ramus, resulting in sharp, posteriorly-pointed barb on one clasper, low rounded protuberance on other. Internal border of horizontal ramus with single tooth approximately equidistant between junction of rami and two terminal denticles, and slightly proximal to emargination on external border.

FEMALE: Shell of biunguis female subelliptical with greatest height near midlength; length 310-320 μ (average, 320 μ); height, 170-200 μ (average, 180 μ).

Relationships. — The peniferum of *Ankylocythere ephydra* resembles that of *A. talulus* (Hoff) and *A. hyba* Hobbs and Walton, but differs from them in possessing a distinct projection on the external border of the horizontal ramus. In some specimens, particularly those from McMinn Co., Tenn., the clasper grades closer to that of *A. telmoecea* (Crawford), showing a rounded protuberance rather than a sharp barb.

Type-locality. — Creek 4.5 mi E of Pulaski on rt 64, Giles Co., Tennessee, U.S.A. (types, USNM: holotype, 138443, allotype, 138444). Hosts, *Orconectes juvenilis*, *O. rusticus mirus*, *Cambarus tenebrosus*, and *C.* sp.

Distribution (*in addition to type-locality*). —

United States

ALABAMA — LIMESTONE CO. — 6 mi E of Athens on rt 72 (hosts, *Orconectes compressus* and *O. spinosus*); $3\frac{1}{4}$ mi W of Athens on rt 72 (host, *O. spinosus*).

TENNESSEE — GILES CO. — Walker Spring, 1.5 mi E of Lawrence Co. line on rt 64 (hosts, *Orconectes spinosus* and *O. rusticus mirus*). McMINN CO. — 7.5 mi SW of Athens on rt 11 (host, *O. erichsonianus*).

Ankylocythere hamata (Hobbs, 1957)

Plate II, figures 6-10
Plate XLVII

Entocythere hamata Hobbs, 1956 (1957): 433-436 (M, H, R, A, D), figs. 1-6. — Hart, 1962: 122 (C). — Straskraba, 1969: 9 (D, H, S).

Ankylocythere hamata. — Hart, 1962: 127 (C, D, S). — Hobbs and Walton, 1963a: 459 (R). — Hobbs, 1966: 69 (D, K, H), fig. 1. — Ferguson, 1968: 501 (C, D). — Hobbs, 1971: 32 (M, H, R, A, D), fig. 18.

Diagnosis. — MALE: Shell length 280-310 μ (average, 290 μ); height, 170-180 μ (average, 180 μ). Clasping apparatus of male resembling fish hook without barb and with broad, adnate, only slightly produced talon rendering distal fourth of external border of clasping apparatus triscalloped; internal border with one moderately well developed tooth above talon and with irregular serrations proximally along horizontal ramus; apex with two dorsally directed denticles. Ventral portion of peniferum rounded with subacute anteroventral, ventrally directed, projection.

FEMALE: Shell similar to that of male, but larger; length, 320-420 μ (average, 360 μ); height, 200-250 μ (average, 230 μ).

Type-locality. — "Un arroy pequeño cerca del laboratorio de investigacion del Jardin de Atkins en Central Soledad, Cienfuegos, Provincia de las Villas, Cuba" (types, USNM: holotype and allotype, 100938). Host, *Procambarus cubensis cubensis*.

Distribution. — This species is known only from the type-locality.

Ankylocythere harmani Hobbs, 1966

Plate II, figures 11-14
Plate XLV

Ankylocythere harmani Hobbs, 1966: 70 (K, D, H), 71-75 (M, H, R, D), figs. 8, 12. — Hobbs III, 1970: 175 (A), 179 (A). — Hobbs and Hobbs, 1970: 14 (A).

Diagnosis. — MALE: Shell length average 350 μ; height, 210 μ. Eye present. Shell subovate with ventral margin entire; greatest height slightly posterior to midlength; submarginal setae somewhat regularly and comparatively closely spaced from level of eye anteriorly and along posterior fourth.

Copulatory complex without finger guard; ventral portion of peniferum slightly swollen and ventral margin broadly truncate but undulating with slight acute prominence on anteroventral extremity. Penis with spermatic and prostatic elements contiguous. Length of penis approximately one-sixth length of clasping apparatus. Clasping apparatus divisible into vertical and horizontal rami, vertical ramus almost three times as long as latter; internal border of horizontal ramus with single prominent tooth along proximal half and slight elevation just proximal to two terminal denticles; external border of horizontal ramus with talon arising at base; talon slender, more than twice as long as least vertical diameter of ramus, and disposed subparallel to ramus with distalmost portion deflected anteroventrally.

TRIUNGUIS FEMALE: Slightly larger than male (average length, 370 μ; average height, 230 μ) but with no features distinguishing it from females of other species of the genus. Amiculum and associated structures absent.

Relationship. — *Ankylocythere harmani* has its closest affinities with *A. copiosa,* from which it differs most conspicuously by the undulating ventral margin of the peniferum rather than a bifid one. It is also similar to *A. hobbsi* in possessing a long vertical ramus of the clasping apparatus, but differs in lacking a thickened area in the distal portion of the horizontal ramus.

Type-locality. — Swamp, 1 mi S of Pigeon, Iberville Parish, Louisiana, U.S.A. (types, USNM: holotype, 123532; allotype, 123533; paratype, 123534). Host, *Cambarus diogenes* × *ludovicianus.*

Distribution (in addition to type-locality). —

United States

ALABAMA — FAYETTE CO. — 5 mi W of Fayette on rt 96 (hosts, *Cambarus (Depressicambarus)* sp. and *Orconectes* sp.).

ARKANSAS — NEVADA CO. — near Falcon (Hobbs, 1966) (host, *Cambarus diogenes* subsp.).

LOUISIANA — LINCOLN PARISH — lawn at Louisiana Tech. at Ruston (host, *Fallicambarus dissitus*).

MISSISSIPPI — MONROE CO. — Muldon (host, *Cambarus diogenes* subsp.).

TEXAS — JEFFERSON CO. — ditch about ½ mi W of Nome on rt 90 at Liberty Co. line (hosts, *Procambarus* sp. and *P. hinei*). LIBERTY CO. — 7 mi W of Dayton on rt 90 (hosts, *Procambarus hinei, P. a. acutus* and *Cambarellus puer*).

Ankylocythere heterodonta (Rioja, 1940)

Plate II, figures 15-19
Plate XLVI

Entocythere "nueva especie" Rioja, 1940c: 57.

Entocythere heterodonta Rioja, 1940a: 471 (*nomen nudum*). — Rioja, 1940b: 593-609 (M, R, B, D, H), 33 figs. — Rioja, 1941a: 177 (H), 178-189 (M), 9 figs. — Rioja, 1941b: 193 (C, H), 194 (F), 195-197 (R). — Ferguson, 1944: 715, 717 (C, R). — Hoff, 1944b: 330 (R), 332 (K), 352 (R), 356 (R). — Rioja, 1949: 315-321 (R, in part), 327, 328 (R), fig. 1. — Tressler, 1947: 705 (C). — Tressler, 1954: 138 (D). — Hobbs, 1956 (1957): 431 (D, R). — Tressler, 1959: 731 (K), fig. 28.191. — Crawford, 1959: 178 (R). — Hart, 1962: 121, 122 (C).

Entocythere (Cytherites) heterodonta. — Rioja, 1942b: 688 (K), 694-696 (R), fig. 21. — Hoff, 1943: 280 (R). — Rioja, 1943a: 554, 562, 564, 566 (R), 566 (H, D). —Rioja, 1943b: 567-584 (M, R), 10 figs.

Entocythere talirotundi. — Rioja, 1949: 316-320 (M, H, R, D), figs. 2-8. (Type-locality, Lago Lerma and Lagunas de Zempoala, Mexico, Types no longer extant.). — Hobbs, 1956 (1957): 431, 435, 436 (R). — Hart, 1962: 122 (C).

Entocythere talirotunda. — Crawford, 1959: 173 (R).

Ankylocythere heterodonta. — Hart, 1962: 127 (C, D, S). — Hart, 1964: 246 (R). — Hobbs, 1966: 67, 69 (S), 67 (M), 69, 71 (K, D, H), figs. 2, 3 — Hobbs, Holt, and Walton, 1967: 6 (C), 14 (B, H). — McGregor and Kesling, 1969: 227, 228 (B, C). — Hart and Hart, 1969b: 160, 164 (M). — Hobbs and Walton, 1970: 852 (M). — Hobbs, 1971: 32-34 (M, H, R, A, D), fig. 19.

Ankylocythere talirotunda. — Hart, 1962: 127-128 (C, D, S). — Hobbs and Walton, 1963a: 459 (R). — Hart, 1964: 246 (R). — Hobbs, 1966: 67, 69 (S), 69 (K, D, H). — Ferguson, 1968: 501 (C, D).

Diagnosis. — MALE: Shell length 340-380 μ (average, 360 μ); height, 200-210 μ (average, 210 μ). Clasping apparatus of male with well developed, broadly rounded talon or with talon reduced to small rounded excrescence, if strongly developed never projecting distally so much as distance equivalent to its own width; internal border with single tooth situated above or proximal to base of talon and sometimes irregularly serrate in vicinity of junction of horizontal and vertical rami; vertical ramus seldom longer than horizontal ramus, apex of horizontal ramus with two upturned denticles. Ventral portion of peniferum entire and tapering.

FEMALE: Shell similar to that of male, but larger; length, 390-430 μ (average, 420 μ); height, 220-270 μ (average, 240 μ).

Type-locality. — Tlalpan, Mexico, D.F., Mexico. (Restricted by Hart, 1962: 127) (types, not extant). Host, *Cambarellus montezumae.*

Distribution (in addition to type-locality). —

Mexico

DISTRITO FEDERAL — Chapultepec (Rioja, 1941b: 197) (host, *Cambarellus montezumae*); Xochimilco (Rioja, 1940a: 608) (host, *C. montezumae*).

MEXICO — Laguna de Lerma (Rioja, 1940b: 608) (host, *Cambarellus montezumae*); Laguna Zumpango (host, *C. montezumae*); Río San Joaquin, km 125 carretera Mexico a Toluca (host, *C. montezumae*).

HIDALGO — Ixmiquilpan (host, *C. montezumae*); Pachuca, estanque en el km 135, de la carretera Mexico a Tuxpan, (host, *C. montezumae*); San Miguel Regla (Rioja, 1949: 316) (host, *C. montezumae*).

JALISCO — Lagunas de Magdalena, near Magdalena (host, *C. chapalanus*); Tizapan (host, *C. chapalanus*); 4 mi N of Jocotepec (host, *C. montezumae*); 13 mi N of Chapala (host, *C. chapalanus*).

MICHOACAN — Jacona (host, *Procambarus digueti*); Lago de Patzcuaro (Rioja, 1940b: 608) (host, *C. patzcuarensis*); Zamora (host, *C. chapalanus*).

MORELOS — Lagunas de Zempoala (Rioja, 1941b: 197; not "Cempoala" as indicated by Rioja) (host, *C. zempoalensis*).

NAYARIT — Acaponeta (host, *Cambarellus* sp.).

PUEBLA — Cholula (host, *C. montezumae*); Necaxa (Rioja, 1943a: 566) (host, *Procambarus hoffmanni* (= *Paracambarus* sp.).

Ankylocythere hobbsi (Hoff, 1944)

Plate III, figures 1-3
Plate XLV

Entocythere hobbsi Hoff, 1944b: 330 (R), 332 (K), 352-356 (M, H, R, D), figs. 26-33. — Rioja, 1949: 328 (R). — Tressler, 1947: 705 (C). — Tressler, 1959: 732 (K), fig. 28.194. — Crawford, 1959: 150 (B), 151 (K), 173 (R), 178 (A), 180-181 (A). — Hart, 1959: 193 (D), 194 (K), 202, 203 (D, H), figs. 17-19. — Walton and Hobbs, 1959: 115 (K).

Ankylocythere hobbsi. — Hart, 1962: 127 (C, D, S). — Crawford, 1965: 149 (R). — Hobbs, 1966: 70 (K, D, H), 72 (R), fig. 11.

Diagnosis. — MALE: 340-400 μ in length (average, 370 μ); 210-250 μ in height (average, 230 μ). Distal portion of peniferum weakly bifid. Clasping apparatus extending distally beyond peniferum; bearing well-developed talon on external border of horizontal ramus, extending distally subparallel to horizontal ramus with one well-developed tooth near mid-length; distal extremity of horizontal ramus with two or three denticles; juncture between horizontal and vertical rami indistinct.

FEMALE: Shell similar to, but larger than, that of male; 400-430 μ in length (average, 410 μ), 250-260 μ in height (average, 250 μ).

Type-locality. — Burrows near Fitzgerald, Ben Hill Co., Georgia, U.S.A. (types, USNM: holotype, allotype, and paratype, 81562). Host, *Procambarus advena.*

Distribution (in addition to type-locality). —

United States

FLORIDA — ALACHUA CO. — burrows along Hogtown Creek, 2 mi from Gainesville city limits on 9th Street (host, *Procambarus advena*); 3 mi N and ½ mi W of Gainesville on rt 441 (host, *P. advena*); 4 mi N of Gainesville (host, *P. advena*). BAY CO. — 3 mi W of Youngstown in wiregrass flatweeds (host, *P. rogersi* intergrades); 6.3 mi W of Beacon Hill on rt 98 (host, *P. rogersi* intergrades). DUVAL CO. — burrows, 1 mi N of jct of Edgewood Ave. and Lake City Rd on Waycross Rd (hosts, *P. advena* and *P. paeninsulanus*). FLAGLER CO. — 10.8 mi ? of San Mateo, 3.9 mi N of rt 28 (host, *P. paeninsulanus*). FRANKLIN CO. — 1.9 mi S of Sumatra on rt 12 (host, *P. rogersi* intergrades); 150 ft from Bay in ditch, near jct of rts 12 and 319 (hosts, *P. rogersi* intergrades and *P. kilbyi*). GADSDEN CO. — Hart 1959 ref. (hosts, *P. leonensis, P. paeninsulanus,* and *P. spiculifer*). ORANGE CO. — Hoff 1944b ref. (host, *P. alleni*). WASHINGTON CO. — small stream flowing from swampy pond near Ebro (host, *P. pycnogonopodus*).

GEORGIA — COOK CO. — 5.6 mi N of Hahira on rt 41 (host, *P. advena*). ECHOLS CO. — 3.6 mi NE of state line on rt 89 (host, *P. advena*).

LAURENS CO. — 3.7 mi W of Glenwood on rt 280 (hosts, *P. howellae, P. advena,* and *Faxonella clypeata*). McINTOSH CO. — 16.9 mi N of Darien on rt 17 (hosts, *P. advena* and *P. barbatus*). THOMAS CO. — 9.5 mi SE of Meigs on rt 19 (hosts, *P. advena* and *P. paeninsulanus*).

NORTH CAROLINA — ALAMANCE CO. — Haw River at Altamahaw (host, *P. blandingii*).

SOUTH CAROLINA — HAMPTON CO. — Hoff 1944b ref. (host, *P. lunzi*). RICHLAND CO. — ditch on rt 48, 9.1 mi SE of Univ. of South Carolina stadium (hosts, *Fallicambarus uhleri* and *Procambarus troglodytes*); Colonel's Ck on rt 262, 14.6 mi E of Columbia (hosts, *P. enoplosternum* and *P. acuminatus*).

Ankylocythere hyba Hobbs and Walton, 1963

Plate III, figures 4-7
Plate XLV

Ankylocythere hyba Hobbs and Walton, 1963: 457-459 (M, H, R, A, D), figs. 4-6. — Hobbs, 1966: 69 (K, D, H), fig. 4. — Ferguson, 1968: 501 (C, D). — Hobbs III, 1970: 180 (A).

Diagnosis. — MALE: Length 310-330 μ (average, 320 μ); height, average, 170 μ. Copulatory complex with peniferum comparatively slender in lateral aspects, ventral margin sloping anteriorly and forming with anterior margin a subangular tip. Penis situated far distal to base of clasping apparatus and entirely within distal third of area between dorsal margin of spermatic loop and ventral extremity of peniferum. Dorsal finger with basal portion comparatively broad and short; ventral finger slender with angular bend at end of proximal fourth; principal axes of fingers subparallel. Clasping apparatus L-shaped and clearly divisible into vertical and horizontal rami; vertical ramus slightly longer than horizontal one and with entire margins. Internal border of horizontal ramus with single tooth near midlength; external border with two excrescences, larger proximal one, corresponding to the talon of related species, situated slightly distal to level of tooth on internal border, and smaller prominence immediately distal to larger one. Distal extremity of ramus with two teeth.

FEMALE: Not definitely correlated with male.

Relationships. — In the reduction of the talon, *Ankylocythere hyba* approaches *A. heterodonta, A. talulus, A. hamata,* and *A. telmoecea*. *A. telmoecea* possesses only a single prominence on the external border of the horizontal ramus of the clasping apparatus, whereas the other four species possess at least two. In *A. hamata* there are three prominences. In *A. talulus,* the two prominences are of about equal size, while in *A. talirotunda* and *A. hyba,* the proximal one is always larger. In the former, however, the proximal prominence lies at a level opposite the single tooth on the internal border and in *A. hyba,* it is situated at a level distal to this tooth. Furthermore, in *A. hyba* the two rami of the clasping apparatus are joined at an approximate right angle, whereas in *A. talirotunda,* the angle is always distinctly acute. The peniferum of *A. hyba* resembles that of *A. talirotunda* more closely than it does those of the other species.

Type-locality. — Hurricane and Tumbling creeks, SE of Waverly, Humphreys Co., Tennessee, U.S.A. (types, USNM: holotype, 108016; paratype, 108017). Hosts, *Orconectes placidus, O. compressus, O. r.* subsp., and *Cambarus striatus* Hay.

Distribution (in addition to type-locality). —

United States

ALABAMA — WALKER CO. — 15.7 mi S of Double Springs on rt 195 (hosts, *Cambarus (Depressicambarus)* sp. and *Orconectes* sp.).

FLORIDA — CALHOUN CO. — 6 mi N of Blountstown and 27.3 mi SE of Marianna (host, *Procambarus r. rogersi*).

TENNESSEE — HICKMAN CO. —locality unknown (hosts, *Orconectes compressus, O. rusticus* subsp., and *O. placidus*). HUMPHREYS CO. — Tumbling Ck E of Bold Spring (hosts, *O. compressus, O. placidus, O. rusticus,* and *Procambarus* sp.); Hurricane Ck 10.2 mi E of Waverly city limits on rt 70 (host, *O. compressus*). LAWRENCE CO. — Peter Cave Ck (host, *O. compressus*). LINCOLN CO. — Swan Ck, about 11 mi W of Fayetteville on rt 64 (hosts, *O. erichsonianus* and *O. mirus*). MARION CO. — creek on Marion-Sequatchie Co. line on rt 28 (host, *O. erichsonianus*). SEQUATCHIE CO. — S of city limits of Dunlap on rt 28 (host, *O. erichsonianus*).

Ankylocythere maya Hobbs, 1971

Plate III, figures 8-10
Plate XLVI

Ankylocythere maya Hobbs, 1971: 34-35 (M, H, R, A, D), figs. 20, 21.

Diagnosis. — MALE: Shell length 270-320 μ (average, 290 μ); height, 150-220 μ (average, 170 μ). Eye completely bleached in all available specimens. Shell ovate with greatest height at midlength; ventral margin gently curved and lacking concavity. Submarginal setae present except dorsally; in addition, irregular row of setae anterodorsally somewhat removed from shell margin and others scattered over shell surface. Copulatory complex with peniferum curved anteroventrally, subtruncate distally with small, subacute projection at anteroventral angle. Penis much longer than half anteroposterior diameter of peniferum at level of penis and curved with proximal and distal ends disposed at angle of approximately 130 degrees. Dorsal and ventral fingers only moderately heavy and both terminating in simple setiform tips; ventral finger with one conspicuous bend slightly distal to end of proximal fourth and gently curved distally. Clasping apparatus with small, low, almost completely adnate talon on external border of horizontal ramus, internal border with single tooth above or proximal to base of talon and with irregular serrations at and distal to junction of horizontal and vertical rami; apex with two upturned denticles. Ventral portion of peniferum subtruncate with anteroventral angle bearing small subacute projection directed anteriorly or ventrally.

FEMALE: Shell similar to that of male, but slightly larger: length 290-320 μ (average, 310 μ); height, 190-220 μ (average, 210 μ).

Type-locality. — Eight km SSW of Simojovel, Chiapas, Mexico (types, USNM: holotype and allotype, 128822). Host, *Procambarus* (*Austrocambarus*) sp.

Distribution (*in addition to type-locality*). —

Mexico

CHIAPAS — Hacienda San José del Arco, Comitán (host, *Procambarus* (*A.*) sp.); Pojsoch, Región de los lagos de Monte Bello (host, *Procambarus* (*A.*) sp.); Finca La Esperanza, 46 km NE of Tuxtla, Gutiérrez (host, *Procambarus* (*A.*) sp.).

Ankylocythere sinuosa (Rioja, 1942)

Plate III, figures 11-13
Plate XLVI

Entocythere cambaria. — Hobbs, 1941: 4 (B, H) (in part).

Entocythere (*Cytherites*) *heterodonta sinuosa* Rioja, 1942a: 203-204 (M, H, R, D), figs. 5, 6. — Rioja, 1953: 287, 291 (C). — Hart, 1962.

Entocythere (*Cytherites*) *sinuosa.* — Rioja, 1942b: 688 (K), 689 (R), 695-696 (M, H, R, A, D), fig. 20. — Rioja, 1943a: 564 (R). — Rioja, 1943b: 576 (R).

Entocythere sinuosa. — Hoff, 1944b: 330 (R, S), 332 (K), 356 (R). — Rioja, 1949: 321 (R, H), 322 (R, H, in part), figs. 13, 14. — Rioja, 1951: 170 (B). — Rioja, 1953: 291-292 (B, C, D). — Tressler, 1954: 138 (D). — Hobbs, 1956 (1957): 431 (D, R). — Tressler, 1959: 731 (K), fig. 28.190. — Crawford, 1959: 178 (R).

Ankylocythere sinuosa. — Hart, 1962: 127 (C, D, S). — Hart, 1964: 246 (R). — Crawford, 1965: 149 (R). — Reddell, 1965: 156 (D). — Hobbs, 1966: 70 (K, D, H), fig. 18. — Ferguson, 1968: 501 (C, D). — Hobbs and Walton, 1968: 246 (A, D). — Baker, 1969: 293 (H, B). — Young, 1971: 399-409 (B). — Hobbs, 1971: 34-35 (M, H, R, A, D), fig. 22. — Reddell and Mitchell, 1971: 142 (C).

Diagnosis. — MALE: Shell length 340-370 μ (average, 350 μ); height, 119-220 μ (average, 220 μ). Clasping apparatus with well developed, comparatively narrow talon projecting posteroventrally; internal border with single tooth situated distal to base of talon, otherwise entire; apex with two upturned teeth. Ventral portion of peniferum deeply cleft but without ventrally directed prominences.

FEMALE: Shell similar to that of male, but slightly larger; length, 350-430 μ (average, 390 μ); height, 230-270 μ (average, 240 μ).

Remarks. — Not until some twenty-four years after the description of *Ankylocythere sinuosa* from La Cueva Chica was an extension of its range recorded. Hobbs (1966: 70) reported it from several of the states cited below and from all of the hosts except *P. gonopodocristatus.* Except for a few additional remarks concerning its morphology, and relationships to other entocytherids, no other additions to our knowledge of the species were made until Baker (1969: 293) observed living individuals which were maintained for a period of some ten weeks away from their host, *Procambarus simulans simulans.* Young (1971) carried out further studies on population densities, reproduction, and development in a Texas farm pond (see p. 2).

Type-locality. — La Cueva Chica, 3 km SE of Ciudad Valles and 609 m SE of pueblo de Pujal, San Luis Potosi, Mexico (types, not extant). Host, *Procambarus acutus cuevachicae.*

Distribution (*in addition to type-locality*). —

United States

ALABAMA — BARBOUR CO. — 2 mi W of Eufaula on rt 82 (hosts, *Procambarus verrucosus* and *Cambarus latimanus*). BULLOCK CO. — 5.5 mi SE of Midway on rt 82 (host, *P. verrucosus*). BUTLER CO. — 2.6 mi W of Greenville on rt 10 (hosts,

P. a. acutus and *C. diogenes* subsp.). CONECUH CO. — 3.9 mi NE of Castlebury on rt 31 (hosts, *P. a. acutus* and *P. versutus*). LOWNDES CO. — burrows in clay banks of ck on SE side of Hayneville (hosts, *P. lophotus* and *C. diogenes* subsp.); 3.9 mi SW of jct of rts 80 & 11 on 11, 3.4 mi SE of Hayneville (hosts, *P. lewisi, P. lophotus* and *C.* (*Depressicambarus*) sp.); 32.4 mi S of Montgomery on rt 31 (hosts, *P. acutissimus, P. lewisi, C. latimanus*, and *C. diogenes* subsp.). MACON CO. — 18.7 mi E of Montgomery on rt 80 (hosts, *P. acutissimus* and *Faxonella clypeata*). SUMPTER CO. — ditch 1.7 mi N of Boligee on rt 11 (host, *P. hybus*); locality unknown (host, *P. acutissimus*). TUSKALOOSA CO. — 9 mi S of Tuskaloosa on rt 11 (hosts, *P. viaeviridis, P. a. acutus, Fallicambarus fodiens,* and *C. diogenes* subsp.). WINSTON CO. — 20.9 mi W of Cullman on US 278 (hosts, *C. obstipus* and *Orconectes* sp.).

ARKANSAS — NEWTON CO. — head of ravine 5.6 mi E of Fallsville on rt 16 [host, *Orconectes nais* (?)]. POLK CO. — Iron's Fork Ck near Mena (host, *O. palmeri longimanus*). PULASKI CO. — Ivesville, on rt 10 W of Little Rock (host, *O. meeki*). SCOTT CO. — 9 mi N of Waldron at jct of rts 23 & 71 (host, *O. p. longimanus*); 6.0 mi E of Cauthron on rt 28 (host, *O. p. longimanus*); 23 mi NE of Mena on US 71, Fourche la Fave River (host, *O. p. longimanus*).

CALIFORNIA — LAKE CO. — Clear Lake near Pirate's Cove (host, *Pacifastacus l. leniusculus*).

FLORIDA — BAY CO. — swamp adjoining Econfina River W of Bennet (host, *Procambarus rogersi* intergrades); 3 mi W of Youngstown in wiregrass flatweeds (host, *P. rogersi* intergrades). ESCAMBIA CO. — Drainage ditch at Bluffs Springs [host, *P. capillatus* (?)]. SANTA ROSA CO. — swampy area along Escambia River on rt 62 (hosts, *Cambarus diogenes* subsp., *P. a. acutus, P. bivitatus, P. shermani, P. versutus,* and *Fallicambarus* sp.).

GEORGIA — BLECKLEY CO. — 7.9 mi S of Twiggs-Bleckley Co. line on rt 129 (hosts, *Procambarus howellae* and *Faxonella clypeata*). CRISP CO. — 0.9 mi N of Crisp-Turner Co. line on US 41 in Flint River drainage (hosts, *P. seminolae* and *F. clypeata*). TWIGGS CO. — 0.5 mi E of Twiggs Co. line on rt 96 (hosts, *F. clypeata* and *P. a. acutus*).

HAWAII — OAHU — drainage ditches in sugar cane fields W of Honolulu (host, *Procambarus clarkii*).

INDIANA — CLAY CO. — Birch Creek at jct of rt 59, 0.4 mi N of Ashboro (host, *Orconectes virilis*).

KENTUCKY — HICKMAN CO. — Obion Creek in area of Murphy's Pond (host, *Procambarus acutus acutus*).

LOUISIANA — ACADIA PARISH — locality unknown (host, *Procambarus clarkii*). ALLEN CO. — locality unknown (hosts, *P. a. acutus, P. dupratzi, P. s. simulans, Cambarellus puer,* and *Orconectes difficilis*); Coney Branch (hosts, *P. a. acutus* and *P. dupratzi*). BEAUREGARD PARISH — locality unknown (host, *P. dupratzi*); Little Sugar Ck, rt 113 (hosts, *O. difficilis* and *P. dupratzi*); Hoosier Ck on rt 389 (host, *P. dupratzi*). CATAHOULA PARISH — Greeks Ck (host, *P. natchitochae*). GRANT PARISH — Sugarhouse Bayou (hosts, *P. clarkii* and *P.* sp.); Gray Mill Branch (host, *P. vioscai*); James Ck (hosts, *C. shufeldtii, O. lancifer,* and *P. vioscai*); Cypress Ck (host, *P. vioscai*). LASALLE PARISH — Little Cow Ck (hosts, *O. p. longimanus* and *P. vioscai*). LINCOLN PARISH — lawn at Louisiana Tech. in Ruston (host, *Fallicambarus dissitus*); 2 mi N of Ruston (host, *P. vioscai*); Madden Creek (host, *P. vioscai*); South Choudrant Ck (hosts, *O. p. longimanus, P. natchitochae, P. vioscai*). LIVINGSTON PARISH — roadside ditch 2.9 mi W of Albany on rt 190 (hosts, *C. shufeldtii, F. clypeata, P. a. acutus, P. clarkii, P. planirostris*). OUICHITA PARISH — 1.7 mi W of Calhoun on rt 80 (host, *P. vioscai*); Swartz-Fairbanks Highway (connection between rts 534 & 134) N of Monroe (host, *Faxonella clypeata*). SABINE PARISH — locality unknown (host, *P. dupratzi*); Little Middle Ck (hosts, *O. difficilis* and *P. natchitochae*); Bayou Tora (hosts, *O. p. longimanus, P. dupratzi,* and *P. tulanei?*); Dry Ck (hosts, *P. a. acutus* and *P. dupratzi*); Little Bayou Sara (hosts, *P. a. acutus* and *P.* sp.); Arthur Ck (hosts, *P. a. acutus* and *P. dupratzi*); Unknown Ck (host, *P. dupratzi*); Beech Bayou, rt 1215 (hosts, *P. a. acutus, P. clarkii,* and *P. dupratzi*); Bayou Scie, rt 120 (hosts, *P. dupratzi, P. s. simulans, P. vioscai* and *P.* sp.); Hapson Bayou (hosts, *O. p. longimanus, P. a. acutus,* and *P. dupratzi*); Bayou Dupont (host, *P. natchitochae*). ST. TAMMANY PARISH — 2 mi W of Covington on rt 190 (host, *P. a. acutus*). VERNON PARISH — Billy Goat Bayou (hosts, *C. shufeldti, O. lancifer, P. clarkii, P. natchitochae,* and *P. vioscai*).

MISSISSIPPI — AMITE CO. — 3 mi SW of Smithdale on Old Jackson Trail (host, *P. a. acutus*). BOLIVAR CO. — 3 mi E of Shaw (host, *O. lanci-*

fer). JONES CO. — locality unknown (hosts, *Cambarus* sp. and *Procambarus* (*Pennides*) sp.). LEAKE CO. — 11.9 mi S of Carthage on rt 35 (host, *P. a. acutus*). LOWNDES CO. — 2.3 mi W of Columbus on rt 82 (hosts, *Fallicambarus fodiens, P. a. acutus, P. hybus,* and *P. viaeviridis*); 11.3 mi W of Columbus on rt 82 (hosts, *P. a. acutus* and *P. hybus*). NEWTON CO. — 10.7 mi N of Chunky of rt 80 (host, *P. a. acutus*). OKTIBBEHA CO. — 6.4 mi SW of Mississippi State Univ. campus on rt 12 (host, *P. acutissumus*). RANKIN CO. — 0.5 mi from Rankin-Scott Co. line on rt 80 (host, *P. a. acutus*). SCOTT CO. — 8.6 mi S of Leake-Scott Co. line on rt 35 (host, *P. a. acutus*); 13 mi S of Leake-Scott Co. line on rt 35 (host, *P. a. acutus*).

OKLAHOMA — COMANCHE CO. — Temporary pond on artillary range at Fort Sill (host, *Procambarus simulans simulans*).

TEXAS — BLANCO CO. — locality unknown (host, *Procambarus clarkii*). LIBERTY CO. — 7 mi W of Dayton on rt 90 (hosts, *Camberellus puer, Procambarus acutus acutus, P. hinei*). MENARD CO. — Powell's Cave, 200 feet downstream from well pipe (host, *P. clarkii*); Neel Cave (host, *P. clarkii*). WHEELER CO. — 8 mi E of McLean on US 66 (host, *P. s. simulans*). WILLIAMSON CO. — locality unknown (host, *P. clarkii*).

Mexico

CHIHUAHUA — Laguna Toronto, 27.4 km SE of Cuidad Camargo (host, *Procambarus clarkii*).

HIDALGO — km 267 carretera de México a Tuxpán (Rioja, 1949: 322) (host, none cited).

NUEVO LEON — Arroyo de la Cruz, km 296 carretera Monterrey a Mexico (host, *P. simulans regiomontanus*); Rio San Juan at San Juan (host, *P. s. regiomontanus*).

TAMAULIPAS — Acequia, km 34 carretera a Cuidad Victoria (host, *P. simulans regiomontanus*).

VERACRUZ — Maria de la Torre, km 407 carretera de México a Nautla (host, *P. gonopodocristatus*); Paso Largo, 16 km NE Martinez de la Torre (host, *P. gonopodocristatus*).

Ankylocythere tallapoosa Hart and Hart, 1971

Plate III, figures 14-18
Plate XLV

Ankylocythere tallapoosa Hart and Hart, 1971: 107-108 (M, H, R, A, D), figs. 2a-e.

Diagnosis. — MALE: Shell length 430-450 μ (average, 390 μ) height, 220-270 μ (average, 240 μ). Compulatory complex with peniferum extending to level of external border of horizontal ramus of clasping apparatus. Peniferum of uniform width, with slight bend $\frac{1}{5}$ peniferum length from end, at level of penis. Distal end of peniferum appearing bifid, with two points joined by membrane. Cephalodistal margin folded mesially.

Clasping apparatus with horizontal and vertical rami of subequal length and disposed at right angles to each other. Internal border of horizontal ramus with one tooth approximately $\frac{2}{3}$ ramus length from distal end, second tooth near two terminal denticles. External border of horizontal ramus with short talon, its length less than width of horizontal ramus. Talon occasionally reduced to slight protuberance posterior to semicircular concavity.

FEMALE: Shell length 490-530 μ (average, 510 μ); height, 280-320 μ (average, 310 μ). Shells otherwise similar to those of males.

Relationships. — *Ankylocythere tallapoosa* is unique in having the minor tooth on the horizontal ramus of the clasping apparatus grouped distally with the apical denticles. On the only other *Ankylocythere* with two teeth, *A. bidentata* (Rioja), the minor tooth lies subjacent to the major one. The peniferum resembles those of several other members of the genus in being pseudobifid — that is, terminating in two distinct points which are joined by a barely discernible membrane.

Type-locality. — Route 27 at Tallapoosa River, Haralson Co., Georgia, U.S.A. (types, USNM: holotype and allotype, 138445; paratype, 138446). Hosts, *Procambarus spiculifer* and *Cambarus halli.*

Distribution (in addition to type-locality). —

United States

ALABAMA — LEE CO. — trib crossing US 85, 2½ mi ENE of Auburn exit (host, *Procambarus verrucosus*).

GEORGIA — CARROLL CO. — 9.0 mi S of Bremen on rt 27A, trib to Little Tallapoosa R (host, *Cambarus latimanus, C. halli*). HARALSON CO. — 1.3 mi S of Tallapoosa R in ck parallel to rt 27 (host, *C. halli*). PAULDING CO. — rt 101 at Tallapoosa R (hosts, *C. latimanus, C. halli*).

Ankylocythere talulus (Hoff, 1944)

Plate III, figures 19-20
Plate XLV

Entocythere talulus Hoff, 1944b: 332 (K), 349, 352 (M, H, R, D), figs. 22-25. — Tressler, 1947: 705 (C). — Rioja, 1949: 320 (R), 328 (R). — Hart, 1959: 194 (K), 203 (D, H), fig. 13. — Walton and Hobbs, 1959: 115 (K). — Crawford, 1959: 173 (R). — Tressler, 1959: 731 (K), fig. 28.189.

Ankylocythere talulus. — Hart, 1962: 128 (C, D, S). — Hobbs and Walton, 1963a: 459 (R). — Hobbs, 1966: 69 (K, D, H), fig. 5.

Diagnosis. — MALE: Shell length 330-340 μ (average, 330 μ); height, 190-210 μ (average 200 μ). Peniferum with distinct anteroventrally directed terminal protuberance. Clasping apparatus with horizontal and vertical rami subequal in length; internal border of horizontal ramus bearing single tooth at approximately mid-length; external border with reduced talon anterior to position of tooth on internal border; horizontal ramus terminating in two denticles.

Type-locality. — Under boards in marsh at NW 20th St. and 27th Ave., Miami, Dade Co., Florida, U.S.A. (types, USNM: holotype, allotype, and paratype, 81561). Host, *Procambarus alleni.*

Distribution. — Known only from type-locality.

Ankylocythere telmoecea (Crawford, 1959)

Plate IV, figures 1-5
Plate XLVI

Entocythere telmoecea Crawford, 1959: 167-173, 178, 183 (M, H, R, D), figs. 24-30, 150 (B), 151 (H, K), 180-181 (A). — Crawford, 1961: 237. — Hart, 1962: 123, 128 (C).

Ankylocythere telmoecea. — Hart, 1962: 128 (C, D, S). — Hobbs and Walton, 1963a: 459 (R). — Hobbs, 1966: 68 (K, D, H), fig. 9. — Hobbs, Holt, and Walton, 1967: 77. — Hart and Hart, 1971: 106-111. — Ferguson, 1968: 501 (C, D).

Diagnosis. — MALE: Shell subreniform in lateral aspect with dorsal margin slightly arched; ventral margin with very slight concavity just anterior to mid-length; shell highest at mid-length; both anterior and posterior margins without angular prominences and setae sparsely scattered on lateral surfaces. Conspicuous submarginal row of setae along anterodorsal margin with few irregularly scattered setae on ventral and posterior margins. Prominent eye located about one-sixth of shell length from anterior end. Shell length 350-380 μ (average 360 μ); height, 180-210 μ (average, 190 μ).

Peniferum terminating bluntly, with slight apical bifurcation. Clasping apparatus with horizontal and vertical rami subequal in length; horizontal ramus with single vestigial talon on external border; single tooth at mid-length of internal border of horizontal ramus; crenulations at internal angle of rami may be mistaken for additional teeth; terminal portion of horizontal ramus with two denticles.

FEMALE: Shell similar to that of male, but slightly larger; length, 390-410 μ (average, 400 μ); height, 210-240 μ (average, 230 μ). The female (triunguis) genitalia are in the form of a short, thick, heavily sclerotized, hollow tube (*pieza quitinosa femenina,* Rioja 1943b: 575) which is directed ventrally. This tubular process has been observed in a few copulating biunguis females; however, it is generally absent in the biunguis forms.

Type-locality. — Gill's Creek on South Carolina rt 48, 2.7 mi SSE of Univ. of South Carolina stadium, Richland Co., South Carolina, U.S.A. (types, USNM: holotype, 103570; allotype, 103571; morphotype, 103572; paratype, 103573). Host, *Procambarus troglodytes.*

Distribution (in addition to type-locality). —

United States

FLORIDA — FLAGLER CO. — 10.8 mi E of San Mateo, 3.9 mi N of rt 28 (host, *Procambarus paeninsulanus*).

GEORGIA — BURKE CO. — Rocky Creek 5 mi W of Waynesboro (host, *Cambarus pubescens*). CHATHAM CO. — Savannah (host, *Procambarus troglodytes*). HEARD CO. — 5.3 mi W of Newman on rt 34 (hosts, *C. latimanus* and *P. spiculifer*). TWIGGS CO. — 4.7 mi E of Twiggs-Wilkinson Co. line on rt 96 (hosts, *C. diogenes* subsp., *C. latimanus, Faxonella clypeata,* and *P. enoplosternum*); 8.1 mi E of Irwinton on rt 57 (hosts, *C. diogenes* subsp., *F. clypeata, P. enoplosternum, P. howellae, P. spiculifer*).

NORTH CAROLINA — RANDOLPH CO. — 2.8 mi NE of Randolph Co. line on rt 49 (hosts, *Cambarus acuminatus* and *Procambarus acutus acutus*).

SOUTH CAROLINA — BAMBERG CO. — 5.4 mi W of Olar (host, *Procambarus troglodytes*). COLLETON CO. — 6.6 mi W of Walterboro on rt 64 (host, *P. troglodytes*). LANCASTER CO. — spring draining into ditch along Potter Road 2 mi N of Pleasant Hill (hosts, *Cambarus* sp. and *P.* (*Ortmannicus*) sp.); ck on W. C. Marshall property, 2 mi N of Pleasant Hill on Potter Rd (hosts, *C.* sp. and

P. (*Ortmannicus*) sp.). ORANGEBURG CO. — Edisto River, at police camp 4 mi S of Orangeburg on rt 49 (host, *P. troglodytes*).

Ankylocythere tiphophila (Crawford, 1959)

Plate IV, figures 6-8
Plate XLV

Entocythere tiphophila Crawford, 1959: 173-178 (M, H, R, D), 150 (B), 151 (H, K), 180-181 (A), figs. 31-37.— Hart, 1962: 123, 128 (C).

Ankylocythere tiphophila.— Hart, 1962: 128 (C, D, S). — Hart, 1964: 245 (R).— Crawford, 1965: 149 (R).— Hobbs, 1966: 71 (K, D, H), fig. 16.— Ferguson, 1968: 501 (C, D).

Ankylocythere tiphophyla.— Hobbs, Holt, and Walton, 1967: 77 (erroneous spelling).

Diagnosis. — MALE: Shell subovate with posterior end slightly more angular than anterior and with maximum height just caudal to mid-length: dorsal margin arched with ventral one slightly convex; both anterior and posterior ends evenly rounded. Setae unevenly scattered along free margin of shell and sparsely dispersed on lateral surface. Eye located about one-fifth shell length from anterior end. Shell length 330-350 μ (average, 350 μ); height, 180-210 μ (average, 200 μ).

Distal portion of peniferum entire and terminating in acute projection. Clasping apparatus divisible into vertical and horizontal rami; former only slightly arched, particularly on flexor margin; vertical ramus at an angle of slightly less than 90 to horizontal ramus; latter slightly more than half as long as former, bearing prominent tooth on flexor surface just proximal to mid-length; conspicuous talon borne on extensor surface opposite tooth on flexor surface; cancavity present on extensor surface of horizontal ramus distal to talon and tooth tapering to distal end; three denticles on distal end directed dorsally.

FEMALE: Shell similar in shape to that of male; but slightly larger; length, 360-390 μ (average, 370 μ); height, 230-240μ (average, 240 μ). Shell of triunguis female subovate with dorsal margin more strongly arched than that of male; ventral margin bearing concavity anterior to mid-length; postero-dorsal margin gently sloping with posteroventral edge evenly rounded; anterior margin smoothly rounded; both free margin of shell and lateral surface sparsely and irregularly setiferous. Eye located about one-fourth of valve length from anterior end.

Type-locality. — Ditch on rt 48, 9.1 mi SE of Univ. of South Carolina Stadium, Richland Co., South Carolina, U.S.A. (type, USNM: holotype, 103574; allotype, 103575). Hosts, *Fallicambarus uhleri* and *Procambarus troglodytes*.

Distribution (in addition to type-locality). —

United States

FLORIDA — SANTA ROSA CO. — flooded McCaskill's Mill Creek (hosts, *Fallicambarus* sp., *Procambarus acutus acutus, P. shermani*).

GEORGIA — MONTGOMERY CO. — 8.3 mi S of Truetler Co. line on US 221 (host, *Faxonella clypeata*). TELFAIR CO. — N city limits of Lumber City on US 341 (hosts, *F. clypeata* and *P. howellae*). WILKINSON CO.— 8.1 mi E of Irwinton on Ga 57 (hosts, *Cambarus diogenes* subsp., *F. clypeata, P. enoplosternum, P. howellae,* and *P. spiculifer*). WAKE CO. — 1 mi S of Durham Co. line on rt 55 (host, *C. d. diogenes*).

OHIO — MIAMI CO. — Sec. 35 Union Township (host, *Orconectes rusticus*). STARK CO. — Beech Creek, 3.8 mi W of Alliance on US 62 (host, *O. obscurus*).

Ankylocythere toltecae Hobbs, 1971

Plate IV, figures 9-12
Plate XLVI

Ankylocythere toltecae Hobbs, 1971: 36-38 (M, H, R, A, D), fig. 23.

Diagnosis. — MALE: Shell length 200-310 μ (average, 300 μ); height, 170-190 μ (average, 180 μ). Clasping apparatus without distinct talon but with two obliquely disposed emarginations delimiting low round lobe on horizontal ramus; internal border with single tooth situated above or slightly proximal to lobe on external border; apex with two upturned denticles. Ventral portion of peniferum shallowly excavate with anteroventral subacute projection.

TRIUNGUIS FEMALE. — Shell length 290-350 μ (average, 300 μ); height, 180-220 μ (average, 200 μ). Eye, if present, situated approximately one-fifth of shell length from anterior end. Shell ovate with greatest height posterior to midlength and with ventral margin shallowly concave slightly anterior to midlength. Submarginal setae as in male, setae on remaining surface of valve less numerous.

Genital complex consisting of simple, short, slightly sclerotized cylindroid projection.

Type-locality. — Cueva del Agua, 75 km W of Ciudad Valles, San Luis Potosi, México (types, USNM: holotype, 128823; allotype, 128824). Host, *Procambarus villalobosi.*

Distribution (*in addition to type-locality*). —

Mexico

HIDALGO — Cueva de El Tenango, about 6.4 km SW of Municipio de Chapulhuacan (host, *Procambarus toltecae*).

SAN LUIS POTOSÍ — Sotano de Huitzmolotitla, about 8.3 km N of Xilitla (host, *P. toltecae*).

TAMAULIPAS — Le Cueva de San Nicolas, about 2 km SE of El Panchon (host, *Procambarus toltecae*).

Ankylocythere tridentata Hart, 1964

Plate IV, figures 13-15
Plate XLV

Ankylocythere tridentata Hart, 1964: 245-246 (M, R, A, D), figs. 3, 4. — Hobbs, 1966: 69 (K, D, H), fig. 17. — Ferguson, 1968: 501 (C, D).

Diagnosis. — MALE: Shell subelliptical in outline with greatest shell height at about midlength. Marginal setae apparently absent. Eye spot reduced or absent. Two known specimens measure 360 μ in length and 200 μ in height (Maryland) and 360 μ in length and 240 μ in height (Indiana). Copulatory complex with clasping apparatus extending beyond peniferum. Clasping apparatus with distinct horizontal and vertical rami, subequal in length. Internal border of horizontal ramus provided with one tooth; external border provided with three teeth, most distal of which not so well developed as other two; distal extremity with three denticles. Peniferum relatively simple; distal extremity bifurcate with points apposed to one another. Penis small, situated in distal portion of peniferum.

FEMALE: Not known.

Relationships: Within the genus *Ankylocythere,* this species appears to have its closest affinities with *A. heterodonta* (Rioja, 1940b) and *A. sinuosa* Rioja, 1942a). These species are known only from Mexico. *A. heterodonta* shows evidence of possible relationships with *A. tridentata* based on the number of teeth on the external border of the horizontal ramus of the clasping apparatus; *A. sinuosa* shows evidence of possible relationships in that it possesses a bifurcate peniferum similar to that found in *A. tridentata.* The bifurcate peniferum, on the other hand, also suggests possible relationships with certain members of the genus *Uncinocythere.*

Type-locality. — Plummers Island, Montgomery Co., Maryland U.S.A. (type, USNM: holotype, 110942). Host, *Cambarus d. diogenes* Girard.

Distribution (*in addition to type-locality*).—

United States

INDIANA — MARTIN CO. — 4.9 mi N of Loogootee on US 231 (host, *Orconectes immunis*).

Ankylocythere villalobosi Hobbs, 1971

Plate IV, figures 16-21
Plate XLVI

Ankylocythere villalobosi Hobbs, 1971: 38-39 (M, H, R, A, D), figs. 24-25.

Diagnosis. — MALE: Shell length 320-350 μ (average, 340 μ); height, 190-210 μ (average, 190 μ). Eye present, situated between a fifth and a quarter of shell length from anterior margin. Shell subovate with greatest height at approximate midlength; ventral margin almost straight or slightly convex. Submarginal setae present except dorsally, other setae widely scattered over surface of shell. Clasping apparatus with talon well developed, more than twice as long as least width; internal border with single tooth situated above talon and with or without irregular serrations proximal to tooth and sometimes extending onto vertical ramus; apex with two upturned teeth. Ventral portion of peniferum tapering or curved, if evenly curved, anteroventral extremity with slender, short projection.

Type-locality. — Palizada, Campeche, Mexico (types, USNM: holotype and paratype, 128825; allotype and paratype, 128826). Host, *Procambarus llamasi.*

Distribution (*in addition to type-locality*). —

Mexico

CAMPECHE — Santa Rita (host, *P. llamasi*); Estacion de Haro, Escarcega (host, *P. llamasi*).

CHIAPAS — Villa Margaritas (host, *Procambarus pilosimanus*); Gruta de Zapaluta, 30 km S of Comitan (host, *P. pilosimanus*).

ASCETOCYTHERE HART, 1962

Asceta Group Hobbs and Walton, 1962: 42 (R) (in part). *Ascetocythere* Hart, 1962: 125 (K), 128 (M, S, C), type-species, *Entocythere asceta* Hobbs and Walton, 1962.—Hart, 1963: 51 (R, M, D).—Hobbs III, 1965: 160 (K).—Hobbs and Hart, 1966: 36 (K), 37-38 (M, R, D).—Hobbs, Holt, and Walton, 1967: 8 (H), 35-36 (M), 68 (R, E).—Hart and Hart, 1969a: 188-189 (D, E, R).—Hart and Hart, 1969b: 158 (M, R).—Hobbs III, 1969a: 170 (R).—Hobbs and Hobbs, 1970: 2 (K), 3 (M).

Diagnosis.—Terminal tooth of mandible pectinate. Male copulatory complex without finger guard; peniferum extending ventrally much beyond clasping apparatus, elongate, comparatively slender, with subterminal bulbous enlargement bearing one to several projections; penis consisting of separate dorsal spermatic and ventral prostatic elements basally and directed anteriorly or anteroventrally; both elements always shorter than clasping apparatus; accessory groove lacking. Clasping apparatus well developed and may or may not be clearly divisible into vertical and horizontal rami; external border of horizontal ramus entire, internal border with two, three, or no teeth along apical half, if present, often grouped far distally with three apic denticles. Triunguis female lacking pectinate prc cess on distal podomere of second antenna; genit complex consisting of genital papilla but lacking J-shaped rod and amiculum. (Slightly modified from Hobbs and Hart, 1966: 37-38.)

Type species.—*Entocythere asceta* Hobbs and Walton, 1962: 43; by designation, Hart, 1962: 128.

Distribution.—United States: Mountainous areas from northeastern Tennessee to southern Pennsylvania. Also Prince Georges Co., Maryland. Hosts.—Freshwater crayfishes of the family Astacidae.

Figure 21.—Distribution of the genus *Ascetocythere*.

Key to the Species of the Genus Ascetocythere

(Modified from Hobbs and Hobbs, 1970, from Hobbs and Hart, 1966)

1	Internal border of distal half of clasping apparatus with one or more teeth situated proximal to distal denticles, or internal border unarmed (Asceta Group)	2
1′	Internal border of clasping apparatus with teeth and distal denticles grouped distally (Coryphodes Group)	10
2(1)	Bulbous portion of peniferum with well-developed cephalic process	3
2′	Bulbous portion of peniferum lacking cephalic process	8
3(2)	Ventral portion of peniferum with angular flange	4
3′	Ventral portion of peniferum lacking angular flange	5
4(3)	Ventral extremity of peniferum with finger-like projection extending ventrally beyond angular flange	*A. asceta*
4′	Ventral extremity of peniferum lacking finger-like projection extending ventrally beyond angular flange	*A. ozalea*
5(3′)	Ventral portion of peniferum with posterior process	6
5′	Ventral portion of peniferum lacking posterior process	7

6(5) Anterior process of peniferum much shorter than digitiform process and situated on level above base of penis .... *A. sclera*
6′ Anterior process of peniferum as long as digitiform process and situated on level below base of penis .... *A. holti*
7(5′) External border of clasping apparatus with proximal bend at end of proximal fifth of apparatus; penis reaching posterior surface of digitiform process .... *A. lita*
7′ External border of clasping apparatus with proximal bend at end of proximal third of apparatus; penis extending along anterior side of digitiform process .... *A. batchi*
8(2′) Ventral portion of peniferum lacking angular flange .... *A. didactylata*
8′ Ventral portion of peniferum with angular flange 9
9(8′) Ventral portion of peniferum with distinct curved digitiform process extending posteroventrally from flange .... *A. hoffmani*
9′ Ventral portion of peniferum lacking distinct digitiform process .... *A. hyperoche*
10(1′) Anterior surface of subterminal or terminal enlargement of peniferum with anteroventrally projecting, club-like prominence and two additional prominences .... *A. cosmeta*
10′ Anterior surface of terminal enlargement of peniferum with single prominent snout-like projection 11
11(10′) Snout-like prominence less than half as long as longitudinal diameter of peniferum at level of prominence .... *A. myxoides*
11′ Snout-like prominence more than two-thirds as long as longitudinal diameter of peniferum at level of prominence .... *A. coryphodes*

Asceta Group Hobbs and Hart, 1966

Asceta Group Hobbs and Hart, 1966: 39 (D, M, R). — Hobbs and Walton, 1968: 238 (R). — Hobbs and Walton, 1970: 857 (R). — Hobbs and Hobbs, 1970: 3 (K).

Diagnosis. — Major angle of clasping apparatus situated proximal to midlength: internal border with 0 to 3 teeth, if none, low elevation aways present slightly distal to midlength and another immediately proximal to three terminal denticles; distal portion of peniferum armed with finger-like projections or angular prominences.

Remarks. — Nine species are assignable to this group: *A. asceta, A. batchi, A. didactylata, A. hoffmani, A. holti, A. hyperoche, A. lita, A. ozalea,* and *A. sclera.* Their combined range is located in the central and southwestern part of the range of the genus as a whole, southern West Virginia and southern Virginia to northeastern Tennessee.

Ascetocythere asceta (Hobbs and Walton, 1962)

Plate V, figures 1-5
Plate XLVII
Page 239

Entocythere asceta Hobbs and Walton, 1962: 42-45 (M, H, R, D, K), figs. 1-4. — Hart, 1962: 123, 128 (C). — Hobbs and Hart, 1966: 35.

Ascetocythere asceta. — Hart, 1962: 128 (C, D, S), fig. 2. — Hobbs III, 1965: 159 (M, R), 162 (R). — Hobbs and Hart, 1966: 38 (K), 39 (M, H, R, D), 40-42, 53, figs. 2-7. — Hobbs, Holt, and Walton, 1967: 7 (B), 8 (H), 12 (B, H), 33 (F), 34 (K), 36 (M, D, H, A, S), 68-69 (D, E), 23, 41, 42, 51, 55, 58, 61, 64 (A, H), 72-75, figs. 7.c, 8.a. — Ferguson, 1968: 501 (C, D). — Hobbs and Walton, 1968: 238 (R). — Hart and Hart, 1969a: 188 (H). — Hobbs and Hobbs, 1970: 3 (K), 4 (K). — Walton and Hobbs, 1971: 87-88, 90-93, 95-96, 102 (B).

Diagnosis. — Peniferum with four prominences extending from subterminal bulbous expansion — anterior process delicate, club-shaped, sometimes difficult to see, extending anteroventrally from cephalic surface; ventral process large, finger-like, with distinct angular flange near mid-length, and directed ventrally; posterior process small, rounded, and situated at posterior base of ventral process; prominent penis guide (acute projection) lying between flange and anterior process; penis reaching surface anterodorsal to flange. Two rami of clasping apparatus not sharply defined; vertical ramus less than $\frac{1}{3}$ total length of apparatus; internal border of horizontal ramus with two teeth along distal third, distal tooth approximately mid-way between proximal tooth and three apical denticles; external border entire.

Relationships. — This ostracod has its closest affinities with *A. ozalea, A. hyperoche,* and *A. hoffmani.* This relationship is best seen in the angular flange on the ventral process of the peniferum. *A. asceta* may be distinguished from them by the elongate, ventrally directed, finger-like extension of the ventral process.

Type-locality. — Crayfish burrows on golf course of Mountain Lake Hotel, off Route 700, about 9 mi NE of Pembroke, Giles Co., Virginia, U.S.A. (types, USNM: holotype, 108174; allotype, 108175; paratypes, 108176 and 108177). Host, *Cambarus dubius* (=*C. carolinus* Hobbs and Walton, 1962, not Erichson).

Distribution (in addition to type-locality). —

United States

VIRGINIA — CRAIG CO. — Burrows along Middle Ck, 1.2 mi N of Maggie on rt 636 (host, *Cambarus dubius*). GILES CO. — Mountain Lake Biological Station grounds (host, *C. dubius*).

WEST VIRGINIA — MONROE CO. — burrows along trib of Rays Branch, 2 mi SE of Waiteville (host, *Cambarus dubius*).

Ascetocythere batchi Hobbs and Walton, 1968

Plate V, figures 6-9
Plate XLVII

Ascetocythere batchi Hobbs and Walton, 1968: 237, 238 (M, H, R, A, D), figs. 1a-e. — Hobbs and Walton, 1970: 857 (R). — Hobbs and Hobbs, 1970: 3 (K), 4 (R).

Diagnosis. — MALE: Shell length 400-410 μ (average, 410 μ); height, 220-230 μ (average, 230 μ). Submarginal setae widely separated but somewhat evenly distributed, very few dorsally.

Peniferum with three prominences extending from subterminal bulbous expansion — anterior process delicate, subspatulate, and extending anteroventrally for distance equivalent to length of anteroposterior plane of peniferal bulb at distal base of process; ventral process heavier and slightly longer than anterior one, finger-like, and disposed ventrally; third process forming penis guide, with almost closed groove to receive two elements of penis, situated subparallel to ventral process along its anteromesial surface and terminating about one-third length of ventral process from its ventral extremity. Clasping apparatus not clearly divisible into vertical and horizontal rami; internal border somewhat gently curved between broad base and proximal tooth located at base of distal third and with smaller tooth slightly proximal-to-midway between proximal tooth and three apical teeth; external border of proximal half curved so that extensions of its extremities separated by angle of approximately 120 degrees; distal half with proximal subangular bend of approximately 130 degrees; distal third of apparatus much slenderer than proximal two-thirds.

TRIUNGUIS FEMALE. — Eyes unpigmented, or perhaps, absent. Shell similar in shape to that of male but with posterior end more produced. Length of single female 0.41 mm; height 0.22 mm.

Genital area with anteroventrally produced cuticular plate bearing subspherical amorphous mass apically.

Relationships. — *Ascetocythere batchi* is clearly a member of the Asceta Group and has its closest affinities with *A. sclera;* it differs from *sclera* in lacking a posterior process on the peniferum and possessing a much longer anterior process. The clasping apparatus more nearly resembles those of *A. hoffmani, A. hyperoche,* and *A. asceta,* but each of these species has a flange on the ventral process of the peniferum that is lacking in *A. batchi.*

Type-locality. — Tributary of Cumberland River, Pine Mountain State Park, Bell Co., Kentucky, U.S.A. (types, USNM: holotype and allotype, 123321). Host, *Cambarus parvoculus.*

Distribution. — Known only from type-locality.

Ascetocythere didactylata Hobbs and Hart, 1966

Plate VI, figures 4-6
Plate XLVII

Ascetocythere didactylata Hobbs and Hart, 1966: 38 (K), 43-44 (M, H, R, D), figs. 17, 18. — Ferguson, 1968: 501 (C, D). — Hobbs and Walton, 1970: 857 (R). — Hobbs and Hobbs, 1970: 3 (K).

Diagnosis. — MALE: Eye without pigment. Shell subovate with ventral margin weakly concave anterior to midlength; greatest height ⅔ length from anterior margin. Length, 430 μ; height, 240 μ. Submarginal setae present except dorsally, somewhat more widely spaced ventrally than anteriorly and posteriorly.

Peniferum with only two prominent extensions from subterminal bulbous expansion-anterior process absent; ventral process digiform and directed ventrally with distal portions of penes approaching surface on its posteroproximal surface; posterior process thumb-like, approximately ½ as long as anterior process. Vertical ramus of clasping apparatus less than ⅓ total length of apparatus; internal border of horizontal ramus with three teeth along distal third, teeth progressively smaller distally; external border entire; apex with three denticles as large as distal two teeth on internal border.

FEMALE. — Unknown.

Relationships. — *Ascetocythere didactylata* has its closest affinities with *A. sclera*, from which it differs chiefly in the peniferum lacking an anterior process and a penial sheath.

Type-locality. — Tributary of Powell River, 5.1 mi N of Tazewell on Route 25E, Claiborne Co., Tennessee, U.S.A. (types, USNM: holotype, 113448). Hosts, *Cambarus parvoculus, C. bartonii cavatus,* and *C.* sp.

Distribution. — Known only from type-locality.

Ascetocythere hoffmani Hobbs and Hart, 1966

Plate VI, figures 7-10
Plate XLVII

Ascetocythere hoffmani Hobbs and Hart, 1966: 38 (K), 40 (M, H, R, D), figs. 8, 9. — Ferguson, 1968: 501 (C, D). — Hobbs and Walton, 1968: 238 (R). — Hobbs and Hobbs, 1970: 3 (K).

Diagnosis. — MALE: Shell length 400-430 μ (average, 420 μ); height, 250-260 μ (average, 255 μ). Eye present. Shell subovate with ventral margin exhibiting only slightest concavity short distance anterior to midlength. Submarginal setae present except dorsally, setae relatively close together and regularly spaced anteriorly and posteriorly.

Peniferum with three prominences, extending from subterminal bulbous expansion — anterior process absent; ventral process large with prominent angular flange anteroventrally, finger-like portion directed posteroventrally and somewhat laterally; posterior process emarginate with caudal extremity of emargination angular and cephalic extremity rounded; prominent penis guide forming leaf-like acute prominence under cover of which penes reach surface anterodorsal to flange of ventral process. Vertical ramus of clasping apparatus slightly more than ⅓ total length of apparatus; internal border of horizontal ramus without teeth but sometimes with two low elevations, one at base of distal third of clasping apparatus and distal one just proximal to three apical denticles; external border entire.

FEMALE. — Unknown.

Relationships. — *Ascetocythere hoffmani* has as its closest relatives *A. asceta, A. hyperoche,* and *A. ozalea,* sharing with them an angular flange on the ventral process of the peniferum. It differs from the three in possessing a slender, posteroventrally directed finger-like extension of the ventral process.

Type-locality. — Crayfish burrows at Burkes Garden, Tazewell Co., Virginia, U.S.A. (types, USNM: holotype, 113441). Host, *Cambarus dubius.*

Distribution. — Known only from type-locality.

Ascetocythere holti Hobbs and Walton, 1970

Plate VII, figures 1-4
Plate XLVII

Ascetocythere holti Hobbs and Walton, 1970: 853-857 (M, H, R, A, D), figs. 3a, b. — Hobbs and Hobbs, 1970: 3 (K).

Diagnosis. — MALE: Shell length 370-410 μ (average, 390 μ); height, 210-240 μ (average, 220 μ). Shell ovate in silhouette but slightly concave anteroventrally, greatest height slightly posterior to midlength. Submarginal setae, anteriorly, posteriorly, and ventrally; those situated anteriorly progressively farther from margin dorsally; setae apparently absent dorsally. Eye pigmented.

Copulatory complex with peniferum bearing three prominences extending from subterminal expansion: anterior process flattened, its length about half that of anteroposterior plane of distal portion of peniferum, and directed anteroventrally with distal portion deflected more ventrally; ventral process slightly heavier, subequal in length to anterior process, directed ventrally, and bearing fold along proximoposterior margin serving as penis guide; posterior process, situated immediately posterior to ventral process, slightly undulating, acute, about one-half as long as latter, and also directed ventrally. Penis complex, long and extending ventrally along penis guides on posterior surface of ventral process. Clasping apparatus not clearly divisible into vertical and horizontal rami; internal border gently curved between broad base and tapering distal third (distal elevation acute, almost toothlike in some specimens); distal extremity with three anterodorsally directed denticles; external border also rounded with broadly oblique subangular bend; extensions of principal proximal and distal axes forming angle of approximately 105 degrees; height of distal extremity of tapering apparatus only approximately one-fourth anteroposterior diameter of base of apparatus. Dorsal finger comparatively stout and terminating in bifid seta extending posteroventrally; ventral finger moderately heavy, disposed somewhat subparallel to anterior margin of ventral portion of peniferum, with one subangular bend, and directed posteroventrally.

TRIUNGUIS FEMALE: Shell length 380-400 μ (average, 390 μ); height, 240-250 μ (average, 240 μ). Shell similar in shape to that of male but distinctly higher in posterior third; submarginal setae disposed as in male. Eye pigmented.

Genital complex consisting of sclerotized papilla surrounded by amorphous hyaline material (presumably the spermatophore) with dangling, irregularly shaped, delicate saclike membrane.

Relationships: *Ascetocythere holti* is a member of the Asceta Group of the genus and seems to have its closest affinities with *A. sclera*. It shares with all of the species of the group a clasping apparatus in which the major bend occurs proximal to its midlength, and with *A. sclera, A. didactylata,* and *A. batchi* the absence of a flangelike process projecting from the ventral surface of the peniferum. It differs from *A. didactylata* in possessing three processes on the ventral portion of the peniferum, from *A. sclera* in having a much longer anterior process, an acute, undulating posterior process, and a subangular bend on the posteroventral margin of the peniferum, and from *A. batchi* in possessing a posterior process.

Type-locality. — Stream 8.3 mi W of jct of roads 2451 and 3387, on latter road SE of Oneida, Scott Co., Tennessee, U.S.A. (types, USNM: holotype and allotype, 126974). Hosts, *Cambarus sphenoides,* an unidentified crayfish related to *C. distans,* and *C. bouchardi.*

Distribution. — Known only from the type-locality.

Ascetocythere hyperoche, Hobbs and Hart, 1966

Plate VII, figures 5-6
Plate XLVII

Ascetocythere hyperoche Hobbs and Hart, 1966: 38 (K), 39-40 (R), 41-42 (M, H, R, D), figs. 12, 13. — Ferguson, 1968: 501 (C, D). — Hobbs and Walton, 1968: 238 (R). — Hobbs and Hobbs, 1970: 3 (K).

Diagnosis. — MALE: Shell length 420 μ; height, 250 μ. Shell ovate highest at midlength and with but slightest concavity on ventral margin. Submarginal setae evenly spaced around shell except dorsally where absent. Eye with pigment.

Peniferum with only two prominences extending from subterminal bulbous expansion — no anterior process on cephalic surface; ventral process rudimentary, rounded, situated at posterior base of angular prominence; prominent penis guide (acute projection) lying at anterior base of ventral process; penis reaching anterior surface at base of ventral process. Vertical ramus of clasping apparatus less than ⅓ total length of apparatus; internal border of horizontal ramus with two teeth along distal third, distal tooth only slightly proximal to apical denticles.

Type-locality. — Tributary of Cherokee Reservoir, 0.5 mi W of jct of Routes 11W and 25E, Granger Co., Tennessee, U.S.A. (types, USNM: holotype, 113444). Hosts, *Orconectes juvenilis, Cambarus longirostris,* and *C. bartonii cavatus.*

Distribution. — Known only from type-locality.

Ascetocythere lita Hobbs and Hobbs, 1970

Plate VII, figures 7-11
Plate XLVII

Ascetocythere lita Hobbs and Hobbs, 1970: 3 (K), 4, 5 (M, H, R, A, D), figs. 1a-d.

Diagnosis. — MALE: Shell length, 390-420 μ (average 410 μ); height, 230-260 μ (average, 240 μ). Shell subovate with posterior region much higher than anterior, greatest height slightly posterior to midlength; ventral margin weakly concave anterior to midlength. Submarginal setae, although sparse, present anteriorly, posteriorly, and ventrally; absent dorsally. Eye pigmented.

Copulatory complex with peniferum bearing two prominences extending from subterminal expansion: anterior process straight and rodlike, its length subequal to that of anteroposterior plane of distal portion of peniferum, and directed anteriorly from upper surface of bulbous expansion; ventral process heavier than anterior process, broad at base and tapering anteroventrally, bearing longitudinal penial groove. Penis long, otherwise unremarkable. Clasping apparatus divisible into vertical and horizontal rami, former with subangular bend proximally; internal border of horizontal ramus with three teeth, proximal one most prominent; external border entire, curving anterodorsally from junction of rami to level of proximal tooth on internal border, from there extending anteriorly to base of three apical denticles. Dorsal finger comparatively stout and terminating in simple seta directed ventrally; ventral finger slender, curved anteriorly parallel to anterior process of peniferum.

TRIUNGUIS FEMALE. — Shell length 420 μ; height; 240 μ. Shell of approximately same size and otherwise similar to that of male. Eye pigmented. Genital complex situated posterodorsally and consisting of sclerotized cone-shaped prominence tapering ventrally with orifice at ventral extremity; paired fingerlike projections lateral to cone directed ventrally; small sclerotized rod surrounded by angular mass located anterodorsally to cone.

Relationships — *Ascetocythere lita* has its closest affinities with the members of the Asceta Group and is more closely allied to *A. batchi* than to any other member of the genus. It may be distinguished from all other species of the group by the major bend of the clasping apparatus occurring distal to midlength, one of the characters utilized by Hobbs and Hart (1966: 44) in defining the Coryphodes Group. In light of the fact that *A. lita* obviously is more closely related to *A. asceta* and its allies than to *A. coryphodes* and its relatives, this character must be deleted from the definition of the two groups. Like the members of the Asceta Group, the internal border of the clasping apparatus of *A. lita* has three teeth not grouped with the terminal denticles, and the distal portion of the peniferum is armed with projections.

Type-locality. — Creek and burrows in seepage area, 4.8 mi SW of jct of rts 25W and 75, on rt 25W, SW of Corbin, Whitley Co., Kentucky, U.S.A. (types, USNM: holotype, 126251; allotype, 126252). Hosts, *Cambarus sphenoides* and *C.* sp.

Distribution. — Known only from type-locality.

Ascetocythere ozalea Hobbs and Hart, 1966

Plate VIII, figures 4-7
Plate XLVII

Ascetocythere ozalea Hobbs and Hart, 1966: 38 (K), 40-41 (M, H, R, D), figs. 10, 11. — Ferguson, 1968: 501 (C, D). — Hobbs and Hobbs, 1970: 3 (K).

Diagnosis. — MALE: Shell length, 390-430 μ (average, 420 μ); height, 230-250 μ (average, 240 μ). Shell subovate, highest ⅓ length of shell from posterior end, with ventral margin exhibiting no concavity. Submarginal setae present except dorsally, nowhere abundant Eye unpigmented.

Peniferum with three prominences extending from subterminal bulbous expansion — anterior process elongate, somewhat delicate, and extending anteroventrally from cephalic surface; ventral process, corresponding largely to flange in *A. asceta,* large, angular, directed ventrally and slightly anteriorly; posterior process small, acute, and situated at posterior base of angular projection; prominent penis guide (acute projection) lying between ventral and anterior processes; penis reaching surface anterodorsal to ventral process. Vertical ramus of clasping apparatus less that ⅓ total length of apparatus; internal border of horizontal ramus with two teeth along distal third, distal tooth only slightly proximal to apical denticles.

FEMALE. — Unknown.

Relationships. — *Ascetocythere ozalea* is most closely allied to *A. asceta, A. hoffmani,* and *A. hyperoche*; all of them share an angular flange as a part of the ventral process of the peniferum. It differs from *asceta* and *hoffmani* in lacking the distal finger-like projection of the ventral process and from *hyperoche* in lacking the anterior process and having a more angular ventral process.

Type-locality. — Harden in Weber City, Scott Co., Virginia, U.S.A. (types, USNM: holotype, 113442; paratype, 113443). Host, *Cambarus dubius.*

Distribution (in addition to type-locality). —

United States

VIRGINIA — RUSSELL CO. — Clinch Mountain near Lebanon (host, *Cambarus dubius*).

Ascetocythere sclera Hobbs and Hart, 1966

Plate VIII, figures 8-13
Plate XLVII

Ascetocythere sclera Hobbs and Hart, 1966: 38 (K), 42-43 (M, H, R, D), figs. 15, 16. — Ferguson, 1968: 501 (C, D). — Hobbs and Walton, 1968: 238 (R). — Hart and Hart, 1969a: 189 (R, E). — Hobbs and Walton, 1970: 857 (R). — Hobbs and Hobbs, 1970: 3 (K).

Diagnosis. — MALE: Shell length 390-450 μ (average, 420 μ); height, 230-280 μ (average, 250 μ). Shell subovate, highest short distance posterior to midlength. Submarginal setae somewhat closer together anteriorly and posteriorly than ventrally; absent dorsally. Eye without pigment.

Peniferum with three prominences extending from subterminal bulbous expansion — anterior process small, thumb-like, extending anteroventrally from cephalic surface; ventral process long, slender, finger-like, extending ventrally; posterior process small, thumb-like, extending ventrally immediately posterior to base of ventral process; penes reaching surface through digitiform penial process (sheath) situated anterolateral to ventral process and only slightly shorter than latter; penis reaching anterior surface at base of ventral process. Vertical ramus of clasping apparatus less than ⅓ total length of apparatus; internal border of horizontal ramus with two or three teeth along distal third, distal tooth only slightly proximal to apical denticles.

TRIUNGUIS FEMALE: Shell length 420-460 μ (average, 440 μ); height, 240-280 μ (average, 260 μ). Shell with posterior portion produced into distinct bulge and highest posterior to midlength. Submarginal setae present except dorsally, closer together anteriorly and posteriorly than ventrally. Eye without pigment.

Genital area with ventrally-directed, dorsal finger surrounded, except for dorsal attachment to integument, by amorphous mass with adhering bits of debris. Posterior to this structure a distinct fold dorsally disrupting even contour of posterodorsal margin of one or both valves.

Relationships. — *Ascetocythere sclera* is most closely allied to *A. didactylata,* having the finger-like process extending ventrally from the peniferum; also only these two members of the genus have three teeth (only occasionally in *A. sclera*) along the internal border of the horizontal ramus of the clasping apparatus. *A. sclera* may be distinguished from *A. didactylata* in possessing a penial sheath that projects ventrally beyond the subterminal bulbous portion of the peniferum.

Variations. — There is remarkable uniformity in the specimens from the several localities. The only conspicuous variation occurs in the number of teeth along the internal border of the horizontal ramus of the clasping apparatus. There are two or three teeth along the distal ⅓ of the internal border. No variation observed seems to be regionally restricted.

Type-locality. — Crayfish burrows in Montcoal, Raleigh Co., West Virginia, U.S.A. (types, USNM: holotype, 113445, allotype, 113446; paratype, 113447). Host, *Cambarus dubius.*

Distribution (*in addition to type-locality*). —

United States

VIRGINIA — BUCHANAN CO. — Vansant (host, *Cambarus dubius*). DICKENSON CO. — Haysi (host, *C. robustus*). RUSSELL CO. — Skeggs, dug from field along Lewis Ck, 2.1 mi SW of Co. line on rt 27 (host, *C. dubius*). TAZEWELL CO. — Shrader (host, *C. dubius*).

WEST VIRGINIA — McDOWELL CO. — 8.8 mi S of Baileysville, Barrenshe Ck, trib to Tug R (host, *Orconectes juvenilis*). WYOMING CO. — Rockhouse River near Baileyville (host, *C. dubius*).

Coryphodes Group Hobbs and Hart, 1966

Coryphodes Group Hobbs and Hart, 1966: 44 (D, M, R). — Hobbs and Hobbs, 1970: 3 (K).

Diagnosis. — Angle (bend) of clasping apparatus situated distal to midlength or gently curved throughout; internal border with one or two teeth situated near distal end of clasping apparatus, grouped with terminal denticles; distal portion of peniferum with prominent projection on cephalic surface (*A. coryphodes* and *A. myxoides*) or with clublike and acute projections (*A. cosmeta*).

Remarks. — Only the three species listed immediately above are assignable to the Coryphodes Group; the ranges of the former two mark the northeastern extremity of the range of the genus, and the single known locality for the latter occurs on the western slopes of the Blue Ridge on the southeastern border of the range of the genus.

Ascetocythere coryphodes Hobbs and Hart, 1966

Plate V, figures 10-12
Plate XLVII

Ascetocythere coryphodes Hobbs and Hart, 1966: 38 (K), 44-45 (M, H, R, D), figs. 19, 20. — Ferguson, 1968: 501 (C, D). — Hobbs and Hobbs, 1970: 3 (K), 4 (R).

Diagnosis. — MALE: Shell length 390-430 μ (average, 400 μ); height, 230-240 μ (average, <240 μ). Shell subovate with ventral margin bearing distinct concavity anterior to midlength; greatest height slightly posterior to midlength. Submarginal setae present except dorsally and subequally spaced. Eye pigmented.

Peniferum with two prominences extending from terminal bulbous expansion — anterior process much elongated, almost as long as anterior-posterior diameter of peniferum at level of process; ventral process absent, posterior process with two or three small anteriorly-projecting, acute processes only one of which reaching beyond distal extremity of bulbous expansion; penis guide consisting of two spinous elements directed somewhat posteriorly and situated laterally rather than projecting anteriorly or anteroventrally; penis not reaching anterior surface. Vertical ramus of clasping apparatus at least twice as long as horizontal ramus, although two rami not clearly delimited; internal border of horizontal ramus with two teeth situated very close to three terminal denticles; vertical diameter of distal portion of horizontal ramus slightly increased in area of teeth.

FEMALE. — Unknown.

Relationships. — *Ascetocythere coryphodes* has its closest relationships with *A. myxoides* and *A. cosmeta.* The similarity exists in the clasping apparatus, in which all of the prominences are grouped distally on the comparatively short horizontal ramus. *A. coryphodes* differs from *A. myxoides* in possessing a much longer and slenderer anterior projection from the peniferum, and from *A. cosmeta* in lacking conspicuous ventral projections from the same structure.

Type-locality. — Brooks Run, 17.2 mi W of Elkins on rt 33, Randolph Co., West Virginia, U.S.A. (types, USNM: holotype, 113449; paratype, 113450). Host, *Cambarus b. bartonii.*

Distribution (in addition to type-locality). —

United States

PENNSYLVANIA — ALLEGHENY CO. — Nine Mile Run, Pittsburgh (host, *Cambarus monongalensis*).

Ascetocythere cosmeta Hobbs and Hart, 1966

Plate VI, figures 1-3
Plate XLVII

Ascetocythere cosmeta Hobbs and Hart, 1966: 38 (K), 46-47 (M, H, R, D), figs. 24, 25. — Ferguson, 1968: 501 (C, D). — Hobbs and Hobbs, 1970: 3, 4 (K).

Diagnosis. — MALE: Shell length, 430 μ; height, 250 μ. Shell ovate with ventral margin very shallowly excavate anterior to midlength; greatest height slightly posterior to midlength. Submarginal setae present except dorsally, closer together anteriorly and posteriorly than ventrally. Eye pigmented.

Peniferum with three prominences extending from subterminal bulbous expansion — anterior process club-shaped in outline and directed anteroventrally; ventral process rather heavy but terminating in acute recurved spine; posterior process broad and not sharply delimited from bulbous portion of peniferum but anteroventral angle acute; penis guide not evident; penis not reaching anterior surface, but, emerging ventrally with ventral and posterior processes. Vertical ramus only slightly longer than horizontal ramus although two rami not clearly delimited; internal border of horizontal ramus with two teeth situated immediately proximal to three terminal denticles; vertical diameter of distal portion of horizontal ramus not appreciably increased.

FEMALE. — Unknown.

Relationships. — This ostracod has its closest affinities with the two members of the Coryphodes Group, *A. coryphodes* and *A. myxoides,* but may be distinguished from either by the conspicuous ventrally directed processes from the peniferum. These processes also point to affinities with the members of the Asceta Group, and as indicated above, *A. cosmeta* combines part of the diagnostic characters of both groups.

Type-locality. — Seepage area along stream, 4 mi N of jct of rt 89 and the Blue Ridge Parkway, Grayson Co., Virginia, U.S.A. (type, USNM: holotype, 113453). Host, *Cambarus dubius.*

Distribution. — Known only from type-locality.

Ascetocythere myxoides Hobbs and Hart, 1966

Plate VIII, figures 1-3
Plate XLVII

Ascetocythere myxoides Hobbs and Hart, 1966: 38 (K), 45-46 (M, H, R, D), figs. 21-23. — Ferguson, 1968: 501 (C, D). — Hobbs and Hobbs, 1970: 3, 4 (K).

Diagnosis. — MALE: Shell length 380-400 μ (average, 390 μ); height, 220-230 μ (average, 220 μ). Shell subovate with ventral margin bearing slight concavity anterior to midlength; greatest height short distance posterior to midlength. Submarginal setae surrounding shell, widely spaced dorsally. Eye pigmented.

Peniferum with only one conspicuous prominence extending from terminal bulbous expansion — anterior process resembling that of inverted human nose, about $\frac{1}{2}$ as long as anterior-posterior diameter of peniferum at level of process; ventral process absent and posterior process projecting only slightly beyond ventral margin of peniferum; penis guide conspicuous; penis not reaching anterior surface. Vertical ramus of clasping apparatus approximately 1.5 times as long as horizontal ramus with single tooth situated very close to three terminal denticles, vertical diameter of distal portion of horizontal ramus slightly increased at level of tooth.

FEMALE. — Unknown.

Relationships. — *Ascetocythere myxoides* has its closest affinity with *A. coryphodes* but may be distinguished from it by the shorter anterior process of the distal portion of the peniferum and by the clasping apparatus which has four distal prominences instead of five.

Type-locality. — Cheat Mountain, Monongahela National Forest, Randolph Co., West Virginia, U.S.A. (types, USNM: holotype, 113451; paratype, 113452). Host, *Cambarus monongalensis.*

Distribution (in addition to type-locality). —

United States

MARYLAND — PRINCE GEORGES CO. — Paint Branch (host, *Cambarus dubius*).

CYMOCYTHERE HART, 1962

Asceta Group Hobbs and Walton, 1962: 42 (in part).
Cymocythere Hart, 1962: 125 (K), 128-129 (in part) (M, S, C), type-species, *Entocythere cyma* Hobbs and Walton, 1960. — Hart, 1963: 51 (R, D). — Hobbs III, 1965: 160 (K), 163 (R). — Hobbs and Hart, 1966: 49-50 (D, M). — Hobbs and Walton, 1968: 247 (K). — Hart and Hart, 1969a: 187 (D, E, R). — Hart and Hart, 1969b: 158 (M, R). — Hobbs and Hobbs, 1970: 2 (K).

Diagnosis. — Terminal tooth of mandible with cusps. Copulatory complex of male with well developed bifid finger guard; peniferum extending ventrally much beyond clasping apparatus, elongate, entirely corneous with subterminal bulbous enlargement containing penis; penis less than ⅓ length of clasping apparatus and with contiguous spermatic and prostatic elements; accessory groove lacking; clasping apparatus well developed, external border entire, internal border with or without prominences, apex with three to five teeth or scallops.

Type-species. — *Entocythere cyma* Hobbs and Walton, 1960: 18 (by original designation, Hart, 1962: 129).

Distribution. — United States: Mountainous areas from central Alabama to southwestern Virginia — including Georgia, South Carolina, North Carolina, and Tennessee.

Host. — Freshwater crayfishes of the family Astacidae.

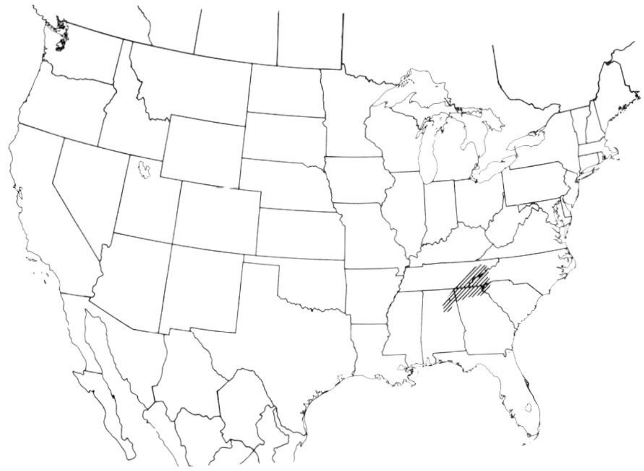

Figure 22. — Distribution of the genus *Cymocythere.*

Key to the Species of the Genus Cymocythere

1	Clasping apparatus with teeth on internal border of horizontal ramus	2
1′	Clasping apparatus without teeth on internal border of horizontal ramus	*C. gonia*
2(1)	Vertical ramus longer than horizontal ramus	*C. clavata*
2′	Vertical ramus shorter than horizontal ramus	*C. cyma*

Cymocythere clavata Crawford, 1965

Plate IV, figures 1-6
Plate XLVII

Cymocythere clavata Crawford, 1965: 149-151 (M, R, A, D), figs. 4, 5, 8, 9. — Hobbs and Hart, 1966: 50-51 (M, H, R, D), figs. 34, 35. — Ferguson, 1968: 501 (C, D).

Diagnosis. — MALE: Shell length, 360-390 μ; height, 210-270 μ. Ventral extremity of peniferum truncate, scarcely tapering from subterminal bulbous area containing penis. Clasping apparatus not clearly divisible into vertical and horizontal rami, proximal and distal extremities directed approximately at right angle to each other; internal border of horizontal portion with three prominences; distal extremity with three scallops.

TRIUNGUIS FEMALE: Shell averaging 380 μ in length, 350 μ in height.

Relationships. — *Cymocythere clavata,* while distinctly related to the other two members of the genus, is not so closely related to them as they are to each other. Both the structure of the clasping apparatus and the disposition of the penis are unique in the genus.

Type-locality. — East fork of Chattooga River at jct with Indian Camp Ck on grounds of U.S. Fisheries Station, 21 mi NW of Walhalla, Oconee Co., South Carolina, U.S.A. (types, USNM: holotype, 113479; allotype, 113480). Hosts, *Cambarus asperimanus* and *C. b. bartonii.*

Distribution (in addition to type-locality). —

United States

SOUTH CAROLINA — OCONEE CO. — 14.1 mi NW of Westminster on rt 76 (host, *Cambarus* sp.).

Cymocythere cyma (Hobbs and Walton, 1960)

Plate IX, figures 7-11
Plate XLVII

Entocythere cyma Hobbs and Walton, 1960: 18-19 (M, H, R, A, D), figs. 11-16. — Hobbs and Walton, 1962: 43 (K), 44-45 (R). — Hart, 1962: 123 (C).

Cymocythere cyma. — Hart, 1962: 129 (C, D, S). — Crawford, 1965: 150 (R). — Hobbs III, 1965: 163 (R). — Hobbs and Hart, 1966: 50 (M, H, R, D), figs. 32, 33. — Ferguson, 1968: 501 (C, D).

Diagnosis. — MALE: Shell length 380-400 μ (average, 390 μ); height, 210-220 μ (average, 220 μ). Peniferum terminating anteroventrally in broadly tapering, rounded prominence from subterminal bulbous area containing penis. Clasping apparatus strongly curved with vertical and horizontal rami bent to form angle of 70 to 75 degrees; internal border of horizontal ramus entire; distal extremity with five denticles.

Relationships. — *Cymocythere cyma* has its closest affinities with *C. gonia*; evidence for this kinship is in the similarity of the clasping apparatus and the similarly disposed penis which seems to be mesial rather than ventral, as in *C. clavata.* It may be distinguished from *C. gonia,* however, by the rounded ventral portion of the peniferum.

Type-locality. — Four Mile Ck, 1.8 mi SW of Benton, Polk Co., Tennessee, U.S.A. (types, USNM: holotype, 105962; paratype, 108183). Host, *Cambarus longirostris.*

Distribution (in addition to type-locality). —

ALABAMA — TALLADAGA CO. — 5.3 mi S of Munford on rt 21 (host, *Cambarus (Depressicambarus)* sp).

GEORGIA — GORDON CO. — 1.8 mi N of Calhoun on rt 41 (hosts, *Cambarus (Depressicambarus)* sp., *C. diogenes* subsp., *Procambarus spiculifer*).

TENNESSEE — McMINN CO. — 7.5 mi SW of Athen on US 11 (host, *Orconectes erichsonianus*); Oostanaula Ck, trib to Hiwassee River, near Athens (hosts, *Cambarus b. bartonii, C. diogenes, O. erichsonianus*). MEIGS CO. — Goodfields Ck 1 mi N of Decatur on rt 58 (host, *C. diogenes* subsp.). POLK CO. — 0.2 mi W of Ocoee (hosts, *C. (Depressicambarus)* sp. and *O. erichsonianus*).

Cymocythere gonia Hobbs and Hart, 1966

Plate IX, figures 12-14
Plate XLVII

Cymocythere gonia Hobbs and Hart, 1966: 51, 52 (M, H, R, D). — Ferguson, 1968: 501 (C, D).

Diagnosis. — MALE: Shell length 400 μ; height, 250 μ. Shell highest posterior to midlength; ventral margin without concavity. Submarginal setae mostly evenly spaced around shell except dorsally where absent, slightly closer together posteriorly than elsewhere. Eye pigmented.

Peniferum tapering from level of base of clasping apparatus; however, expanding into subterminal bulbous area containing horizontally situated penis with tip directed mesially; terminal portion of peniferum with angle posteroventrally on thickened area bearing anteriorly directed tooth mesially. Finger guard moderately well developed and bifid. Clasping apparatus divisible into horizontal and vertical rami with internal and external borders of both entire, rami joined in broad arc with proximal and distal ends directed at angle of 40 to 50 degrees; dorsodistal extremity with five denticles. Ventral finger subparallel to clasping apparatus; dorsal finger comparatively large, in this respect resembling *C. cyma.*

FEMALE. — Unknown.

Relationships. — *Cymocythere gonia* has its closest affinities with *C. cyma.* The strong similarities are evident primarily in the strongly curved clasping apparatus with five terminal denticles and the mesially disposed penis. The two may be separated by the angular, as opposed to rounded, posteroventral portion of the peniferum in *C. gonia.*

Type-locality. — Stream, 6 mi SE of Thornhill on rt 11W, Grainger Co., Tennessee, U.S.A. (type, USNM: holotype, 113455). Hosts, *Orcenectes juvenilis, Cambarus longirostris,* and *Cambarus bartonii cavatus.*

Distribution. — Known only from type-locality.

DACTYLOCYTHERE HART, 1962

Runki Group Crawford, 1959: 157 (M). — Crawford, 1961: 236, 241, 244 (R). — Hart and Hobbs, 1961a: 25 (D). — Hart and Hobbs, 1961b: 173, 174 (K, M, R in part). — Hobbs and Walton, 1961: 381 (R). — Hobbs and Walton, 1962: 42, 46 (R).

Runki Subgroup. — Hart and Hobbs, 1961b: 174, 177 (R).

Dactylocythere Hart, 1962: 124 (K), 129 (M, S, C), type-species, *Entocythere runki* Hobbs, 1955. — Hart, 1963: 51 (R, D). — Hart, 1965b: 255 (H). — Hobbs III, 1965: 160 (K), 163 (R). — Hart and Hart, 1966a: 2, 3, 5 (R). — Hobbs, 1967: 1 (D), 3 (R), 4-5 (S, K). — Hobbs, Holt, and Walton, 1967: 41 (M). — Hart and Hart, 1969a: 185-187 (D, E, R). — Hart and Hart, 1969b: 158 (M, R). — Hart and Hart, 1970: 282 (R). — Hobbs and Hobbs, 1970: 2, 5 (K), 5 (M). — Hart and Hart, 1971: 582, (R).

Diagnosis. — Terminal tooth of mandible pectinate. Male copulatory complex with finger guard; peniferum, except in *D. leptophylax,* never extending ventrally beyond clasping apparatus and never with bulbous enlargement or conspicuous terminal or subterminal processes nor terminating in barbed point; penis complex directed anteroventrally, simple, with spermatic and prostatic elements contiguous through-out their lengths and shorter than clasping apparatus; accessory groove conspicuous except in *D. leptophylax.* Clasping apparatus well developed, often divisible into vertical and horizontal rami; external border with or without teeth. Triunguis female with second antenna lacking accessory pectinate process on distal podomere; genital complex including J-shaped rod and amiculum except in *D. coloholca* (see under the species).

Type-species. — *Entocythere runki* Hobbs, 1955: 330 (by designation, Hart, 1962: 129).

Distribution. — United States: Indiana, southward to northern Florida and eastward to southern New Jersey. Canada: Ontario. Hosts. — Freshwater crayfishes of the family Astacidae.

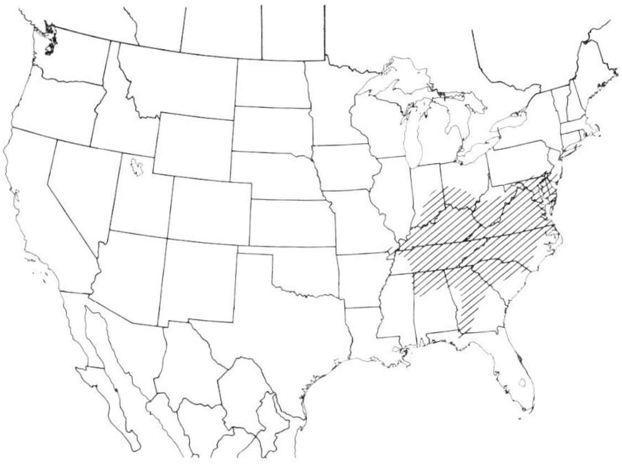

Figure 23. — Distribution of the genus *Dactylocythere.*

Key to the Species of the Genus Dactylocythere

(Modified from Hobbs and Hobbs, 1970)

1 Ventral portion of peniferum heavily sclerotized, terminating in single acute projection; resembling clasping apparatus on cursory inspection .... *D. banana*
1′ Ventral portion of peniferum not heavily sclerotized; not resembling clasping apparatus 2
2(1′) Ventral portion of peniferum with two or more small acute projections; finger guard very slender and tridentate; clasping apparatus almost U-shaped with subterminal flare; accessory groove reduced or obsolete .... *D. leptophylax*
2′ Ventral portion of peniferum without projections; finger guard seldom slender, if tridentate, comparatively stocky; clasping apparatus variable but never with subterminal flare; accessory groove well developed 3
3(2′) Apex of clasping apparatus without denticles 4
3′ Apex of clasping apparatus with denticles 5
4(3) Clasping apparatus scythelike, sometimes with single tooth on internal border of horizontal ramus, otherwise unadorned .... *D. falcata*
4′ Clasping apparatus not scythelike, without teeth but with two or three grooves appearing to encircle distal portion of horizontal ramus .... *D. striophylax*
5(3′) Shell with posteroventral projection 6
5′ Shell without posteroventral projection (*D. exoura* has a small prominence on shell) .. 8
6(5) Shell with anteroventral protuberance .... *D. amphiakis*
6′ Shell without anteroventral protuberance 7
7(6′) Accessory groove extending dorsally clearly above ventral level of spermatic loop 32
7′ Accessory groove extending dorsally only to ventral level of spermatic loop .. *D. pughae*
8(5′) Accessory groove extending dorsally much beyond level of dorsal extremity of spermatic loop 9
8′ Accessory groove never extending dorsally more than slightly beyond dorsal extremity of spermatic loop, sometimes not reaching loop 13
9(8) External border of vertical ramus of clasping apparatus with angular shoulder .... *D. macroholca*
9′ External border of vertical ramus of clasping apparatus without angular shoulder 10
10(9′) Vertical ramus of clasping apparatus almost straight .... *D. mecoscapha*
10′ Vertical ramus of clasping apparatus distinctly arched 11
11(10′) Proximal and distal portions of clasping apparatus subparallel .... *D. prionata*
11′ Proximal and distal portions of clasping apparatus directed at angles of 50° to 70° to each other 12
12(11′) Three proximal teeth on internal border of horizontal ramus of clasping apparatus subequal in size and equally spaced .... *D. arcuata*
12′ Proximal tooth on internal border of horizontal ramus of clasping apparatus much larger than more distal teeth 36
13(8′) Dorsal margin of accessory groove not reaching ventral portion of spermatic loop 14
13′ Dorsal margin of accessory groove reaching almost to, or slightly beyond, dorsal extremity of loop of spermatic duct 17
14(13) External border of clasping apparatus with two subangular bends 31
14′ External border of clasping apparatus with only one angular or subangular bend 15
15(14′) Finger guard short, truncate distally, with posterior margin strongly arched, reaching ventrally only to level of base of terminal seta on dorsal finger .... *D. coloholca*
15′ Finger guard long, tapering, with posterior margin straight or only slightly arched, reaching ventrally to or beyond tip of terminal seta on dorsal finger 16
16(15′) Finger guard heavy, tapering from base, posterior margin straight .... *D. exoura*
16′ Finger guard slender, tapering irregularly with distal portion abruptly narrower than basal and middle portions .... *D. prinsi*
17(13′) External border of vertical ramus of clasping apparatus with distinct angular shoulder 18
17′ External border of vertical ramus of clasping apparatus without shoulder 20
18(17) Internal border of horizontal ramus of clasping apparatus with two teeth .... *D. jeanae*
18′ Internal border of horizontal ramus of clasping apparatus with three or more teeth 19
19(18′) Ventral portion of peniferum terminating in posteriorly directed spine; internal border of clasping apparatus with four teeth, distal extremity with two .... *D. crawfordi*

19′ Ventral portion of peniferum subtruncate and directed posteroventrally; internal border of clasping apparatus with three teeth, distal extremity with three .... *D. phoxa*
20(17′) Internal border of horizontal ramus of clasping apparatus with or without one prominent tooth and occasionally with undulations distal to tooth 21
20′ Internal border of horizontal ramus of clasping apparatus with more than one reduced or prominent tooth 24
21(20) Ventral margin of peniferum gently curved or with posteroventral extremity subangulate 33
21′ Ventral margin of peniferum rounded without angle at posteroventral extremity 22
22(21′) Clasping apparatus L-shaped with vertical ramus nearly straight or concave posteriorly; finger guard with anteroventral tubercle most prominent .... *D. myura*
22′ Clasping apparatus C-shaped with vertical ramus convex posteriorly; finger guard with posteroventral tubercle most prominent 23
23(22′) Finger guard terminating in two tubercles .... *D. chalaza*
23′ Finger guard terminating in three tubercles .... *D. spinata*
24(20′) Finger guard with three distinct distal tubercles 25
24′ Finger guard with less than three distinct distal tubercles 26
25(24) Anteroventral extremity of peniferum rounded; distal portion of internal border of clasping apparatus serrate .... *D. suteri*
25′ Anteroventral extremity of peniferum angular; distal portion of internal border of clasping apparatus with teeth but not serrate 34
26(24′) External border of clasping apparatus subangular at junction of vertical and horizontal rami 27
26′ External border of clasping apparatus curved at junction of indistinctly delineated horizontal and vertical rami 28
27(26) Peniferum with prominent hump on posterior margin paralleling accessory groove; posterior margin of finger guard strongly concave, distal end with single emargination .... *D. xystroides*
27′ Peniferum without prominent hump on posterior margin paralleling accessory groove; posterior margin of finger guard only slightly concave, distal end with two emarginations 35
28(26′) Internal border of clasping apparatus with only two teeth .... *D. ungulata*
28′ Internal border of clasping apparatus with three or more teeth 29
29(28′) Finger guard with distinctly curved posterior margin .... *D. runki*
29′ Finger guard straight, or almost so 30
30(29′) Finger guard with longitudinal ridge; extremities of clasping apparatus divergent .... *D. cooperorum*
30′ Finger guard without longitudinal ridge; extremities of clasping apparatus subparallel .... *D. steevesi*
31(14) Finger guard terminating in three lobes .... *D. charadra*
31′ Finger guard terminating abruptly; truncate .... *D. brachystrix*
32(7) Clasping apparatus with strongly developed horizontal ramus; internal border of horizontal ramus with three or four teeth .... *D. daphnioides*
32′ Clasping apparatus with slender horizontal ramus, bearing one well defined tooth at mid-length .... *D. koloura*
33(21) Clasping apparatus with two teeth on internal border of horizontal ramus .... *D. enoploholca*
33′ Clasping apparatus with single tooth at approximately mid-length of internal border of horizontal ramus .... *D. megadactylus*
34(25′) Clasping apparatus with four indistinct teeth on internal border of clasping apparatus .... *D. chelomata*
34′ Clasping apparatus with three or four distinct and irregular teeth on distal half of internal border .... *D. sandbergi*
35(27′) Clasping apparatus heavily developed, with one or two distinct teeth on internal border of horizontal ramus .... *D. pachysphyrata*
35′ Clasping apparatus more delicate than above, with two indistinct teeth on internal border of horizontal ramus .... *D. speira*
36(12′) Finger guard with proximal portion bulbous .... *D. amicula*
36′ Finger guard subequal in width throughout .... *D. susanae*

Dactylocythere amicula Hart and Hart, 1966

Plate X, figures 1-4
Plate XLVIII

Dactylocythere amicula Hart and Hart, 1966a: 1-3 (M, R, D, A), figs. 1, 2. — Hobbs, 1967: 5 (K). — Ferguson, 1968: 501 (C, D). — Hobbs and Walton, 1970: 853 (M, R). — Hobbs and Hobbs, 1970: 5-7 (K), 10 (R).

Diagnosis. — MALE: Shell length 550-570 μ (average, 560 μ); height, 320-360 μ (average, 330 μ). Shell subelliptical in outline with greatest height ⅓ shell length from posterior end. Setae scarce, scattered on anterior and posterior margins. Eye pigmented, situated approximately ⅕ shell length from anterior end.

Copulatory complex with peniferum large, corneous, and extending ventrally to about level of external border of horizontal ramus of clasping apparatus. Distal portion of peniferum terminating in two points; upper one tapering and sharply pointed, ventral one thicker and more blunt. Accessory groove present, well developed, and extending proximally for ¾ length of peniferum — well beyond level of spermatic duct loop. Penis small (averaging 16 μ in length), approximately horizontal, and situated on level with proximal portion of clasping apparatus. Finger guard well developed and heavily sclerotized. Clasping apparatus robust and clearly divisible into horizontal and vertical rami; rami subequal in length, forming slightly acute angle with one another, and with their thickest portions at their juncture. External border of vertical ramus concave in proximal half, convex in distal half; internal border concave. Horizontal ramus smoothly convex on external border; internal border concave and bearing one large tooth approximately at mid-length and two somewhat indistinct teeth distal to it; distal extremity of horizontal ramus bearing three dorsally directed denticles. Indistinct teeth on internal border marking three horizontal lines or grooves extending proximally along mid-portion of ramus.

TRIUNGUIS FEMALE: Shell length 550-560 μ (average, 560 μ); height, 350-370 μ (average, 360 μ). Shell similar to that of male, but slightly greater in average height, with greatest height in posterior ⅕ of shell. Amiculum greatly hypertrophied as compared to those of other *Dactylocythere* species, extending posteriorly beyond valves for more than 100 μ, forming a striking fan- or crest-shaped structure with a crenulated free margin.

Biunguis females unknown.

Relationships. — This ostracod appears to have its closest relationships with *Dactylocythere runki.* This is evident in the clasping apparatus, both of which have similar tooth arrangements on the inner margin of the horizontal ramus, and both of which have the same horizontal to vertical ramus length ratios (1:1). The shapes of the male and female shells are also similar. However, in no other *Dactylocythere* does the amiculum reach the size found in *D. amicula.*

Type-locality. — Buckhorn Ck at Camp Robertson, near Noble, 5.2 mi NE of rt 15 on unnumbered dirt road, Breathitt Co., Kentucky, U.S.A. (types, USNM: holotype, 113330; allotype, 113331; paratype, 113332). Hosts, *Cambarus* (*Jugicambarus*) sp. and *Orconectes juvenilis.*

Distribution (*in addition to type-locality*). —

United States

KENTUCKY — MASON CO. — 3.4 mi W of Rectorville on rt 10 (host, *C. laevis*). ROWAN CO. — East Fork of Triple Ck, 3.5 mi E of Moorehead on rt 60 (hosts, *C. b. bartonii, C. robustus, C.* (*Depressicambarus*) sp., and *O. juvenilis*).

Dactylocythere amphiakis Hart and Hart, 1966

Plate X, figures 5-10
Plate XLVIII

Dactylocythere amphiakis Hart and Hart, 1966a: 3-5 (M, R, D, A), figs. 3-7. — Ferguson, 1968: 501 (C, D). — Hobbs, 1967: 5 (K). — Hart and Hart, 1969a: 185 (M, R). —Hart and Hart, 1969b: 154 (M). — Hobbs and Hobbs, 1970: 5-7 (K).

Diagnosis. — MALE: Shell length 570-700 μ (average, 600 μ); height, 300-450 μ (average, 350 μ). Shell subelliptical in outline with greatest height ⅓ shell length from posterior end pointed protuberances present on posteroventral and anteroventral margins; posterior protuberance directed caudally, anterior protuberances directed dorsally. Setae scarce; scattered on anterodorsal and posterior margins. Eyes present, situated approximately ⅕ shell length from anterior end.

Copulatory complex with peniferum not extending beyond external border of horizontal ramus of clasp-apparatus; distal extremity blunt, with two corneous ventrally directed points separated by bulging membranous area. Accessory groove present and extending proximally above level of spermatic duct loop.

Penis situated in distal ⅕ of peniferum; main body almost horizontal, with tip curved and directed ventrally. Clasping apparatus with distinct horizontal and vertical rami; horizontal ramus slightly longer than vertical. Internal border of horizontal ramus with three evenly-spaced vestigial teeth, distal extremity directed dorsally and bearing four denticles; external border smoothly convex. External border of vertical ramus concave in proximal ⅓, convex in distal ⅔. Thickest portion of clasping apparatus at juncture of two rami. Dorsal and ventral fingers present; finger-guard well developed, bifid distally.

FEMALE: Shell of biunguis females similar to, but averaging slightly smaller than, that of males. Length 510-570 μ (average, 550 μ); height, 290-320 μ (average, 300 μ).

Shell of triunguis females similar to male, but averaging slightly larger. Length 620-700 μ (average, 660 μ); height, 360-420 μ (average, 390 μ). Amiculum similar to that of *D. chalaza.*

Relationships. — Although the general configuration of the clasping apparatus is similar to that of *D. chalaza,* this ostracod does not appear to bear close relationships with any of the described species of *Dactylocythere.* At first glance it might be thought that a close relationship exists between *D. amphiakis* and *D. daphnioides,* but the copulatory apparatus of the two species are distinctly different, and the overall shell shapes are dissimilar.

Variations. — The teeth on the internal surface of the horizontal ramus of the clasping apparatus vary somewhat in size and position; in some specimens they are quite distinct, in others they are so poorly developed as to be almost non-existent, and not all are evenly spaced. The terminal denticles of the horizontal ramus vary in number from three to five, and the more distal of the two points of the peniferum is, in some specimens, directed anteroventrally instead of ventrally.

Type-locality. — Springs in Rich Pond Area, Warren Co., Kentucky, U.S.A. (types, USNM: holotype and paratype, 113333; allotype, 113334; paratype, 113335 and 113336). Host, *Cambarus* sp.

Distribution (in addition to type-locality). —

United States

KENTUCKY — CHRISTIAN CO. — Sinking Fork of Little R 7.1 mi W of Hopkinsville on rts 68 & 80 (host, *Orconectes tricuspis*). TRIGG CO. — 1 mi SW of Cadiz in spring (host, *Cambarus tenebrosus*); spring used for cattle feeding 1.5 mi W of Cadiz near limestone bluff (host, *C. tenebrosus*). WARREN CO. — intermittent springs in field 3 mi SE of rts 31 & 242 in Rich Pond area about 8 mi S of Bowling Green on rts 31 & 242 (host, *C. tenebrosus*).

TENNESSEE — DICKSON CO. — Montgomery Bell State Park (host, *Orconectes* sp.).

Dactylocythere arcuata (Hart and Hobbs, 1961)

Plate XI, figures 1-3
Plate XLVIII

Entocythere arcuata Hart and Hobbs, 1961b: 173 (K), 177 (M, H, R, A, D), figs. 9-11.
Dactylocythere arcuata. — Hart, 1962: 129 (C, D, S). — Hobbs, 1967: 5 (K). — Ferguson, 1968: 501 (C, D). — Hobbs and Hobbs, 1970: 5-7 (K).

Diagnosis. — MALE: Shell length 450-480 μ (average, 470 μ); height, 250-280 μ (average, 270 μ). Shell subelliptical, but distinctly higher posteriorly than anteriorly, and with a slight ventral emargination just anterior to midlength. Maximum height, short distance posterior to midlength. Setae widely spaced on anterior, anterodorsal, and anteroventral surfaces. Posterolateral surface almost hirsute. Eye pigmented.

Copulatory complex with peniferum not extending distally beyond external border of clasping apparatus; distal extremity truncate. Accessory groove extending proximal to level of loop of spermatic duct. Finger guard slender and truncate distally. Dorsal finger with two terminal setae. Clasping apparatus with vertical and horizontal rami not clearly defined. External borders of both rami entire; internal border of horizontal ramus with three evenly spaced strong teeth, distal extremity with three denticles.

FEMALE. — Unknown.

Relationships. — *Dactylocythere arcuata* has its closest affinities with *D. ungulata, D. prionata,* and *D. steevesi* — but may be distinguished from them by the structure of the clasping apparatus.

Type-locality. — Saltpeter Cave, (Sauta Cave), Jackson Co., Alabama, U.S.A. (Erroneously given by Hart and Hobbs, 1961, as Franklin Co., Alabama. The mistake was pointed out by Dr. Thomas C. Barr, Jr.) (types, ANSP). Host, *Cambarus tenebrosus.*

Distribution (in addition to type-locality). —

United States

ALABAMA — JACKSON CO. — Russell Cave (host, *C. tenebrosus*).

Dactylocythere banana Hart and Hart, 1971

Plate XI, figures 4-6
Plate XLVIII

Dactylocythere banana Hart and Hart, 1971: 108-109 (M, H, R, A, D), figs. 3a-c.

Diagnosis. — MALE: Shell length 440-480 μ (average, 450 μ); height, 250-280 μ (average, 270 μ). Shell subelliptical, with greatest height in posterior half. Eye pigmented, situated ⅕ shell length from anterior end.

Copulatory complex with peniferum tapering distally to recurved sclerotized point. Penis situated within distal half of peniferum, with tip directed ventrally. Terminal position of penis apparently introduced into sclerotized structure in anterior portion of peniferum — possibly homologous with "penis guide" found in some members of genus Uncinocythere. Delicate membrane connecting sclerotized guide-structure with distal portion of peniferum.

Clasping apparatus thick, heavily sclerotized, evenly curved throughout entire length, not divided into distinct horizontal and vertical rami. External margin evenly convex and smooth. Internal margin evenly concave, with single small tooth situated ⅓ length of entire clasping apparatus from distal end, apparatus terminating in four denticles. In some specimens two teeth present on internal margin of clasping apparatus — one about ⅓ length of clasping apparatus from distal end, another immediately proximal of terminal denticles. Teeth inconspicuous, occasionally not detected because of angle at which clasper is viewed.

Finger guard heavy and well-developed; trifid distally.

FEMALE. — Only available specimen crushed; no distinguishing characteristics noted.

Relationships. — Superficially, the copulatory complex of *Dactylocythere banana* resembles that of *Rhadinocythere serrata* but can be distinguished from it by the presence of a well-developed finger guard. The finger guard, the tapering distal portion of the peniferum, and the position of the penis indicate placement in the genus *Dactylocythere.* The even curve of the relatively heavy clasping apparatus and the unique heavily sclerotized single-element terminal portion of the peniferum indicate that it is not closely related to other species within the genus.

Type-locality. — Jackson River at junction of rts 220 and 84, 26 mi N of Warm Springs, Highland Co., Virginia, U.S.A. (types, USNM: holotype, 138447; paratype, 138448). Host, *Cambarus longulus* Girard.

Distribution (in addition to type-locality). —

United States

VIRGINIA — GREENE CO. — Middle R, 3.5 mi N of Stanardsville on rt 230 (host, *Cambarus b. bartonii*).

Dactylocythere brachystrix Hobbs and Walton, 1966

Plate XI, figures 7-11
Plate XLVII

Dactylocythere brachystrix Hobbs and Walton, 1966: 2-3 (M, H, R, A, D), figs. 1a-d. — Hobbs, 1967: 5 (K). — Hobbs and Walton, 1968: 239 (R). — Hobbs and Walton, 1970: 853 (M, R). — Hobbs and Hobbs, 1970: 5-7 (K).

Diagnosis. — MALE: Shell length 460-480 μ (average, 470 μ); height, 260-290 μ (average, 270 μ). Shell with highest portion posterior to midlength and with dorsal and ventral margins more gently tapered anteriorly than posteriorly; ventral margin of shell entire; submarginal setae anteriorly; ventrally, and posteriorly somewhat evenly spaced, absent dorsally. Eye pigmented.

Copulatory complex with posterior margin of peniferum expanded into three humps; both anteroventral and posteroventral extremities directed anteroventrally. Penis small, its length approximately one-third the anterior-posterior dimension of distal portion of peniferum, and situated at level of base of ventral two-fifths of peniferum. Accessory groove extending dorsally almost to ventral extremity of spermatic duct. Dorsal finger slender with its terminal seta extending to ventral margin of finger guard; ventral finger gently curved throughout its length. Finger guard prominent with its anterior margin straight and its ventral margin emarginate with three small prominences. Clasping apparatus with vertical and horizontal rami somewhat distinctly delimited; vertical ramus with a bend of about 45° just proximal to midlength and joining horizontal ramus (external borders) at an angle of about 100°; internal border of horizontal ramus with three evenly spaced teeth and apex with three denticles; external borders of both rami entire.

TRIUNGUIS FEMALE: Shell length 440-510 μ (average, 480 μ); height, 270-320 μ (average, 290 μ). Shell distinctly larger than that of male, with its greatest height slightly posterior to midlength and distinctly lower anteriorly than posteriorly; ventral margin entire; marginal setae as in male. Prominent J-shaped rod and ruffled amiculum present.

Relationships. — *Dactylocythere brachystrix* is closely related to *D. daphnioides, D. runki, D. chalaza,* and *D. pachysphrata,* but differs from all of them in possessing an accessory groove on the peniferum which extends dorsally only to the ventral extremity of the loop of the spermatic duct. In the other species, the groove reaches the dorsal extremity of the loop.

Type-locality. — Otter Creek, about 3.5 mi upstream from Obed River, Cumberland Co., Tennessee, U.S.A. (types, USNM: holotype and allotype, 111251; paratype, 111252). Host, *Cambarus* sp.

Distribution (in addition to type-locality). —

United States

TENNESSEE — GRAINGER CO. — Indian Ck 11.4 mi SE of Tazewell on rt 25E (host, *C. b. bartonii*). LAWRENCE CO. — 1.4 mi E of Lawrenceburg on rt 64 (host, *Orconectes spinosus*).

Dactylocythere chalaza (Hobbs and Walton, 1962)

Plate XII, figures 1-6
Plate XLVII

Entocythere chalaza Hobbs and Walton, 1962: 42 (K), 45-47 (M, H, R, D), figs. 6-9. — Hart, 1962:, 123, 129 (C).

Dactylocythere chalaza. — Hart, 1962: 129 (C, D, S). — Hobbs and Walton, 1963a: 460, 461 (R). — Hart and Hart, 1966a: 4 (R). — Hobbs and Walton, 1966: 4 (R). — Hobbs, 1967: 3, 4 (K). — Hobbs, Holt, and Walton, 1967: 7 (B), 8 (H), 12 (B, H), 34 (K), 36 (A), 41, 42 (M, D, H, A, S), 55, 69 (D, E), 23, 51, 58, 61, 64 (A, H), 72-75 figs. 7e, i; 9c. — Ferguson, 1968: 501 (C, D). — Hobbs and Walton, 1970: 853 (M, R), 858 (A), 862 (R). — Hobbs III, 1970: 7. — Hobbs and Hobbs, 1970: 5-7 (K). — Walton and Hobbs, 1971: 87-103 (B). — Hart and Hart, 1971: 118.

Diagnosis. — MALE: Shell length 430-470 μ (average, 450 μ); height, 250-280 μ (average, 260 μ). Shell subovate with margins entire; maximum height slightly posterior to midlength. Submarginal setae rather close together anterodorsally and posteriorly, but sparse ventrally; none on dorsal margin. Eye pigmented.

Copulatory complex with peniferum usually heavily sclerotized and robust and with a conspicuous hump, or swelling, on posterodistal portion. Apical end of peniferum directed anteriorly. Penis of moderate size and situated at level of finger guard. Accessory groove present and extending proximally to loop of spermatic duct. Dorsal finger slender and with a single long seta; ventral finger curved throughout so that apex is directed anterodorsally. Finger guard prominent, strongly arched anteriorly, bifid distally with the larger posterior element directed posteroventrally. Clasping apparatus divisible into vertical and horizontal rami. Vertical ramus strongly curved and gradually broadens distally so that its maximum width occurs at junction with horizontal ramus, the thickest portion of the clasping apparatus. Horizontal ramus with external border entire and slightly convex; internal border with one prominent angular elevation slightly proximal to midlength; apex of clasping apparatus without distinct denticles but marked by two longitudinal grooves which extend posteriorly on lateral surface of horizontal ramus.

FEMALE: Shell of triunguis female essentially similar to that of male but frequently with a shallow ventral emargination just anterior to midlength, and often with a posteroventral lobular expansion. Length 430-470 μ (average, 460 μ); height, 250-280 μ (average, 270 μ). Prominent J-shaped rod and ruffled amiculum present.

Relationships. — *D, chalaza* is closely allied to *D. daphnioides, D. runki, D. mecoscapha, D. suteri,* and *D. chelomata.* It shares with all of these the presence of a finger guard, at least one tooth on the internal border of the clasping apparatus, and an amiculum in the female. The males of *D. chalaza* may be distinguished by the thickened angle of the clasping apparatus and the single "tooth" on its internal border.

Type-locality. — Crayfish burrows on golf course of Mountain Lake Hotel, off rt 700, about 9 mi NE of Pembroke, Giles Co., Virginia, U.S.A. (types, USNM: holotype, 108179; allotype, 108180; paratype, 108181). Hosts, *Cambarus b. bartonii, C. longulus,* and *C. acuminatus.*

Distribution (*in addition to type-locailty*). —

United States

TENNESSEE — SULLIVAN CO. — 6.9 mi N of Elizabethton on rt 19E in clear, sluggish stream (hosts, *Cambarus b. bartonii* and *C. longirostris*). UNICOI CO. — Indian Ck at jct with Devil Fork (hosts, *C. longirostris* and *C. robustus*).

VIRGINIA — CRAIG CO. — crayfish burrows along Middle Ck, 1.2 mi N of Maggie on rt 636 (host, *Cambarus dubius*). Mountain Lake Biological Station grounds (host, *C. dubius*). MONTGOMERY CO. — small trib to North Fork of Roanoke R, 0.2 mi NE of jct of rts 785 and 723, on 785 (hosts, *C. acuminatus, C. bartonii*, and *C. longulus*).

WEST VIRGINIA — MONROE CO. — burrows along trib of Rays Branch 2.0 mi SE of Waiteville (host, *Cambarus dubius*).

Dactylocythere charadra Hobbs III, 1971

Plate XII, figures 7-10
Plate XLVII

Dactylocythere charadra Hobbs III, 1971: 142-144 (M, H, R, A, D), figs. 2g-l.

Diagnosis. — MALE: Shell length 480-490 μ (average, 485 μ); height, 270-280 μ (average, 275 μ). Eyes pigmented, located approximately one-fifth shell length from anterior margin. Shell subovate with greatest height posterior to midlength; margins entire; submarginal setae present in limited numbers anteriorly, ventrally, and posteriorly, absent dorsally. Sternal spine short, directed posteroventrally.

Copulatory complex possessing finger guard tapering from broad base and flaring distally in 3 tubercles, posterior margin markedly concave; peniferum elongate, posteroventral portion gently rounded; accessory groove short, not reaching ventralmost level of spermatic loop; peniferal groove very long, extending dorsally beyond level of dorsalmost portion of spermatic loop; apex of peniferal groove narrower than least diameter of vertical ramus of clasping apparatus; penis horizontally situated in distal one-fourth of peniferum and slightly curved ventrally at distal end, less than half width of peniferum; clasping apparatus extending ventrally beyond peniferum; and clearly divisible into horizontal and vertical rami, proximal and distal portions directed at angle of approximately 80 degrees; external border of vertical ramus with subangular bend distally, otherwise external borders of both rami and internal border of vertical ramus entire; internal border of horizontal ramus with prominent, acute, proximal tooth and two more distal rudimentary ones; apex of ramus terminating in three acute denticles; both dorsal and ventral fingers slender and terminating in single seta.

TRIUNGUIS FEMALE: Shell similar to but slightly larger than that of male. Shell length 480-510 μ (average, 490 μ); height, 290-330 μ (average, 310 μ). Eye pigmented, located approximately one-seventh shell length from anterior margin. Shell subovate with greatest height posterior to midlength; ventral margin with shallow concavity anterior to midlength, otherwise entire; submarginal setae like those of male. Genital complex located posterodorsally, composed of distinct J-shaped rod and amiculum; amiculum consisting of amorphous mass suspended from lower third of J-shaped rod and supported by U-shaped ribs; anteriormost pair of supporting ribs with rami extending anteriorly; dorsal portion of J-shaped rod bifid, appearing cracked and angled anterodorsally, with no apparent genital papilla.

Relationships. — *Dactylocythere charadra* is most closely allied to *D. chalaza* from which it differs most conspicuously by the much elongated peniferal groove; it can further be distinguished from *D. chalaza* by the presence of two reduced teeth located distally on the internal border of the horizontal ramus of the clasping apparatus; *D. charadra* possesses a sternal spine suggesting a not too remote relationship to other members of the genus that possess this character, *D. amicula, D. brachytrix, D. chalaza, D. chelomata, D. daphnoides, D. exoura, D. pachysphyrata, D. runki, D. spinata,* and *D. xystroides.* Although closely allied, *D. charadra* may be distinguished from these species by the distinct characters of the accessory groove, finger guard, shape of peniferum, and clasping apparatus.

Type-locality. — Sinking Creek in "Hoss Cove" near Johnson City, Washington Co., Tennessee, U.S.A. (types, USNM: holotype and allotype, 135622). Host, *Cambarus* sp.

Distribution (*in addition to type-locality*). —

United States

TENNESSEE — SULLIVAN CO. — clear, sluggish stream 6.9 mi N of Elizabethton on rt 19E (hosts, *Cambarus longirostris* and *C. b. bartonii*). UNICOI CO. — Indian Ck at jct with Devil Fort (hosts, *C. robustus* and *C. longirostris*).

Dactylocythere chelomata (Crawford, 1961)

Plate XII, figures 11-15
Plate XLVIII

Entocythere chelomata Crawford, 1961: 242-244 (M, H, R, D), figs. 15-21. — Hobbs and Walton, 1962: 46 (R). — Hart, 1962: 123 (C).

Dactylocythere chelomata. — Hart, 1962: 130 (C, D, S). — Hart and Hart, 1966a: 6 (R). — Hobbs, 1967: 5 (K), 8 (R). — Ferguson, 1968: 501 (C, D). — Hobbs and Walton, 1970: 853 (M, R). — Hobbs and Hobbs, 1970: 5-7 (K).

Diagnosis. — MALE: Shell length 420-470 μ (average, 450 μ); height, 250-270 μ (average, 260 μ). Shell oval in lateral aspect with margins smoothly rounded. Posterior end higher than anterior with maximum height just posterior to midlength. Prominent eye located one-fifth of valve length from anterior end. Setae sparsely distributed on lateral surface and dispersed as usual along free margin. Vertical ramus of clasping apparatus arched, proximally clavate, and bearing prominence on posterior margin; horizontal ramus bearing four teeth on distal portion of internal margin and three prominent terminal denticles. Distal portion of peniferum acute and directed anteriorly; distinct accessory groove in mid-posterior portion of peniferum.

TRIUNGUIS FEMALE: Shell length 430-480 μ (average, 450 μ); height, 250-300 μ (average, 280 μ). Shell subovate, with concavities on the posterodorsal and ventral margins, the latter being slightly anterior to midlength. Pubescence similar to that on male valve.

Relationships. — The closest relatives of this species seem to be *D. runki* and *D. suteri,* differing from the former in the slenderness of the horizontal ramus of the clasping apparatus, and the more proximal serrations on the finger margin of *D. runki* are much more prominent and evenly spaced than are those of *D. chelomata.* The over-all outline of the clasping apparatus of *D. suteri* is similar to *D. chelomata.* The proximal end of the vertical ramus of the clasping apparatus of *D. suteri* appears to be evenly rounded, while that of *D. chelomata* is conspicuously knobbed.

Type-locality. — West fork of Little Pigeon River, 1 mi S of Gatlinburg on rt 441, Sevier Co., Tennessee, U.S.A. (types, USNM: holotype, 107461; allotype, 107463; paratype, 107462; morphotype, 107464). Hosts, *Cambarus longirostris, C. b. bartonii,* and *Orconectes juvenilis.*

Distribution (in addition to type-locality). —

United States

NORTH CAROLINA — CHEROKEE CO. — Valley R at Andrews (hosts, *Cambarus bartonii* and *C. longirostris*). MACON CO. — Sugar Fork of Cullasaga R NW of Highlands (host, *C. b. bartonii*).

Dactylocythere coloholca Hobbs and Hobbs, 1970

Plate XIII, figures 1-5
Plate XLVIII

Dactylocythere coloholca Hobbs and Hobbs, 1970: 6 (K), 7, 8 (M, H, R, A, D), figs. 2a-d.

Diagnosis. — MALE: Shell length 430-480 μ (average, 460 μ); height 250-270 μ (average, 270 μ). Shell oblong-ovate with posterior region vaulted and posteroventral margin subangulate; margins entire. Submarginal setae moderate in number and evenly spaced anteriorly, posteriorly, and ventrally. Sternal spine lacking. Eye pigmented, located approximately one-third shell length from anterior margin.

Copulatory complex with finger guard strongly concave posteriorly and slightly concave distally; peniferum heavy, tapering ventrally but with ventral surface truncate; accessory groove short, extending posterodorsally but not nearly reaching level of ventral extremity of loop of spermatic duct; peniferal groove opening ventrally, its apical width approximately two-thirds that of diameter of vertical ramus of clasping apparatus; penis L-shaped and situated in ventral third of peniferum; clasping apparatus, extending ventral to peniferum, divisible into vertical and horizontal rami, forming angle of less than 90 degrees, external borders of both rami and internal border of vertical ramus entire, internal border of horizontal ramus with three teeth, apex of ramus with three denticles; both dorsal and ventral fingers slender, latter more than twice length of former and gently curved throughout its length.

TRIUNGUIS FEMALE: Shell length 490-540 μ (average, 520 μ); height, 320-340 μ (average, 320 μ). Shell distinctly larger than that of male, highly vaulted posteriorly, subangular posteroventrally, and with slight ventral concavity anterior to mid-length. Submarginal setae as in male. Genital complex situated posterodorsally and consisting of posteroventrally directed rod surrounded by granular mass together with small ventrally directed papilla situated immediately posterodorsal to rod; J-shaped rod and amiculum absent. In lacking latter, triunguis female of this species unique in genus. Eye pigmented, located approximately one-fifth shell length from anterior margin.

Relationships. — *Dactylocythere coloholca* appears to have its closest affinities with *D. prinsi,* which resembles it in lacking a sternal spine and possessing a comparatively slender clasping apparatus in which the vertical and horizontal rami are clearly divisible, and in having a very short accessory groove in the peniferum. Except for the latter characteristic, *D. myura* also appears to be a near relative. *Dactylocythere coloholca* may be distinguished from *D. prinsi* by possessing a shorter finger guard and by the opening of the peniferal groove being directed ventrally, instead of posteriorly. Furthermore, the females differ from all other members of the genus in lacking a J-shaped rod and an amiculum.

Type-locality. — Creek and burrows in seepage area 4.8 mi SW of jct of rts 25W and 75, on rt 25W, SW of Corbin, Whitley Co., Kentucky, U.S.A. (types, USNM: holotype, 126253; allotype, 126254). Host, *Cambarus sphenoides.*

Distribution. — Known only from type-locality.

Dactylocythere cooperorum Hobbs and Walton, 1968

Plate XIII, figures 6-9
Plate XLVIII

Dactylocythere cooperorum Hobbs and Walton, 1968: 238-242 (M, H, R, A, D), figs. 1f-k. — Hobbs and Hobbs, 1970: 5-7 (K).

Diagnosis. — MALE: Shell length 430-440 μ (average, 435 μ); height, 220-240 μ (average, 230 μ). Shell ovate with posterior portion higher than anterior and slightly bulging posteroventrally. Submarginal setae comparatively close together posteriorly, somewhat more widely spaced dorsally and ventrally; anteriorly setae not numerous, but two or three somewhat removed from margin. Eye pigmented, subcylindrical to oblong-ovate.

Peniferum with concavity near mid-length of posterior margin, ventral portion tapering with apex directed ventrally; accessory groove extending posterodorsally to level of lower extremity of spermatic loop; penis L-shaped, directed anterodorsally with apex extending mesioventrally. Finger guard long, comparatively slender, entire distally, and with one or two longitudinal grooves. Dorsal finger slender and ending in single acute setiform tip; ventral finger gently curved throughout with no irregular bends. Clasping apparatus somewhat C-shaped without clear delineation between vertical and horizontal rami, but somewhat thickened in area of their junction; external border entire, never bearing angular shoulder proximally; internal border of distal half with four serrations, proximal one apparently strongest; narrow distal extremity of apparatus with three small denticles.

TRIUNGUIS FEMALE: Shell length 440-470 μ (average, 450 μ); height, 260-270 μ (average, 260 μ). Shell more highly vaulted posteriorly than in male, and not bulging posteroventrally. Submarginal setae distributed much as in male, but absent dorsally posterior to eye. Genital area with J-shaped rod almost forming circle, and with amiculum assuming 180 degree arc around its posteroventral margin; flared portion of amiculum short and not reaching posterior margin of shell. Eye pigmented.

Relationships. — The closest relative of *Dactylocythere cooperorum* is *D. brachystrix.* The similarities may be seen in the short accessory groove of the peniferum, the sinuate posterior margin of the peniferum, and in the arrangement of the teeth on the internal border of the clasping apparatus. It differs strikingly from *D. brachystrix,* however, in lacking a subangular bend at the junction of the horizontal and vertical rami of the clasping apparatus, in possessing a simple, slender finger guard, in the conformation of the distal extremity of the peniferum, and in the more tightly coiled J-shaped rod in the genital area of the female. The clasping apparatus is unique.

Type-locality. — Three miles WSW of Donansburg on rt 88, Little Barren River, Green Co., Kentucky, U.S.A. (types, USNM: holotype and allotype, 123322; paratype, 123323). Host *Cambarus cornutus.*

Distribution (*in addition to type-locality*). —

United States

KENTUCKY — ADAIR CO. — Russell Creek 1 mi ESE of Columbia on KY 80 (host, *Cambarus cornutus*). ALLEN CO. — Trammel Fork at KY 240, Allen Springs (host, *C. cornutus*). CASEY CO. — South Fork Ck, 2 mi ENE of Dunville on US 127 (host, *C. cornutus*). METCALF CO. — East Fork of Little Barren R, 2 mi E of Sulphur Well at US 68 (host, *C. cornutus*). WARREN CO. — West Fork Ck at KY 240, Boyce (host, *C. cornutus*).

Dactylocythere crawfordi Hart, 1965

Plate XIII, figures 10-13
Plate XLIX

Dactylocythere crawfordi Hart, 1965b: 255-257 (M, R, D), figs. 1, 2. — Ferguson, 1968: 501 (C, D). — Hobbs and Walton, 1968: 242, 243 (R). — Hobbs and Hobbs, 1970: 5-7 (K).

Diagnosis. — MALE: Shell length 440-500 μ (average, 470 μ); height, 260-300 μ (average, 290 μ). Shell subelliptical in outline with greatest height immediately posterior to mid-point; setae scarce, situated on anterior and posterior ends. Eyes present, situated one-fourth shell length from anterior end.

Copulatory complex with peniferum heavily sclerotized and with conspicuous concavity on posterodistal margin; apical end of peniferum terminating in two projections, with the more ventral projection somewhat scoop-shaped and extending beyond the more distal one; accessory groove present and extending proximally to level of loop of spermatic duct; penis small and situated on level with proximal portion of clasping apparatus; finger guard prominent, straight, and heavily sclerotized; clasping apparatus robust and divisible into vertical and horizontal rami, junction of two not sharply defined; external border of vertical ramus convex, with marked break in continuity at about mid-length; internal surface concave, with swelling in proximal half; horizontal ramus about 1.5 times longer than vertical ramus; external border smoothly convex; internal border concave and bearing three evenly spaced teeth; distal extremity with three denticles; three longitudinal lines or grooves extending proximally from one terminal denticle and two horizontal ramus teeth, respectively.

BIUNGUIS FEMALE: Shell length 430-480 μ (average, 450 μ); height, 260-310 μ (average, 280 μ). Shell essentially identical with that of male but slightly smaller.

TRIUNGUIS FEMALE: Shell length 460-510 μ (average, 490 μ); height, 280-330 μ (average, 300 μ). Shell essentially identical with that of male, but larger; prominent J-shaped rod and ruffled amiculum present.

Relationships. — This ostracod appears to have its closest affinities with *Dactylocythere runki.* It differs from *D. runki* in the relative lengths of the horizontal and vertical rami (*D. runki* 1:1; *D. crawfordi* 1.5:1), in the transition between the two rami (*D. runki,* relatively abrupt; *D. crawfordi,* smoothly curved), and in the termination of the peniferum (*D. runki,* relatively abrupt; *D. crawfordi,* ventral projection scoop-shaped and extending beyond the dorsal one).

Type-locality. — Macacheek Creek on rt 287, Logan Co., Ohio, U.S.A. (types, USNM: holotype, 110946; paratypes, 110947 and 110948). Host, *Cambarus d. diogenes.*

Distribution (in addition to type-locality). —

United States

INDIANA — MARION CO. — Irvington (host, *Cambarus diogenes* subsp.). WHITE CO. — locality unknown (host, *C. diogenes* subsp.).

OHIO — CLINTON CO. — Woodland pool 2 mi E of Westboro, Jefferson Township (host, *Cambarus diogenes* subsp.). FRANKLIN CO. — Lockbourne (host, *C. diogenes* subsp.). JACKSON CO. — 6.8 mi S of Rocky Hill on rt 35 (hosts, *C. laevis* and *Orconectes sanborni*).

Dactylocythere daphnioides (Hobbs, 1955)

Plate XIV, figures 1-5
Plate XLVIII

Entocythere daphnioides Hobbs, 1955: 325-330 (M, H, R, D), 331-332 (R), figs. 1-9. — Ferguson, 1958: 198 (C). — Crawford, 1959: 157, 162, 167 (R). — Walton and Hobbs, 1959: 115 (R). — Hobbs and Walton, 1960: 19, 22, 23 (R). — Crawford, 1961: 244 (A). — Hart and Hobbs, 1961b: 175 (R), 177 (R). — Hobbs and Walton, 1961: 381 (R). — Hart, 1962: 122 (C). — Hobbs and Walton, 1962: 46 (R).

Dactylocythere daphnioides. — Hart, 1962: 130 (C, D, S). — Hobbs and Walton, 1963a: 460 (R). — Hart and Hart, 1966a: 5 (R). — Hobbs and Walton, 1966: 3, 4 (R). — Hobbs, 1967: 5 (K), 9 (R). — Hobbs, Holt, and Walton, 1967: 34 (K), 42 (M, D, H, A, S), 69 (D, E), 27, 50, 51, 54, 55, 57, 59, 62 (A, H), figs. 7f, k, 10a. — Hart and Hart, 1969a: 185 (M, R). — Hobbs and Walton, 1970: 853 (M, R). — Hobbs and Hobbs, 1970: 5-7 (K).

Diagnosis. — MALE: Shell length 570-630 μ (average, 590 μ); height, 320-350 μ (average, 330 μ). Shell oblong-ovate with anterior end distinctly narrower than posterior, and the posteroventral margin of each valve with a conspicuous prominence which may or may not be symmetrically situated. Setae on shell sparsely and irregularly dispersed near margins with very few or none along higher dorsal margin. Scattered flecks of dark brown pigment along dorsal portion of shell behind the eyes present in some specimens. Prominent eye located about one-fourth of shell length from anterior margin.

Copulatory complex with peniferum terminating in bifid projections directed anteriorly; accessory groove present in posterodistal half of peniferum. Clasping apparatus heavy; horizontal and vertical rami forming right or acute angle with one another; internal margin of horizontal ramus with three teeth; external margin entire; three terminal denticles present.

TRIUNGUIS FEMALE: Shell similar to that of male except larger; shell length 620-680 μ (average, 640 μ); height, 370-400 μ (average, 380 μ).

Type-locality. — East fork of Greenbriar River at rt 250, E of Durbin, Pocahontas Co., West Virginia, U.S.A. (types, USNM: holotype, 98406; allotype, 98407; morphotype, 98408; paratype, 98409). Hosts, *Cambarus chasmodactylus* and *C. robustus.*

Distribution (*in addition to type-locality*). —

United States

KENTUCKY — WAYNE CO. — 10 mi SW of Monticello on KY 90 (hosts, *Orconectes juvenilis* and *O. rusticus placidus*).

MISSOURI — IRON CO. — 1 mi SW of Middlebrook, Middlebrook Ck (hosts, *Orconectes hylas* and *O. quadruncus*).

NORTH CAROLINA — ALLEGHANY CO. — 16.9 mi E of jct of parkway & rt 18 (hosts, *Cambarus bartonii* and *C. robustus*); 15.9 mi S of Galax on Blue Ridge Parkway (*C. robustus, C. chasmodactylus, C.* sp.); 3.9 mi W of jct of Blue Ridge Parkway and rt 18 on rt 18 (*C. robustus, C.* sp.). ASHE CO. — 13.7 mi N of Watauga Co. line on rt 16 (*C. robustus, C. chasmodactylus*); Buffalo Ck between rt 88 and Creston (*C. chasmodactylus*); 2.9 mi N of Jefferson on rt 16 (*C. robustus, C. b. bartonii*); 5.1 mi S of Virginia line on rt 16 (*C. b. bartonii, C. robustus*); Three Top Ck, near Creston, off rt 88 (*C. chasmodactylus*). AVERY CO. — Ck at Cranberry (*C. b. bartonii, C. robustus*). TRANSYLVANIA CO. — Looking-glass Ck (host, *C. bartonii*). WATAUGA CO. — 3.5 mi S of Boone (*C. robustus, C. chasmodactylus*); 2 mi S of Vilas on rt 194 (*C. robustus*).

TENNESSEE — CARTER CO. — Doe Run, trib to Watauga R near Elizabethton (host, *Cambarus bartonii*). CLAIBORNE CO. — Sycamore Ck 3.1 mi SE of Tazewell on rt 25E (hosts, *C. bartonii* and *C. longirostris*). CLAY CO. — Big Tree Ck, Hermitage Spring (hosts, *Orconectes compressus* and *O. putmani*). FENTRESS CO. — Little Crab Ck (host, *C.* (*Puncticambarus*) sp. and *C. striatus*). GRAINGER CO. — Indian Ck, 11.4 mi SE of Tazewell on rt 25E (host, *C. bartonii*). SEVIER CO. — West Fork of Little Pigeon R, 1 mi S of Gatlinburg (hosts, *C. bartonii, C. longirostris,* and *O.* sp.).

VIRGINIA — BLAND CO. — Clear Fork Ck near Rocky Gap (*C. chasmodactylus, C. sciotensis*); Wolf Ck near Rocky Gap (*C. sciotensis*). CARROLL CO. — 1.3 mi S of Woodlawn on Rt 221 (*C. robustus, C. chasmodactylus*); 2.3 mi S of Dug Spur, trib to Big Reed Island Ck (*C. b. bartonii, C. robustus*). DICKENSON CO. — Trib of Pound R, 13.8 mi E of Pound on rt 83 (*C. veteranus, C.* sp.). FLOYD CO. — Pine Ck, 2.5 mi S of Check (*C. b. bartonii, C. sciotensis*); 7.9 mi S of Floyd on rt 221 (*C. robustus, C. sciotensis*); 15.5 mi SW of Floyd (*C. chasmodactylus, C. robustus*). GILES CO. — Little Stony Ck, 2.3 mi N of Pembroke (*C. sciotensis, C. b. bartonii*); East R at Glen Lynn (*C. sciotensis, C. chasmodactylus*). GRAYSON CO. — 3.8 mi S of Galax on rt 221 (*C. robustus, C.* sp.); 2 mi NE of N.C. State line on rt 16 (*C. b. bartonii, C. robustus*); 2 mi W of Mouth of Wilson on rt 16 (*C. robustus, C. chasmodactylus*); 4.5 mi SW of Independence on rt 58 (*C. robustus, C. chasmodactylus*); 11.9 mi SW of New River Bridge on rt 58 (*C. robustus, C. b. bartonii*); 0.8 mi S of jct of Parkway and rt 89 (*C. robustus, C.* sp.). MONTGOMERY CO. — Norris Run at mouth (*C. sciotensis*). TAZEWELL CO. — 11.2 mi E of Tazewell on rt 460 (*C. sciotensis*). WYTHE CO. — 1.6 mi SW of Wytheville on rt 11 (Reed Creek) (*C. chasmodactylus*); Reed Ck at Carter Wayside north of Wytheville (*C. chasmodactylus*).

WEST VIRGINIA — SUMMERS CO. — Greenbriar R, 7 mi SE of Hinton at jct of rts 3 & 12 (host, *Orconectes s. sanborni*). WIRT CO. — Cranenest Ck at Peewee (host, *O. s. erismophorus*).

Dactylocythere enoploholca Hobbs and Walton, 1970

Plate XIV, figures 6-9
Plate XLVIII

Dactylocythere enoploholca Hobbs and Walton, 1970: 857-859 (M, H, R, A, D), figs. 3c, d. — Hobbs and Hobbs, 1970: 5-7 (K).

Diagnosis. — MALE: Shell length 440-490 μ (average, 450 μ); height, 250-280 μ (average, 270 μ). Shell ovate with greatest height distinctly posterior to midlength. Submarginal setae present except dorsally between level of posterior margin of eye and that of dorsal portion of peniferum. Sternal spine lacking. Eye pigmented, situated approximately one-eighth shell length from anterior margin.

Copulatory complex with finger guard broad at base and tapering to form narrow distal portion, ventral margin emarginate with posteriorly directed acute tip; peniferum moderately heavy with subtruncate ventral margin and posteroventral "heel"; accessory groove reaching level slightly dorsal to dorsal margin of spermatic loop, with its dorsalmost portion folded and possessing irregular margin; peniferal groove opening anteriorly, its apical width approximately two-thirds that of diameter of vertical ramus of clasping apparatus above rounded shoulder (see below); penis somewhat L-shaped, situated in ventral fourth of peniferum; clasping apparatus, extending ventrally beyond ventral margin of peniferum, with two major bends, but not clearly divisible into vertical and horizontal rami and with major axes of extremities forming angle of approximately, 70 degrees; external borders of both rami entire but that of vertical ramus with rounded shoulder at midlength (level of proximal bend); internal border of horizontal ramus with one large tooth near midlength and with three low elevations immediately proximal to three apical denticles. Both dorsal and ventral fingers moderately slender, latter more than twice length of former, gently curved from base and suddenly curved posteriorly at base of distal third.

TRIUNGUIS FEMALE: Shell length 460-550 μ (average, 500 μ); height, 280-340 μ (average, 300 μ). Shell distinctly larger than that of most males, more highly vaulted posteriorly, and with shallow ventral excavation anterior to mid-length. Submarginal setae present except dorsally between level of eye and genital compex. Eye pigmented and situated as in male.

Genital complex consisting of prominent, but short, J-shaped rod and amiculum, latter protruding little, if at all, between valves.

Relationships. — *Dactylocythere enoploholca* has as its closest relatives *D. chalaza, D. pachysphyrata,* and *D. spinata.* In all four species, the accessory groove extends dorsally approximately to the level of the spermatic loop; the aperture of the peniferal groove is directed anteriorly; and the clasping apparatus is curved at almost the same angle and bears only one major tooth (two in *D. pachysphyrata*) on the internal border of the clasping apparatus. *Dactylocythere enoploholca* differs from the other three, however, in having a heel-like prominence on the posteroventral margin of the peniferum, a prominent rounded shoulder on the external border of the vertical ramus of the clasping apparatus, and in lacking a sternal spine.

Type-locality. — South fork of Holston River at jct of rts 600 and 762, Washington Co., Virginia, U.S.A. (types, USNM: holotype and allotype, 126973). Hosts, *Cambarus longirostris* and *C.* sp.

Distribution. — Known only from type-locality.

Dactylocythere exoura Hart and Hart, 1966

Plate XIV, figures 10-11
Plate XLVIII

Dactylocythere exoura Hart and Hart, 1966a: 5-7 (M, R, D, A), figs. 8, 9. — Hobbs, 1967: 5 (K). — Ferguson, 1968: 501 (C, D). — Hobbs and Walton, 1968: 242, 243 (R). — Hobbs and Walton, 1970: 853 (M, R). — Hobbs and Hobbs, 1970: 5-7 (K).

Diagnosis. — MALE: Shell length 400-460 μ (average, 420 μ); height, 250-290 μ (average, 270 μ). Shell subelliptical in outline, with greatest height at mid-length. Setae very scarce, with scattered ones on posterior shell margin. Posterior shell margin of males and biunguis females nearly always with small pointed protuberance at mid-length or slightly lower; protuberance absent on triunguis females. Eyes present; situated approximately $\frac{1}{5}$ shell length from anterior end.

Copulatory complex with peniferum extending ventrally almost to external border of horizontal ramus of clasping apparatus. Distal portion of peniferum terminating in two apposed points, separated by indistinct membranous bulge directed anteriorly. Accessory groove present, but much reduced in length as compared to that of other members of genus; proximal end reaching barely beyond proximal end of vertical ramus of clasping apparatus. Penis small, directed horizontally, and situated on level slightly distal to proximal end of ramus of clasping apparatus. Finger guard present, well developed, and terminating in single point. Clasping apparatus with horizontal and vertical rami forming right angle; horizontal ramus 1.2 times as long as vertical ramus. External border of horizontal ramus slightly convex, entire; internal border with three small teeth in distal $\frac{1}{4}$, one somewhat larger but less acute tooth situated at mid-length, and another, similar one just distal to juncture of two rami; three very small terminal denticles present. External border of vertical ramus concave proximally, convex distally; internal border convex.

FEMALE: Shell of triunguis female with typical amiculum; shell somewhat larger than biunguis females. Shell length 400-420 μ (average, 410 μ); height, 250-260 μ (average, 255 μ).

Relationships. — This ostracod appears to have its closest affinities with *D. chelomata.* The horizontal and vertical rami of the clasping apparatus of these two ostracods have similar proportions, but in *exoura* both rami are straighter.

Type-locality. — Rattlesnake Creek, trib. of Eagle Creek, 4.3 mi NE of Holbrook, Grant Co., Kentucky, U.S.A. (types, USNM: holotype, allotype, and paratype, 113337). Hosts, *Cambarus d. diogenes, C. ornatus, C. ortmanni,* and *Orconectes rusticus.*

Distribution (in addition to type-locality). —

United States

KENTUCKY — OLDHAM CO. — Harrod's Ck, 0.9 mi E of rt 42, 6.7 mi SW of Sligo (hosts, *C. diogenes* subsp. and *O. rusticus*).

Dactylocythere falcata (Hobbs and Walton, 1961)

Plate XV, figures 1-5
Plate XLIX

Entocythere falcata Hobbs and Walton, 1961: 379-382 (M, H, R, D), figs. 2, 3, 7, 8. — Hart, 1962: 123, 130 (C).

Dactylocythere falcata. — Hart, 1962: 130 (C, D, S). — Hobbs, 1967: 4 (K). — Hobbs, Holt, and Walton, 1967: 34 (K), 35 (D), 42-43 (M, D, H, A, S), 69 (D, E), 18, 20, 25, 41, 45, 50, 52, 54, 55, 61, 63 (A, H), 72, 73, 78, figs. 7j, 10b. — Ferguson, 1968: 501 (C, D). — Hart and Hart, 1969a: 187 (B). — Hobbs and Walton, 1970: 858 (A). — Hobbs and Hobbs, 1970: 5 (K).

Diagnosis. — MALE: Shell length 420-450 μ (average, 430 μ); height, 230-250 μ (average, 250 μ). Shell with conspicuous excavation ventrally; maximum height at anterior portion of posterior quarter; marginal setae relatively evenly dispersed except dorsally where there are few.

Copulatory complex with extremity of peniferum broad and truncate; distal portion strongly curved, the axis of its two extremities forming an angle of 125 degrees; dorsal finger slender; its basal portion much shorter than finger guard and directed anteroventrally; ventral finger slender and curved throughout its length with its tip directed anteriorly; accessory groove present, extending proximally with one-third of its length lying proximal to loop of spermatic duct; penis situated at base of distal fourth of peniferum and with a short curved tip; finger guard stout and straight, with posterodistal angle strongly excised; narrow distal portion bearing a shallow notch at its extremity; clasping apparatus, extending distally to distal extremity of peniferum, scythe-like with external and internal borders entire.

FEMALE: Shell length of biunguis female 380-410 μ (average, 400 μ); height, 210-230 μ (average, 220 μ). Shell of biunguis female not so strongly arched as triunguis female, and amiculum replaced by a small, broadly conical sclerotized projection.

Shell length of triunguis female 430-480 μ (average, 460 μ); height, 250-270 μ (average, 260 μ). Shell of triunguis female with shallow concavity on ventral margin slightly anterior to mid-length; posterior border of shell sloping anteroventrally, posterodorsal margin with a shallow to deep emargination immediately dorsal to amiculum; prominent amiculum and J-shaped rod always present.

Type-locality. — Creek 3 mi S of Charleston, Bradley Co., Tennessee, U.S.A. (types, USNM: holotype, allotype, and morphotype, 105966; paratype, 105967). Host, *Cambarus longirostris.*

Distribution (in addition to type-locality). —

United States

ALABAMA — ETOWAH CO. — 10.4 mi E of Etowah Co. line on rt 278 (hosts, *Cambarus (Puncticambarus)* sp. and *C. (Depressicambarus)* sp.). ST. CLAIR CO. — 4.1 mi E of Blount-St. Clair Co. line on rt 231 (hosts, *Orconectes spinosus, C. diogenes* subsp., *C. jordani,* and *C. (Puncticambarus)* sp.).

GEORGIA — CHATOOGA CO. — Lyerly Ck on S city limits of Lyerly on rt 114 (host, *Orconectes spinosus*); 5.2 mi S of Summerville on US 27 (host, *Cambarus (Puncticambarus)* sp.). FLOYD CO. — 23 mi NW of Rome (host, *C. (Puncticambarus)* sp.). MURRAY CO. — 3 mi W of Chatsworth on rt 52 (host, *C. (Puncticambarus)* sp.). WALKER CO. — locality unknown (host, *O. erichsonianus*). WHITFIELD CO. — 9.3 mi S of Dalton (hosts, *O. spinosus, C. (Puncticambarus)* sp., and *C. (Depressicambarus)* sp.).

MISSOURI — WAYNE CO. — Black R Bridge 7 mi S of Piedmont on rt 34 (host, *Orconectes hylas*).

TENNESSEE — ANDERSON CO. — Vess property between Clinton and Norris (host, *Cambarus diogenes* subsp.); 3.7 mi NE of Norris on rt 71 (hosts, *C. longirostris* ?, *C. bartonii,* and *C.* sp.). CARTER CO. — 3.2 mi W of Roan Mountain on rt 19E, Roaring Ck (hosts, *C. longirostris* and *C. b. bartonii*). CLAIBORNE CO. — Caney Valley Ck, trib to Norris Lake 6.1 mi SE of Tazewell on rt 25E (host, *C. b. bartonii*); Sycamore Ck 3.1 mi SE of Tazewell on rt 25E (hosts, *C. longirostris* and *C. b. bartonii*); 5.1 mi N of Tazewell on rt 25E (hosts, *C. parvoculus* and *C. b. bartonii*); spring, trib to Old Town Ck, trib to Powell R, at Tazewell (host, *C. b. bartonii*); 3.1 mi SW of New Tazewell on rt 33 (host, *C. b. bartonii*). FRANKLIN CO. — Trib to Elk R 11.4 mi W of Winchester on rt 64 (hosts, *O. spinosus* and *O. mirus*). GRAINGER CO. — 1.1 mi NE of Thorn Hill on rt 25E, Indian Ck (host, *C. b. bartonii*); trib to Holston R 4.9 mi SW of Rutledge on rt 11W (host, *C. b. bartonii*); 0.9 mi SW of Rutledge on rt 11W (hosts, *O. erichsonianus* and *C. b. bartonii*); 5.3 mi NE of Rutledge on rt 11W (host, *C. b. bartonii*). HAWKINS CO. — Stream 10.8 mi W of Rogersville on rt 11W (host, *C. b. bartonii*). LAWRENCE CO. —Shoal Ck 1.4 mi E of Lawrenceburg on rt 64 (host, *O. spinosus*); Crowson Ck 1.8 mi W of Lawrenceburg on rt 64 (host, *C. longirostris*); 1.4 mi E of Lawrenceburg on rt 64 (host, *C. longirostris*). MARION CO. — Ck on Marion-Sequatchie Co. line on rt 28 (host, *O. erichsonianus*). McMINN CO. — Oostanaula Ck, trib to Hiwassee River near Athens (hosts, *O. erichsonianus, C.* sp., *C. diogenes* subsp., and *C. b. bartonii*); Matlock Spring Ck 7 mi N of Athens, TN (host, *C. b. bartonii*). MONROE CO. — Ck 7 mi N of Madisonville (host, *C. b. bartonii*). MONTGOMERY CO. — Louise Ck 13.9 mi S of Clarksville (hosts, *O. placidus* and *O.* sp.). POLK CO. — 11.4 mi S of Benton (hosts, *C.* (*Depressicambarus*) sp. and *C.* (*Puncticambarus*) sp.); Greasy Ck 4 mi from jct of rts 64 and 30, on 30 (host, *C. b. bartonii*). ROANE CO. — 2 mi S of Harriman on rt 27 (hosts, *C. longirostris, C. b. bartonii,* and *O.* sp.). RHEA CO. — 1 mi W of Dayton on rt 30 (host, *O. forceps*). SEQUATCHIE CO. — S city limits of Dunlap on rt 28 (host, *O. erichsonianus*). SULLIVAN CO. — clear sluggish stream, 15 feet wide, 6.9 mi N of Elizabethton on rt 19E (hosts, *C. longirostris* and *C. b. bartonii*); 6 mi NE of Kingsport on rt 11 (hosts, *C. dubius, C. b. bartonii,* and *C. longirostris*); S Fork of Holston R 4 mi SE of Kingsport (hosts, *C. longirostris* and *C. b. bartonii*). UNION CO. — Suckstone Ck, trib to Bull Run 8.6 mi SW of Maynardsville on rt 33 (hosts, *C. longirostris* and *C. b. bartonii*).

VIRGINIA — LEE CO. — small stream 4 mi W of Ewing (hosts, *Cambarus parvoculus* and *C. b. bartonii*). MECKLENBERG CO. — Pole Branch on rt 58 between rt 1 and Boydon (hosts, *Procambarus a. acutus* and *C. acuminatus*). PATRICK CO. — 1.6 mi S of Patrick-Franklin Co. line on rt 40 (hosts, *C. longulus* and *C. acuminatus*). SCOTT CO. — Big Moccasin Ck at Snowflake on rt 71 (host, *O. erichsonianus*); Stock Ck 1.8 mi NW of Natural Tunnel on rt 58 (host, *O. juvenilis*).

Canada

ONTARIO — DURHAM CO. — 4 mi E of Port Hope on rt 2, flows into Lake Ontario in about 300 yards (host, *O. propinquus*). (This distribution record is questionable.)

Dactylocythere jeanae Hobbs, 1967

Plate XV, figures 6-10
Plate XLIX

Dactylocythere jeanae Hobbs, 1967: 3 (A), 5 (K), 6-8 (M, H, R, A, D), figs. 1g, h, i. — Hart and Hart, 1969a: 187, 188 (M, R, D, H). — Hobbs and Walton, 1970: 853 (M, R), 860 (R). — Hobbs and Hobbs, 1970: 6 (K). — Peters, 1971: 100.

Diagnosis. — MALE: Shell length 460-510 μ (average, 490 μ); height, 270-290 μ (average, 280 μ). Shell highest posterior to midlength, much higher posteriorly than anteriorly; ventral margin with very shallow excavation just anterior to midlength; submarginal setae present except dorsally and closer together anteriorly and posteriorly than ventrally. Pigmented eye present.

Peniferum with thickened, rounded, ventral extremity; accessory groove extending posterodorsally to level of dorsal extremity of spermatic loop; penis L-shaped with bulbous area in horizontal position and subequal in length to slender, terminal vertical portion. Finger guard essentially bifid with anterior ramus longer than posterior. Dorsal finger comparatively heavy with apical part appearing bifid (probably spatulate); ventral finger slender and gently curved to base of terminal "segment." Clasping apparatus with extensions of two rami forming angle of 60° to 80° and without thickening at their junction; vertical ramus with angular prominence just proximal to midlength of posterior surface, anterior border entire; horizontal ramus with external (ventral) border entire and internal border with two teeth on distal third; apex with three dorsodistally directed denticles.

TRIUNGUIS FEMALE: Shell length 490-560 μ (average, 500 μ); height, 260-340 μ (average, 300 μ). Shell highest posterior to midlength and much higher posteriorly than anteriorly, ventral margin with distinct concavity anterior to midlength. Submarginal setae present except dorsally from level of eye to amiculum. Amiculum and J-shaped rod present; however, latter often resembles inverted L more than usual J.

Relationships. — *Dactylocythere jeanae* has its closest affinities with *D. suteri.* Their penifera are distinctly similar and the tapering, L-shaped clasping apparatus without a thickening at the junction of the two rami is found in only two other species of the genus, *D. chelomata* and *D. striophylax.* The structure of the penifera of the latter two, however, suggests that *D. jeanae* is not nearly so closely related to them as to *D. suteri.* The combination of an angular hump on the posterior side of the vertical ramus of the clasping apparatus, two teeth on the internal border of the horizontal ramus, a bifid finger guard, and a rounded anteroventral portion of the peniferum is unique in *D. jeanae.*

Type-locality. — Roadside ditch in Virginia Beach on Route 605, near Stumpy Lake, Princess Anne Co., Virginia, U.S.A. (types, USNM: holotype and allotype, 113475; paratype, 113475; paratype, 113476). Hosts, *Cambarus d. diogenes* and *Procambarus a. acutus.*

Distribution (*in addition to type-locality*). —

United States

DISTRICT OF COLUMBIA — Howard University Reservoir (host, *Cambarus d. diogenes*); "Burville, D.C." near a brook, trib to E branch of Potomac R (host, *C. d. diogenes*).

MARYLAND — MONTGOMERY CO. — Plummer Island in Potomac R (host, *Cambarus d. diogenes*). PRINCE GEORGES CO. — Laurel (host, *C. d. diogenes*); locality unknown (host, *C. d. diogenes*).

NEW JERSEY — CUMBERLAND CO. — 7 mi W of Bridgeton (host, *Cambarus d. diogenes*) (Hobbs 1967: 8).

NORTH CAROLINA — GREENE CO. — 3.5 mi N of Snow Hill on rt 258 (host, *Cambarus d. diogenes*) (Hobbs 1967: 8).

VIRGINIA — AMHERST CO. — Buffalo R at Forks of Buffalo (hosts, *Cambarus longulus* and *C. b. bartonii*). DINWIDDLE CO. — Rowanty Ck 3.3 mi S of Reams Station on rt 301 (host, *Orconectes virginiensis*). FAIRFAX CO. — Meadow near small stream on VA side of Potomac R 1 mi above D.C. (host, *C. d. diogenes*). HANOVER CO. — 1.9 mi E of Doswell on rt 602 (hosts, *Procambarus a. acutus* and *C. d. diogenes*). JAMES CITY CO. — Underground springs near James R (host, *C. d. diogenes*). KING & QUEEN CO. — 6.4 mi E of West Point on rt 14 (host, *C. d. diogenes* and *C. acuminatus*). LOUISA CO. — backwater of South Anna R on rt 22 (host, *C. d. diogenes*) (Hobbs 1967: 7-8). NORFOLK CO. — foot of Little John Drive (host, *C. d. diogenes*). NORTHAMPTON CO. — Seaview (host, *C. d. diogenes*) (Hobbs 1967: 8). PRINCESS ANNE CO. — Acerdale (host, *C. d. diogenes*) (Hobbs 1967: 7-8); Landstown Road off Princess Anne Road, Virginia Beach (host, *C. d. diogenes*) (Hobbs 1967: 8); rt 605 across Stumpy Lake (hosts, *C. diogenes* and *P. a. acutus*). SOUTHAMPTON CO. — 4 mi SE of Sebrell on VA 35 (host, *C. d. diogenes*). SURRY CO. — 9.5 mi N of Surry Court House (host, *C. d. diogenes*).

Dactylocythere koloura Hart and Hart, 1971

Plate XV, figures 11-15
Plate XLVIII

Dactylocythere koloura Hart and Hart, 1971: 109-110 (M, H, R, A, D), figs. 42-c.

Diagnosis. — MALE: Shell length 480-510 μ (average, 490 μ); height, 260-300 μ (average, 290 μ). Shell subovoid, with posteroventral projection; venter weakly convex, curving to contracted anterior end; dorsum evenly convex anteriorly to about ⅕ shell length from posterior end, followed by shallow concavity; posterior margin convex to projection. Greatest height ⅓ shell length from posterior end. Eye pigmented.

Copulatory complex with peniferum large and extending ventrally to about level of external border of horizontal ramus of clasping apparatus. Distal portion of peniferum directed at approximately 70° angle to proximal portion. Accessory groove extending above level of spermatic loop; penis located in ventral ⅕ of peniferum. Main body of penis almost horizontal, with tip curved and directed ventrally.

Clasping apparatus with distinct horizontal and vertical rami, meeting at angle of approximately 70°. Horizontal ramus slightly longer than vertical. Internal border of horizontal ramus with one tooth at midlength and three terminal denticles; external border smoothly convex. External border of vertical ramus concave in proximal ½, convex in distal ½, with shoulder-like angle at midlength. Thickest portion of vertical ramus at midlength.

Finger-guard small and slightly bifid.

TRIUNGUIS FEMALE: Shell length 490-530 μ (average, 510 μ); height, 280-310 μ (average, 290 μ). Shell similar to that of male, but with greater height difference between anterior and posterior ends. Genital complex of triunguis female consisting of rod, which follows conspicuous concavity of shell, and amiculum, which trends anteroventrally from posteroventral end of rod.

Relationships.— The shell resembles that of *Dactylocythere daphnioides* and *D. pughae,* but the clasping apparatus does not resemble either of these. *D. koloura* is the only species of the genus having the combination of an accessory groove extending distinctly dorsal to the spermatic loop and a clasping apparatus with a single tooth on the internal border.

Type-locality. — Montgomery Bell State Park, Dickson Co., Tennessee, U.S.A. (types, USNM: holotype, 138449; allotype, 138450). Host, *Orconectes* species.

Distribution. — Known only from type-locality.

Dactylocythere leptophylax (Crawford, 1961)

Plate XVI, figures 1-5
Plate XLVIII

Entocythere leptophylax Crawford, 1961: 238-242 (M, H, R, A, D), 243 (R), figs. 9-14. — Hart, 1962: 123 (C).

Dactylocythere leptophylax. — Hart, 1962: 130 (C, D, S). — Crawford, 1965: 150 (A). — Hobbs III, 1965: 159 (R), 160 (K), 163 (R). — Hobbs and Hart, 1966: 36 (K, M). — Hobbs, 1967: 4 (K). — Ferguson, 1968: 501 (C, D). — Hobbs and Hobbs, 1970: 5-7 (K).

Diagnosis. — MALE: Shell length 430-460 μ (average, 450 μ); height, 240-290 μ (average, 270 μ). Shell subelliptical, with anterior end narrower than posterior. Margins smoothly rounded except for slight concavity on ventral margin just anterior to midlength. Height greatest slightly posterior to midlength. Setae scattered along free margin of valve but more numerous on anterior end; lateral surface with few widely spaced setae. Eye located about one-sixth of valve length from anterior end.

Finger guard of copulatory complex heavily sclerotized, narrow, and long, with distal end trilobed. Posteroventral portion of peniferum sclerotized, with two small tooth-like projections at ventral extremity. Horizontal and vertical rami of clasping apparatus not clearly delimited. The former approximately ½ as long as the latter, together forming angle of about 60 degrees. The distal ramus is enlarged distally, with six or more subapical serrations, of which the two proximal ones are most prominent. Accessory groove present.

TRIUNGUIS FEMALE: Shell length 460-510 μ (average, 490 μ); height, 300-330 μ (average, 310 μ). Shell higher posteriorly than that of male, and ventral margin has slight concavity just anterior to midlength. The dorsal margin is prominently arched, the anterior end smoothly rounded, and the posterodorsal margin concave.

Genitalia of the triunguis female subtriangular, showing considerable variation. Unlike other members of this group, the triunguis female is without the deeply staining mass associated with the genitalia.

Type-locality. — East fork of Chattooga River at jct with Indian Camp Creek on grounds of U.S. Fisheries Station, 21 mi NW of Walhalla, Oconee Co., South Carolina, U.S.A. (types, USNM: holotype, 197458; allotype, 107460; paratype, 107459). Hosts *Cambarus asperimanus* and *C. b. bartonii.*

Distribution (in addition to type-locality). —

United States

FLORIDA — HOLMES CO. — Clear Run below spring (host, *Procambarus versutus*).

GEORGIA — BANKS CO. — 0.9 mi SW of Homer on rt 98 (hosts, *Procambarus raneyi, Cambarus b. bartonii, C. latimanus,* and *C.* (*Depressicambarus*) sp.). HART CO. — 7.5 mi S of Hartwell on rt 172 (hosts, *C.* (*Depressicambarus*) sp. and *C. b. bartonii*). JACKSON CO. — 2.5 mi N of Jefferson on rt 82 (host, *C. latimanus*). PULASKI CO. — 5 mi W of Hawkinsville on rt 27 (host, *P. spiculifer*). PUTNAM CO. — Rock Eagle State Park at Eatonton (hosts, *C. latimanus* and *P. spiculifer*). UNION CO. — Neil's Gap SE of Blairsville (host, *C. b. bartonii*). WHITE CO. — Duke's Ck 6.6 mi N of Cleveland (host, *C. b. bartonii*); 7.4 mi N of Cleveland on rt 75 (host, *P. spiculifer*).

TENNESSEE — ANDERSON CO. — locality unknown (host, *Orconectes forceps*). POLK CO. — Greasy Ck 4 mi from jct of rts 64 and 30 on 30 (host, *Cambarus b. bartonii*); 5.5 mi W of Ducktown on rt 64 (host, *C. b. bartonii*).

Dactylocythere macroholca Hobbs and Hobbs, 1970

Plate XVI, figures 6-10
Plate XLIX

Dactylocythere macroholca Hobbs and Hobbs, 1970: 6 (K), 9, 10 (M, H, R, A, D), figs. 3a-d.

Diagnosis. — MALE: Shell length 580-620 μ (average, 600 μ); height, 350-410 μ (average, 390 μ). Shell subovate with greatest height posterior to midlength; ventral margin with broad shallow concavity, otherwise entire. Submarginal setae present in limited numbers anteriorly, ventrally, and very sparse posteriorly. Sternal spine lacking. Eye pigmented, located approximately one-fifth shell length from anterior margin.

Copulatory complex possessing finger guard with broad base tapering anteroventrally to fingerlike projection, ending without tubercles; peniferum elongate with greatest width at level of penis; distal apex of peniferum twisted and anteroventral portion with small mesially folded flange; accessory groove very long, extending well beyond dorsal one-seventh of peniferum; apex of peniferal groove approximately one-fourth greatest width of vertical ramus of clasping apparatus; penis small and L-shaped and situated in ventral third of peniferum; clasping apparatus heavy, divisible into horizontal and vertical rami disposed at approximately 90 degrees; internal border of vertical ramus entire, its external border bearing prominent angular shoulder slightly proximal to midlength, and distal half of ramus approximately 1.5 times as broad as width immediately above shoulder; external border of horizontal ramus entire and tapering; internal border bearing four teeth, and apex of ramus with three denticles; both dorsal and ventral fingers slender, dorsal one bifurcate; ventral finger bent at midlength, forming angle of approximately 130 degrees.

TRIUNGUIS FEMALE: Shell length 610 μ; height 400-420 μ (average, 410 μ). Shell obovate, highly vaulted, and highest slightly posterior to midlength; submarginal setae as in male except more abundant posteriorly. Genital complex composed of large J-shaped rod and massive amiculum, latter protruding well beyond posterior margin of shell; two or three posterodorsally directed papillae situated between dorsal portion of J-shaped rod and amiculum. Eye pigmented, located one-third shell length from anterior margin.

Relationships — Only two other entocytherids have such long accessory grooves in the peniferum as is exhibited by *Dactylocythere macroholca*: *D. amicula* and *D. mecoscapha*. Both of these have similarly constructed finger guards and clasping apparatus, and the amiculum of *D. amicula* protrudes from the shell even more conspicuously than does that of *D. macroholca*. Neither, however, has an angular shoulder on the external border of the vertical ramus of the clasping apparatus, and, although the finger guard of *D. mecoscapha* is broad basally, it terminates in two ventrally directed tubercles and the amiculum of the female is not nearly so conspicuous. Of the three, only *D. amicula* has a prominent sternal spine.

Type-locality. — Trammel Creek and tributary, 3.6 mi S of Scottsville on rt 31E, Allen Co., Kentucky, U.S.A. (types, USNM: holotype, 126255; allotype, 126256). Hosts, *Orconectes juvenilis* (Hagen), *Cambarus diogenes* subsp., *Cambarus tenebrosus* Hay, *C. striatus* Hay, and *C.* sp.

Distribution (in addition to type-locality). —

United States

KENTUCKY — BATH CO. — 4.2 mi E of Owingsville on rt 60 (hosts, *O. rusticus, O. juvenilis,* and *C.* (*Depressicambarus*) sp.). MASON CO. — 3.4 mi W of Rectorville on rt 10 (host, *C. laevis*).

Dactylocythere mecoscapha (Hobbs and Walton, 1960)

Plate XVII, figures 1-5
Plate XLVIII

Entocythere mecoscapha Hobbs and Walton, 1960: 19-23 (M, H, R, A, D), figs. 17-20. — Hobbs and Walton, 1961: 381 (R). — Hobbs and Walton, 1962: 46 (R). — Hart, 1962: 123 (C).

Dactylocythere mecoscapha. — Hart, 1962: 130 (C, D, S). — Hobbs, 1967: 4 (K). — Ferguson, 1968: 501 (C, D). — Hobbs and Hobbs, 1970: 5-7 (K), 10 (R).

Diagnosis. — MALE: Shell length 450-500 μ (average, 480 μ); height, 240-300 μ (average, 270 μ). Shell subelliptical, with distinct excavation on ventral margin slightly anterior to midlength; maximum height in posterior half of shell. Marginal setae evenly dispersed except on dorsal side, where few are present.

Copulatory complex with short, straight finger guard bifid distally. Dorsal finger long and directed anteroventrally; ventral finger slender and gently curved throughout its length. Peniferum bifid ventrally; accessory groove long, slender, and extending into proximal third of peniferum. Clasping apparatus angular; external border entire, internal border with three teeth in distal half, distal extremity with three denticles.

FEMALE: Shell length 440-470 μ (average, 460 μ); height, 280-300 μ (average, 300 μ). Shell occasionally with a slight concavity on ventral margin slightly anterior to midlength; posterior margin with one or two undulations, one always present ventral to terminal portion of ruffled skirt. Prominent amiculum and J-shaped rod present.

Relationships. — *Dactylocythere mecoscapha* is related to *D. runki* and *D. daphnioides*. Like them, it possesses both a finger guard and an accessory groove on the copulatory complex. It differs from *D. daphnioides* in lacking a projection on the posteroventral margin of the shell, and from both of them in having the distal portion of the base terminating in an acute projection and having a comparatively longer and more slender accessory groove.

It should be noted that the specimens from Meiggs County are both shorter and not so high as specimens from the type-locality.

Type-locality. — South Chestuee Creek, 0.2 mi W of Ocoee on rt 64, Polk Co., Tennessee, U.S.A. (types, USNM: holotype, 105960; allotype, 105961; paratypes, 108184 and 108185). Hosts, *Cambarus longirostris, C.* (*Depressicambarus*) sp., and *Orconectes erichsonianus*.

Distribution (*in addition to type-locality*). —

United States

ALABAMA — COLBERT CO. — Bear Ck 12.6 mi S of Tuscumbia on US 43 (host, *Cambarus girardianus*). LAUDERDALE CO. — 18.3 mi W of Athens on US 72 (host, *Orconectes spinosus*); Cypress Ck 3 mi N of Florence (hosts, *O. compressus* and *O. spinosus*).

GEORGIA — WALKER CO. — Chickamauga National Military Park near Rossville (hosts, *Procambarus lophotus, Orconectes erichsonianus,* and *Cambarus* (*Depressicambarus*) sp.).

SOUTH CAROLINA — HAMPTON CO. — 1.4 mi SE of Early Branch on rt 28; 13.8 mi SE of Varnville (hosts, *Procambarus lunzi* and *P. barbatus*).

TENNESSEE — FRANKLIN CO. — Trib to Elk River 11.4 mi W of Winchester on rt 64 (hosts, *Orconectes spinosus, O. mirus, Cambarus girardianus,* and *C. striatus*). LAWRENCE CO. — Shoal Ck 1.4 mi E of Lawrenceburg on rt 64 (host, *O. spinosus* and *C. longirostris*); Crowson Ck 1.8 mi W of Lawrenceburg on rt 64 (host, *C. longirostris*). McMINN CO. — 7.5 mi SW of Athens on US 11 (host, *O. erichsonianus*). MEIGGS CO. — 1.9 mi N of Georgetown (host, unknown) (Hobbs and Walton 1960: 22).

Dactylocythere megadactylus Hart and Hart, 1971

Plate XVII, figures 6-10
Plate XLVIII

Dactylocythere megadactylus Hart and Hart, 1971: 110-111 (M, H, R, A, D), figs. 5a-e.

Diagnosis. — MALE: Shell length 380-510 μ (average, 490 μ); height 230-300 μ (average, 280 μ). Shell ovoid, with greatest height $\frac{1}{3}$ shell length from posterior end. Eye pigmented. Sternal spine absent.

Copulatory complex with peniferum large, convex posteriorly in distal $\frac{1}{6}$ of peniferum, concave in dorsal $\frac{1}{6}$, evenly convex in proximal $\frac{2}{3}$. Peniferum extending to level of internal border of clasping apparatus. Distal portion of peniferum appearing bifid with connecting membranous area. Accessory groove extending dorsally slightly beyond dorsal margin of spermatic loop, narrow apically, widening near junction with peniferal groove. Penis horizontal, with tip directed ventrally.

Horizontal ramus of clasping apparatus slightly longer than vertical, with greatest width at junction; angular "heel" present on external border at junction of rami. External border of vertical ramus slightly arched. External border of horizontal ramus evenly convex; internal border with one indistinct tooth and three terminal denticles on holotype. Other specimens with three indistinct teeth marked by horizontal lines or grooves extending proximally along midportion of ramus.

Finger guard large and heavy, bifid terminally; basal posterior portion produced in prominent bulge.

FEMALE: Shell length 540 μ; height, 340 μ (one specimen). Shell of triunguis female subovoid, with greatest height slightly posterior of midlength of shell. Genital complex with amiculum extending along anterior and sides of sinuous genital rod.

Relationships. — This species most resembles *Dactylocythere mecoscapha,* but that species has a much longer accessory groove with more than half length of groove situated dorsal to spermatic loop. The female genital complex also resembles *D. mecoscapha,* but the amiculum of *megadactylus* extends farther ventrally from the genital rod.

Type-locality. — Rt 49, 2.8 mi NE of Randolph Co. line, Randolph Co., North Carolina, U.S.A. (types, USNM: holotype and paratypes, 138451; allotype, 138452; paratypes, 138453 and 138454). Hosts, *Procambarus a. acutus* and *Cambarus acuminatus.*

Distribution (in addition to type-locality). —

United States

NORTH CAROLINA — ROWAN CO. — Trib to Second Creek 5.5 mi E of Cleveland on rt 70 (host, *Cambarus acuminatus*). YADKIN CO. — 5.5 mi S of Elkin on rt 21 (hosts, *C. acuminatus* and *C. bartonii*).

VIRGINIA — BOTETOURT CO. — Stream at Blue Ridge (host, *Cambarus bartonii*).

Dactylocythere myura Hobbs and Walton, 1970

Plate XVII, figures 11-14
Plate XLIX

Dactylocythere myura Hobbs and Walton, 1970: 859-860 (M, H, R, A, D), figs. 3e, f. — Hobbs and Hobbs, 1970: 5-7 (K).

Diagnosis. — MALE: Shell length 460-480 μ (average, 470 μ); height, 250-270 μ (average, 270 μ). Shell elongate ovate with greatest height at midlength. Submarginal setae present anteriorly, posteriorly, and ventrally, but none present dorsally between level of eye and dorsal portion of peniferum. Sternal spine lacking. Eye pigmented, situated approximately one-fifth shell length from anterior margin.

Copulatory complex with finger guard rather heavy, its posterior margin concave and its oblique distal margin with three prominences of which anteriormost extending considerably farther ventrally than posterior one; peniferum moderately heavy with rounded ventral margin lacking tubercles, emarginations, or scallops; accessory groove reaching level of dorsal margin of spermatic loop with simple round dorsal extremity; peniferal groove opening anteriorly, its apical width narrow, no more than one-fourth least diameter of vertical ramus of clasping apparatus; penis L-shaped and situated at base of distal fourth of peniferum; clasping apparatus also L-shaped with vertical ramus slightly bowed anteriorly, extending ventrally beyond peniferum, clearly divisible into vertical and horizontal rami, and major axes forming angle of approximately 85 degrees; external borders of both rami and internal border of vertical ramus entire, that of vertical ramus lacking shoulder; internal border of horizontal ramus usually without teeth but with three low elevations distal to midlength (elevations occasionally subacute), and bearing three or four small dorsally directed denticles. Dorsal finger somewhat heavier than ventral and terminating in bifid tip; ventral finger gently curved throughout its length.

TRIUNGUIS FEMALE: Shell length 480 μ; height, 270-290 μ (average, 280 μ). Eye pigmented and situated slightly more anteriorly than that of male. Shell while scarcely longer than that of male, distinctly more highly vaulted with greatest height posterior to midlength, and with much steeper slope posterodorsally; ventral margin with only faintest indication of shallow excavation anterior to midlength. Submarginal setae distributed as in male.

Genital complex consisting of prominent, long J-shaped rod and long amiculum, latter sometimes slightly protruding between valves.

Relationships. — While *D. myura* is not obviously closely allied to any other species of the genus, the comparatively slender clasping apparatus and the finger guard with three lobes of which the anterior one extends farthest distally are somewhat like those found in *D. suteri.* It may be readily separated from the latter, however, by the more angular clasping apparatus which bears no more than one distinct tooth proximal to the apical denticles, and the distal margin of the finger guard is oblique rather than subtruncate. Its more distant relatives include *D. jeanae* and *D. striphylax;* in neither of these, however, is the finger guard distinctly trilobed distally.

Type-locality. — Crayfish burrows on spring-fed stream bank 3.5 mi SW of Chilhowie, Washington Co. (Holston River drainage), Virginia, U.S.A. (types, USNM: holotype and allotype, 126975). Host, undescribed crayfish related to *Cambarus dubius.*

Distribution (in addition to type-locality). —

United States

VIRGINIA — SMYTH CO. — 1.0 mi SW of Chilhowie off rt 181 (host, *Cambarus dubius*).

Dactylocythere pachysphyrata Hobbs and Walton, 1966

Plate XVIII, figures 1-3
Plate XLIX

Dactylocythere pachysphyrata Hobbs and Walton, 1966: 3-4 (M, H, R, A, D), figs. leg. — Hobbs, 1967: 5 (K). — Hobbs and Walton, 1970: 853 (M, R), 858 (R), 862 (R). — Hobbs and Hobbs, 1970: 5-7 (K).

Diagnosis. — MALE: Shell length 420-440 μ (average, 430 μ); height, 240-250 μ (average, 250 μ). Eye present. Shell highest posterior to midlength and with dorsal and ventral margins more gently tapered anteriorly than posteriorly; submarginal setae limited largely to anterior, anterodorsal, posterior and posterodorsal areas, nowhere abundant, rare ventrally, and absent dorsally.

Copulatory complex with posterior margin of peniferum entire; anteroventral extremity directed ventrally and posteroventral extremity directed anteriorly. Penis small, its length about one-fourth of anterior-posterior dimension of distal portion of peniferum, and situated at level of base of ventral two-fifths of peniferum. Accessory groove extending dorsally to level of dorsalmost extremity of loop of spermatic duct. Dorsal finger moderately stout with its terminal seta extending to ventral margin of finger guard; ventral finger slender, curved throughout its length, somewhat strongly so at base of proximal two-fifths. Finger guard prominent with its anterior margin straight and its ventral margin slightly emarginate but without distinct prominences. Clasping apparatus with vertical and horizontal rami indistinctly delimited by thickened "ankle"; vertical ramus with a bend of about 30° at distal end of proximal two-fifths; internal border of vertical ramus inclined anteroventrally at approximately a 45° angle to main axes; internal border of horizontal ramus with three teeth; apex of clasping apparatus with three denticles; external borders of both rami entire, their posteroventral extensions forming a right angle.

FEMALE: Unknown.

Relationships. — *Dactylocythere pachysphyrata* is related to the same group of species as is *D. brachystrix.* While they share a number of features in common, the most conspicuous is the thickened "ankle" of the clasping apparatus. It is less closely allied to *D. steevesi,* in which the vertical ramus of the clasping apparatus is more uniformly thickened. It may be distinguished from *daphnioides* by lacking a posteroventral extension of the shell; from *chalaza* by possessing more than one tooth on the internal border of the clasping apparatus; and from *runki* by possessing a much more prominently thickened "ankle."

Type-locality. — Otter Creek, about 3.5 mi upstream of Obed River, Cumberland Co., Tennessee, U.S.A. (types, USNM: holotype, 111253; paratype, 111254).

Distribution. — Known only from type-locality.

Dactylocythere phoxa Hobbs, 1967

Plate XVIII, figures 4-7
Plate XLIX

Dactylocythere phoxa Hobbs, 1967: 5 (K), 8-9 (M, R, D, H), figs. 1d, e, f. — Hobbs and Hobbs, 1970: 5-7 (K).

Diagnosis. — MALE: Shell length 460 μ; height, 260 μ (one specimen known). Eye present. Shell highest posterior to midlength, dorsal margin tapering ventrally much more suddenly posteriorly than anteriorly; ventral margin of shell with only a slight concavity just anterior to midlength; submarginal setae evenly spaced except dorsally where absent.

Peniferum with ventrally and slightly anteriorly directed subacute apex; apical portion of peniferal groove (pg) directed more strongly ventrally than that in other members of the genus, width of groove at apex about $\frac{1}{8}$ of maximum anterior-posterior diameter of vertical ramus of clasping apparatus; accessory groove extending posterodorsally to level of dorsal extremity of spermatic loop; penis L-shaped, with bulbous area in horizontal plane and distinctly longer than slender vertical portion. Finger guard simple with distinct groove extending from base to distal end along posterior surface. Dorsal finger comparatively slender with bifid tip; ventral finger gently curved but with long, nearly straight portion just proximal to bulbous area. Clasping apparatus L-shaped with subangular hump at posterior junction of two rami; external borders of both rami and internal border of vertical ramus entire. Internal border of horizontal ramus with comparatively long conical tooth near midlength and three low, but pointed, teeth along distal half; apex of ramus with two teeth.

FEMALE: Shell length 460 μ; height, 280 μ. Eye present. Shell subreniform with proximal emargination ventrally; submarginal setae present except dorsally between levels of eye and amiculum, nowhere conspicuously abundant. Amiculum and J-shaped rod present; amiculum, although partially obscured by adhering debris, fully ruffled and extending beyond posterodorsal margin of shell.

Relationships. — *Dactylocythere phoxa* is most closely related to *D. daphnioides* and *D. runki,* but differs from both in possessing a short vertical ramus of the clasping apparatus, and the proximal tooth on the internal border of the horizontal ramus is conical rather than being the proximal member of a saw-toothed edge. The narrow, almost ventrally directed peniferal groove is not found in any other member of the genus.

Type-locality. — Small stream immediately E of Maxwelton, Greenbrier Co., West Virginia, U.S.A. (types, USNM: holotype, 113477; allotype, 113478). Host, *Cambarus b. bartonii.*

Distribution. — Known only from type-locality.

Dactylocythere prinsi Hobbs and Walton, 1968

Plate XVIII, figures 8-10
Plate XLIX

Dactylocythere prinsi Hobbs and Walton, 1968: 242-243 (M, H, R, A, D), 247 (A), figs. 2e-g. — Hobbs and Hobbs, 1970: 5-7 (K), 8 (R).

DIAGNOSIS. — MALE: Shell length 400 μ; height, 220 μ (one specimen). Eye pigmented, subcircular. Shell oblong-ovate and with very shallow concavity on ventral margin. Submarginal setae present along entire border of shell and rather closely spaced posteriorly.

Peniferum with posteroventral extremity heavily cornified and strongly curved, forming anteriorly directed spine; tip of spine almost in contact with tip of apparently slender, acute, corneous bar on anterior surface; penis long with curved tip directed ventromesially; accessory groove short, extending posterodorsally only to lower level of loop of spermatic duct. Finger guard slender, long, reaching ventrally almost to level of ventral extremity of peniferum, with distal fourth narrower than proximal portion; dorsal finger of moderate size with tip reaching ventrally to about same level as finger guard and peniferum; ventral finger curved near midlength with horizontal and vertical portions situated at 90 degree angle, internal border of horizontal segment with distinct emargination. Clasping apparatus clearly divisible into horizontal and vertical rami situated at approximately 90 degrees to one another; external border entire; internal border of vertical ramus strongly concave, that of horizontal ramus with three evenly spaced teeth; apex of clasping apparatus with three small teeth.

FEMALE: Unknown.

Relationships. — Among the closest relative of *Dactylocythere prinsi* are *D. crawfordi* and *D. exoura.* The similarities are evident in the short, recurved ventral portion of the peniferum, and the arrangement of the teeth on the horizontal ramus of the clasping apparatus. *D. prinsi* may be distinguished from *D. crawfordi,* however, by the longer, slenderer finger guard, in lacking a shoulder on the external border of the longer vertical ramus of the more angular clasping apparatus, and the much narrower accessory groove in the peniferum. It differs from *D. exoura* in possessing a longer, slenderer finger guard and much more prominent teeth on the internal border of the clasping apparatus. The latter is unique in the genus.

Type-locality. — Whitewater River (Savannah R drainage) at rt 107, Jackson Co., North Carolina, U.S.A. (types, USNM: holotype, 123325; paratype, 123326). Host, *Cambarus* sp.

Distribution. — Known only from type-locality.

Dactylocythere prionata (Hart and Hobbs, 1961)

Plate XVIII, figures 11-13
Plate XLIX

Entocythere prionata Hart and Hobbs, 1961b: 174 (K), 178-180 (M, R, H, A, D), figs. 15-17.

Dactylocythere prionata. — Hart, 1962: 130 (C, D, S). — Hart and Hart, 1966a: 5, 8 (A). — Hobbs, 1967: 4 (K). — Ferguson, 1968: 501 (C, D). — Hobbs and Hobbs, 1970: 5-7 (K). — Hobbs and Barr, 1972: 81 (D, H).

Diagnosis. — MALE: Shell length 460 μ; height, 270 μ. Shell subovate with higher portion ⅓ of shell length from posterior margin. A few scattered setae present on cephalic and caudal margins, but absent dorsally and ventrally. Eye unpigmented.

Copulatory complex with tip of peniferum extending ventrally slightly distal to the internal border of the horizontal ramus of clasping apparatus; distal extremity truncate. Accessory groove extends dorsally distinctly above level of loop of spermatic duct. Finger guard moderately heavy; truncate distally. Dorsal finger with a single terminal seta. Clasping apparatus strongly curved, with proximal portion of vertical ramus subparallel to horizontal ramus; internal border of horizontal ramus with three evenly spaced teeth, apical portion with two teeth; external border of horizontal ramus entire.

FEMALE: A single biunguis female was found in this collection, but cannot definitely be assigned to this species.

Relationships. — See discussion of relationships under *D. ungulata.*

Type-locality. — Wind Cave, Pulaski Co., Kentucky, U.S.A. (types, ANSP). Host, *Orconectes australis packardi.*

Distribution (*in addition to type-locality*). —

United States

KENTUCKY — JACKSON CO. — Wind Cave in Kentucky R drainage (host, *Cambarus tenebrosus*). PULASKI CO. — Hydens Cave 1.3 mi NE of Blue John, 6 mi from Sloans Valley (host, *O. a. packardi*). WARREN CO. — intermittent springs in field 3 mi SE of rts 31 and 242 in Rich Pond area about 8 mi S of Bowling Green (host, *C. tenebrosus*).

Dactylocythere pughae (Hobbs and Hobbs, 1970)

Plate XIX, figures 1-5
Plate XLIX

Dactylocythere pughae Hobbs and Hobbs, 1970: 6 (K), 10, 11 (M, H, R, A, D), figs. 4a-d.

Diagnosis. — MALE: Shell length 470-560 μ (average, 530 μ); height, 260-320 μ (average, 300 μ). Shell shaped similarly to those of members of cladoceran genus Daphnia, but with posteroventral prominence, and with shallow ventral concavity anterior to midlength. Marginal setae sparse anteriorly, posteriorly, and ventrally, and absent dorsally. Sternal spine well developed and curved with apex directed anteroventrally. Eye pigmented, located approximately one-fourth shell length from anterior margin.

Copulatory complex with finger guard slender, elongate, undulate, and slightly concave distally; posteroventral portion of peniferum rounded, not forming sharp apex or rounded prominence; accessory groove short, reaching dorsally only to ventralmost level of spermatic loop; apex of peniferal groove wider than least diameter of vertical ramus of clasping apparatus; penis L-shaped, less than half width of peniferum, and situated in ventral third of peniferum; clasping apparatus, extending ventrally beyond peniferum, not clearly divisible into horizontal and vertical rami, and with proximal and distal portions directed at angle of approximately 60 degrees; external border of vertical ramus strongly convex with distinct bend distal to midlength; external borders of both rami and internal border of vertical ramus entire; internal border of horizontal ramus possessing three teeth, proximal one rounded and distal two acute; distal extremely terminating in three denticles; dorsal and ventral fingers slender, each terminating in single seta, with basal third of ventral finger straight and distal two-thirds gently curved.

TRIUNGUIS FEMALE: Shell length 530-600 μ (average, 570 μ); height, 340-380 μ (average, 360 μ). Shell markedly similar to that of the male, but larger. Genital complex situated posterodorsally; J-shaped rod "S-shaped" with horizontal arm long and straight, ends of both arms anteriorly recurved; amiculum suspended from ventral arm. Eye pigmented, situated one-fourth shell length from anterior margin.

Relationships. — *Dactylocythere pughae* shares with *D. daphnioides* a posteroventral projection of the shell, a prominent sternal spine, and a similar clasping apparatus; however, it differs from the latter, among other features, in possessing a shorter accessory groove, a slenderer and undulating finger guard, and the J-shaped rod of the female is greatly produced horizontally.

Type-locality. — Creek and burrows in seepage area, 4.8 mi SW of jct of rts 25W and 75, on rt 25W, SW of Corbin, Whitley Co., Kentucky, U.S.A. (types, USNM: holotype and allotype, 126257). Hosts, *Cambarus sphenoides, Cambarus diogenes* subsp., and *C.* (*Jugicambarus*) sp.

Distribution. — Known only from type-locality.

Dactylocythere runki (Hobbs, 1955)

Plate XIX, figures 6-9
Plate XLIX

Entocythere runki Hobbs, 1955: 330-332 (M, H, R, D), figs. 3, 4, 10. — Ferguson, 1958: 198 (C). — Crawford, 1959: 162, 167 (R). — Hobbs and Walton, 1960: 19, 22, 23 (R). — Crawford, 1961: 244 (R). — Hobbs and Walton, 1961: 381 (R). — Hobbs and Walton, 1962: 46 (R). — Hart, 1962: 112 (C).

Dactylocythere runki. — Hart, 1962: 130 (C, D, S). — Hart, 1965b: 257 (R). — Hart and Hart, 1966a: 2 (R). — Hobbs and Walton, 1966: 3 (R). — Hobbs, 1967: 5 (K), 9 (R). — Hobbs and Walton, 1970: 853 (M, R). — Hobbs and Hobbs, 1970: 47: 5-7 (K).

Diagnosis. — MALE: Shell length 410-490 μ (average, 460 μ); height, 230-290 μ (average, 270 μ). Shell subovate with the anterior portion not as high as posterior, and without a prominence on the posteroventral margin.

Copulatory complex with finger guard truncate. Distal portion of peniferum with two widely spaced anteroventrally directed apices; accessory groove extending dorsally to level of top of spermatic tube loop. Clasping apparatus with heavy vertical ramus; horizontal ramus forming right-angle with vertical ramus; external margin entire; internal margin of horizontal ramus with three teeth, and bearing three terminal denticles.

TRIUNGUIS FEMALE: Shell length 430-510 μ (average, 480 μ); height, 260-320 μ (average, 300 μ). Shell higher posteriorly than that of the male, and the ventral margin is slightly concave just anterior to midlength.

Relationhsip. — *Dactylocythere runki* is closely related to *D. daphnioides.*

Type-locality. — Stream 15.9 mi SW of Galax on Blue Ridge Parkway, Alleghany Co., North Carolina, U.S.A. (types, USNM: holotype, 98410; allotype, 98411; morphotype, 98412). Hosts, *Cambarus robustus* and *C. chasmodactylus.*

Distribution (in addition to type-locality). —

United States

NORTH CAROLINA — ALLEGHANY CO. — 16.9 mi E of jct of Parkway and rt 18 (hosts, *C. robustus* and *C. b. bartonii*); 15.9 mi S of Galax on Blue Ridge Parkway (hosts, *C. robustus, C. chasmodactylus,* and *C.* sp.); 3.9 mi W of junction of rt 18 and Parkway on rt 18 (hosts, *C. robustus* and *C.* sp.). ASHE CO. — Buffalo Ck, between rt 88 and Creston (host, *C. chasmodactylus*); 2.9 mi N of Jefferson on rt 16 (hosts, *C. b. bartonii* and *C. robustus*). AVERY CO. — Stream at Cranberry, trib to Watauga R (hosts, *C. b. bartonii* and *C. robustus*). SWAIN CO. — Wesser Ck to Little Tennessee R, 11.5 mi N of Topton (host, *C. b. bartonii*). WATAUGA CO. — Trib to Watauga R, 2.0 mi S of Vilas on rt 194 (host, *C. robustus*); 3.3 mi S of Trade on rt 88 (host, *C. robustus*); 2.0 mi E of Sugar Grove (host, *C. robustus*).

TENNESSEE — WASHINGTON CO. — Sinking Ck, Hoss Cove, near Johnson City (hosts, *Cambarus robustus, C. longirostris,* and *C. b. bartonii*).

VIRGINIA — CARROLL CO. — 2.3 mi S of Dug Spur, trib of Big Reed Island Ck (hosts, *C. b. bartonii* and *C. robustus*). FLOYD CO. — 1.4 mi NE of Copper Hill on rt 221 (host, *C. b. bartonii*). GRAYSON CO. — 0.8 mi S of junction of Parkway and rt 89 (hosts, *C. robustus* and *C.* sp.).

Dactylocythere sandbergi (Hart and Hart, 1971)

Plate XIX, figures 10-15
Plate XLVIII

Dactylocythere sandbergi Hart and Hart, 1971: 112-113 (M, H, R, A, D), figs. 6a-f.

Diagnosis. — MALE: Shell length 420-440 μ (average, 430 μ); height, 230-260 μ (average, 230 μ). Shell ovoid, greatest height $\frac{1}{3}$ shell length from anterior end. Eye pigmented, situated $\frac{1}{4}$ shell length from anterior end.

Copulatory complex with peniferum extending ventrally to level of internal border of horizontal ramus of clasping apparatus. Distal portion of peniferum terminating in two points with connecting membrane; lower point slender and acute, upper point obtuse. Accessory groove wide, extending slightly above level of spermatic loop, with upward extensions at antero- and posterodorsal extremities of groove. Penis directed anteroventrally, with tip vertical, situated $\frac{1}{5}$ peniferum length from base of peniferum.

Clasping apparatus with length of vertical ramus approximately equal to that of horizontal ramus and meeting it in gentle curve rather than at distinct angle. Vertical ramus sinuous, with external border convex posteriorly in lower $\frac{1}{2}$, concave in upper $\frac{1}{2}$. Internal border convex anteriorly in upper $\frac{1}{3}$, concave in lower $\frac{2}{3}$. External border of horizontal ramus smoothly convex; internal border with one large tooth distal to midlength, followed by deep concavity, two smaller teeth, and three terminal denticles. Clasping apparatus occasionally with three subequal teeth rather than one large and two smaller ones.

Finger guard thick, terminating in three step-like projections, anterior one longest on holotype. In other specimens, three blunt projections sometimes equal in length.

TRIUNGUIS FEMALE: Shell length 420 μ; height, 210 μ. Shell with straight venter and broadly arched dorsum. Greatest height near midlength. Slight concavity posterodorsally in area of amiculum. Eye pigmented, located $\frac{1}{6}$ shell length from anterior end. U-shaped rod with amiculum hanging vertically from it.

Relationship. — In the shell, peniferum, and finger guard, *Dactylocythere sandbergi* most resembles *D. chelomata.* The arrangement of the teeth on the clasper, with the large proximal tooth situated distal to the midlength of the horizontal ramus, is unique.

Type-locality. — Blue Buck Ck, 100 yds upstream from mouth, Hickman Co., Tennessee, U.S.A. (types, USNM: holotype, 138455; allotype and paratype, 138456; paratype, 138457 and 138458). Hosts, *Orconectes compressus* and *O.* species.

Distribution. — Known only from type-locality.

Dactylocythere speira (Hart and Hart, 1971)

Plate XX, figures 1-5
Plate XLIX

Dactylocythere speira Hart and Hart, 1971: 113-114 (M, H, R, A, D), figs. 7a-e.

Diagnosis. — MALE: Shell length 430 μ; height, 240-250 μ (average, 240 μ). Shell subelliptical with greatest height $\frac{1}{3}$ shell length from posterior margin. Eye pigmented and located $\frac{1}{5}$ shell length from anterior end. Sternal spine lacking.

Copulatory complex with peniferum not reaching external border of horizontal ramus of clasping apparatus; distal extremity terminating in two points, with anterodistal portion shorter than posterodistal; posterior more tapering and sharply pointed. Accessory groove not reaching dorsal margin of spermatic loop; penis situated $\frac{1}{4}$ peniferum length from ventral tip.

Clasping apparatus with horizontal and vertical rami approximately equal in length, at slightly acute angle, with thickest portion immediately proximal to juncture. Horizontal ramus smoothly convex on external border, internal border concave with two small teeth along distal half and three terminal denticles.

Finger guard slender, with termination slightly rounded in holotype.

BIUNGUIS FEMALE: Shell length 440-460 μ (average, 450 μ); height, 280-290 μ (average, 290 μ). Shell of biunguis female subovate. Dorsum broadly rounded, venter broadly concave in posterior half, followed by slight convexity rounding into anterior end. Rod apparently paired, lacking dependent amiculum, resembling coiled tube and terminating posteroventrally in two segments. Papilla short, straight, positioned anterior to main body of rod.

Relationships. — *Dactylocythere speira* seems to have affinities with *D. chelomata* in the shape of the clasping apparatus, but the latter has a much longer accessory groove. The short accessory groove is similar to that of *D. exoura* and *D. brachystrix. D. speira* resembles *D. coloholca* in the features of the peniferum and in the shape of the female shell, but the clasping apparatus of *D. speira* is smooth.

Type-locality. — Meadow Creek on Clark Range Road, Putnam Co., Tennessee, U.S.A. (types, USNM: holotype and paratype, 138459; allotype, 138460). Host, *Cambarus* (*Jugicambarus*) sp.

Distribution (*in addition to type-locality*). —

United States

TENNESSEE — CHEATHAM CO. — 1 mi W of Davidson Co. line W of Belleview (host, *Orconectes* sp.).

Dactylocythere spinata (Hobbs and Walton, 1970)

Plate XX, figures 6-9
Plate XLVIII

Dactylocythere spinata Hobbs and Walton, 1970: 853 (M, R), 855, 857 (A), 858 (R), 860 (M, H, R, A, D), figs. 3g, h. — Hobbs and Hobbs, 1970: 5-7 (K).

Diagnosis. — MALE: Shell length 440-480 μ (average, 460 μ); height, 270-290 μ (average, 270 μ). Shell elongate ovate with greatest height some distance posterior to midlength. Marginal setae present anteriorly, posteriorly, and ventrally. Sternal spine prominent, long, directed posteriorly with apical portion only slightly bent ventrally. Eye pigmented, situated slightly more than one-fourth shell length from anterior margin.

Copulatory complex with finger guard rather heavy, its posterior margin deeply convex anteriorly, and its ventral border excavate with anteroventral prominence decidedly smaller than bituberculate, posteroventrally directed posterior prominence, thus ventral margin with three prominences; peniferum moderately heavy with posterior margin slightly undulating, but nowhere angulate or with lobes, and terminating in anteriorly directed acute tip; peniferal groove, only slightly wider at apex than one-half least diameter of vertical ramus of clasping apparatus, and directed anteriorly; penis L-shaped and situated approximately at base of ventral fourth of peniferum; clasping apparatus clearly divisible into vertical and horizontal rami with major axes forming angle of approximately 80 degrees although vertical ramus with proximal portion bent anteriorly, almost paralleling horizontal ramus. External border of both rami and internal border of vertical ramus entire and external border of vertical ramus without conspicuous shoulder; internal border of horizontal ramus with single large tooth near midlength and two or three exceedingly low prominences immediately proximal to three small apical denticles; clasping apparatus thickest in region of junction of two rami, tapering slightly proximally and distally; portion of horizontal ramus distal to major tooth not nearly so thick as that proximal to it.

TRIUNGUIS FEMALE: Shell length 450-490 μ (average, 470 μ); height, 280-320 μ (average, 300 μ). Shell much more highly vaulted posteriorly than in male, with greatest height some distance posterior to midlength and with posterior margin subtruncate; ventral margin with shallow excavation at about midlength. Submarginal setae disposed as in male.

Eye pigmented and located slightly more anteriorly than that of male.

Relationships. — *Dactylocythere spinata* has its closest affinities with *D. chalaza* and its allies (see discussion of relationships of *D. enoploholca* above), and it is more similar to this species than to the other two. It differs from *D. pachysphyrata* in having only a single major tooth on the internal border of the horizontal ramus of the clasping apparatus, a bituberculate posteroventral prominence on the finger guard, and a sternal spine that is directed posteriorly rather than ventrally. In *D. chalaza,* the posteroventral prominence on the finger guard is not bituberculate, the guard is not strongly bowed anteriorly, and the sternal spine is directed posteroventrally rather than posteriorly. It differs from *D. enoploholca* in possessing a sternal spine and in lacking a heel-like prominence on the posteroventral margin of the peniferum.

Type-locality. — Stream 9 mi SW of Caryville on country road to Shea, Campbell Co. (Cumberland River drainage), Tennessee, U.S.A. (types, USNM: holotype and allotype, 126972). Host, *Cambarus distans.*

Distribution (*in addition to type-locality*). —

United States

TENNESSEE — FENTRESS CO. — Laurel Fork, 9.7 mi NE of Jamestown on TN 154 (host, unknown). MORGAN CO. — White Oak Ck on US 27 at Sunbright (host, unknown). SCOTT CO. — Bandy Ck W of Leatherwood Fork (host, unknown).

Dactylocythere steevesi (Hart and Hobbs, 1961)

Plate XX, figures 10-16
Plate XLVIII

Entocythere steevesi Hart and Hobbs, 1961b: 174 (K), 180 (M, H, R, A, D), 176 (A), 177 (A, R), figs. 18-20. — Bedinger and Hobbs, 1965: 94 (H, D) (by implication).

Dactylocythere steevesi. — Hart, 1962: 131 (C, D, S). — Hobbs and Walton, 1963a: 460 (R). — Hart and Hart, 1966a: 5, 9 (A). — Hobbs and Walton, 1966: 4 (R). — Hobbs, 1967: 5 (K). — Ferguson, 1968: 501 (C, D). — Hart and Hart, 1969a: 187 (B). — Hobbs and Hobbs, 1970: 5-7 (K). — Hobbs and Barr, 1972: 81 (D, H).

Diagnosis. — MALE: Shell length 470 μ height, 240 μ. Shell subovate, with greatest height in the posterior third. Setae widely spaced on anterior and posterior margins. Eye spot situated about $\frac{1}{5}$ the length of the shell from the anterior margin.

Copulatory complex with peniferum extending ventrally to a level slightly proximad of external margin of clasping apparatus; distal extremity truncate. Accessory groove extends dorsally slightly about level of loop of the spermatic duct. Finger guard straight, prominent, and rounded distally. Dorsal finger with a single terminal seta. Clasping apparatus (figs. 19-20) strongly curved, with vertical ramus itself curved; proximal subhorizontal portion of vertical ramus comparatively slender, vertical portion thickened to more than twice the dorsoventral thickness of the horizontal ramus; horizontal ramus gently curved with internal border bearing three teeth, the proximal one at base of ramus, and apical portion with three teeth; external border of horizontal ramus entire.

FEMALE: Unknown.

Relationships. — See discussion of relationships of *D. ungulata.*

Type-locality. — Saltpeter Cave (Sauta Cave), Jackson Co., Alabama, U.S.A. (types, ANSP). Host, *Cambarus tenebrosus.*

Distribution (*in addition to type-locality*). —

United States

ALABAMA — JACKSON CO. — Russell Cave (host, *Cambarus tenebrosus*).

KENTUCKY — EDMONSON CO. — pool in Roaring River Passage of Mammoth Cave (host, *Cambarus tenebrosus*). GREEN CO. — Brush Ck Cave

0.8 mi E of Lobb (hosts, *C. tenebrosus* and *Orconectes pellucidus*). HARDIN CO. — ck from cave at Crystal Lake 2 mi N of Muldraugh off rt 31 (host, *C.* sp.). HART CO. — Stream in Hidden River Cave (host, *O. pellucidus*). JEFFERSON CO. — Subterranean limerock spring on Lannert farm (hosts, *C. ornatus* and *C. laevis*). MEADE CO. — Doe Run (host, *C. tenebrosus*). TRIGG CO. — Small creek 2 mi SW of Cadiz (host, *C.* sp.); Spring 1 mi SW of Cadiz (host, *C. laevis*); Spring near limestone bluff 1.5 mi W of Cadiz (host, *C. tenebrosus*). WARREN CO. — Pruett's Saltpeter Cave S of Anna (host, *C. tenebrosus*); Intermittent spring in field 3 mi SE of rts 31 and 242 in Rich Pond area about 8 mi S of Bowling Green (host, *C.* sp.). WAYNE CO. — Horse Hollow Cave (hosts, *C. distans* and *O. australis packardi*).

TENNESSEE — CLAY CO. — Locality unknown (host, *Orconectes placidus*). DAVIDSON CO. — Richland Ck at Bellemeade W of Nashville (hosts, *Orconectes* sp., *O. placidus,* and *Cambarus tenebrosus*). FRANKLIN CO. — Sinking Cove Cave (host, *C. tenebrosus*). PUTNAM CO. — Blind Fish Cave, Calfkiller R Valley (hosts, *O. a. australis* and *C. tenebrosus*).

Dactylocythere striophylax (Crawford, 1959)

Plate XXI, figures 1-5
Plate XLVIII

Entocythere striophylax Crawford, 1959: 150 (B), 151 (H, K), 157-162 (M, H, R, D), 179 (B), 180-181 (A), figs. 10-17. — Hart, 1962: 123 (C).

Dactylocythere striophylax. — Hart, 1962: 131 (C, D, S). — Hobbs, 1967: 4 (K), 8 (R). — Ferguson, 1968: 501 (C, D). — Hobbs and Walton, 1970: 853 (M, R), 860 (R). — Hobbs and Hobbs, 1970: 5-7 (K).

Diagnosis. — MALE: Shell length 420-480 μ (average, 450 μ); height, 240-270 μ (average, 260 μ). Shell subelliptical (oblong-ovate) with anterior end markedly narrower than posterior end and ventral margin with slight concavity just anterior to midlength. Dorsal margin slightly more rounded than ventral one. Setae sparsely scattered along free margin of valve with very few to none along dorsal margin. Lateral surfaces appear to bear no setae. Eye located about one-fourth shell length from anterior margin.

Copulatory complex with peniferum terminating bluntly distally; accessory groove short, occupying posterior portion of peniferum. Finger guard heavy, sclerotized, and trifid. Clasping apparatus gently curved; external margin entire; distal half of internal margin of horizontal ramus bearing three minute teeth and four terminal denticles.

TRIUNGUIS FEMALE: Shell length 450-480 μ (average, 470 μ); height, 290-300 μ (average, 290 μ). Shell larger than that of male and conspicuously dissimilar. Entire dorsal margin strongly arched and bearing prominent concavity on posterodorsal margin; concavity on anteroventral margin also appears more prominent than that in male. Anterior end of valve narrower than posterior and somewhat pointed. Setae sparsely scattered on free margins of valve, with few-to-none on lateral surface and dorsal margin.

Relationships. — This ostracod has its closest affinities with *D. suteri.*

Type-locality. — Cedar Creek, 9.6 mi NNE of Columbia city limits on rt 215, Richland Co., South Carolina, U.S.A. (types, USNM: holotype, 103563; allotype, 103564; morphotype, 103565; paratype, 103566). Hosts, *Procambarus a. acutus, Cambarus b. bartonii, C. latimanus,* and *C. spicatus.*

Distribution (*in addition to type-locality*). —

United States

SOUTH CAROLINA — ABBEVILLE CO. — Intermittent stream on rt 20, 1.4 mi N of Abbeville (host, *Cambarus latimanus*) (Crawford 1961: 238). RICHLAND CO. — 37 collections in Broad River drainage (hosts, *C. spicatus, C. b. bartonii, C. acuminatus, P. a. acutus, enoplosternum,* and *P. troglodytes*); 5 localities in Wateree River drainage (hosts, *C. latimanus* and *P. troglodytes*); 5 localities in Saluda River drainage (hosts, *C. latimanus* and *C. acuminatus*).

Dactylocythere susanae (Hobbs III, 1971)

Plate XXI, figures 6-10
Plate XLVIII

Dactylocythere susanae Hobbs III, 1971: 140-142 (M, H, R, A, D), figs. 2a-f.

Diagnosis. — MALE: Shell length 480-520 μ (average, 500 μ); height, 260-290 μ (average, 270 μ). Shell subovate, with greatest height posterior to midlength; margins entire; submarginal setae present in limited numbers anteriorly, posteriorly, and ventrally. Sternal spine lacking. Eyes pigmented, located approximately one-fourth shell length from anterior margin.

Copulatory complex possessing straight finger guard, terminating distally without tubercles, anterior margin incised short distance proximal to apex; peniferum elongate with distal anteriorly directed flange; posteroventral portion tapering to semi-acute prominence; accessory groove long, extending one-third its length beyond dorsal extremity of spermatic loop; apex of peniferal groove wider than least diameter of vertical ramus of clasping apparatus; penis L-shaped, greater than half width of peniferum, and situated in ventral third of peniferum; clasping apparatus extending ventrally beyond peniferum, nearly C-shaped, not clearly delimited into horizontal and vertical rami, and with proximal and distal portions directed at angle of approximately 75 degrees; external border of vertical ramus strongly convex posteriorly, external and internal border entire; internal border of horizontal ramus bearing three teeth, proximal one most pronounced and rounded, distal one small and subacute, and that between almost obsolete; distal extremity of horizontal ramus terminating in three denticles; dorsal and ventral fingers slender, former terminating distally in two setae, latter in single seta.

TRIUNGUIS FEMALE: Shell length 540-580 μ (average, 560 μ); height, 300-370 μ (average, 330 μ). Shell (Fig. 2e) subovate with entire margins; range of shell size of 10 specimens recorded in Table 1; submarginal setae like those of male. Genital complex with posterodorsal and consisting of coiled J-shaped rod and amiculum; upper portion of J-shaped rod bifid with genital papilla situated between rami; amiculum surrounding coil, supported by U-shaped ribs; rami of ribs extending away from coil; anterodorsalmost pair of supporting ribs with one ramus extending dorsally and other anteriorly. Eyes pigmented, situated one-fourth shell length from anterior margin.

Relationships. — *Dactylocythere susanae* has its closest affinities with *D. steevesi* and *D. ungulata.* All three species possess elongate penifera which taper ventrally to terminate in a flange and a semi-acute prominence at anteroventral angle. Likewise, all possess a rather straight finger guard, which terminates distally without tubercles. *D. susanae* can easily be distinguished from these two species in that it bears an accessory groove which extends dorsally well beyond the level of the spermatic loop. Likewise, the internal border of the horizontal ramus of the clasping apparatus of *D. susanae* bears a large proximal tooth which is greatly reduced in the other two species. *D. susanae* is also related to *D. arcuata.* The ventral portion of the peniferum in both is markedly similar, and the accessory grooves extend dorsally above the spermatic loop. *D. arcuata,* however, has a very slender elongate finger guard and lacks an enlarged proximal tooth on the horizontal ramus of the clasping apparatus.

Type-locality. — Murray Spring Cave, Orange Co., Indiana, U.S.A. (types, USNM: holotype, 135619; allotype, 135620; paratype, 135621). Hosts, *Cambarus laevis.*

Distribution (in addition to type-locality). —

United States

INDIANA — LAWRENCE CO. — Connerly's Cave (host, *Cambarus laevis*) (Hobbs III, 1971: 142). MONROE CO. — Buckner's Cave (host, *C. laevis*) (Hobbs III, 1971: 142); Weaver Springs Cave (host, *C. laevis*) (Hobbs III, 1971: 142); Jack's Defeat Ck, Bloomington, 1.5 mi NE of jct of Woodyard Road and Curry Pike on Woodyard Road (host, *C. laevis*) (Hobbs III, 1971: 142). WASHINGTON CO. — Fredericksburg Cave (host, *C. laevis*) (Hobbs III, 1971: 142).

KENTUCKY — BRECKINRIDGE CO. — Bandy Cave 3.0 mi S of Irvington (hosts, *Cambarus tenebrosus, Orconectes inermis inermis,* and *O. australis packardi*) (Hobbs III, 1971: 142). HART CO. — Cooch Webb Cave 0.4 mi N of Bear Wallow (hosts, *C. tenebrosus, O. i. inermis,* and *O. a. packardi*) (Hobbs III, 1971: 142). PULASKI CO. — Cumberland Crystal Cave at Sloan's Valley (hosts, *C. tenebrosus, O. i. inermis,* and *O. a. packardi*) (Hobbs III, 1971: 142). ROCKCASTLE CO. — Pine Hill Cave at Pine Hill on US 25 (hosts, *C. tenebrosus, O. i. inermis,* and *O. a. packardi*) (Hobbs III, 1971: 142).

Dactylocythere suteri (Crawford, 1959)

Plate XXI, figures 11-15
Plate XLIX

Entocythere suteri Crawford, 1959: 150 (B), 152 (K), 162-167 (M, H, R, D), 178, 180-181 (A), figs. 18-23. — Crawford, 1961: 244 (R). — Hobbs and Walton, 1962: 46 (R). — Hart, 1962: 123 (C).

Dactylocythere suteri. — Hart, 1962: 131 (C, D, S). — Hobbs, 1967: 5 (K), 8 (R). — Hobbs, Holt, and Walton, 1967: 34 (K), 43-44 (M, D, A, H, S), 69 (D, E), 18, 20, 25, 38, 45, 49, 61 (A, H), 72, 73, 77 figs. 7h, 10c. — Ferguson, 1968: 501 (C, D). — Hart and Hart, 1969a: 187 (R). — Hobbs and Walton, 1970: 853, 860 (M, R). — Hobbs and Hobbs, 1970: 7. — Hart and Hart, 1971: 106, 111. — Peters, 1971: 100.

Diagnosis. — MALE: Shell length 450-470 μ (average, 460 μ); height, 260-280 μ (average, 270 μ). Shell oblong-ovate with anterior end narrower than posterior and ventral margin with a slight concavity just anterior to mid-length. Setae scattered on free margin, being numerous on anterior and posterior ends, and sparsely dispersed along ventral margin. Lateral surface with no setae. Eyes located about one-fifth of valve length from anterior end.

Copulatory complex with peniferum terminating bluntly; accessory groove small and terminating on level with dorsal margin of spermatic tube loop. Finger guard large, sclerotized, trifid. Clasping apparatus gently curved; rami junction indistinct; external margin entire; internal margin of horizontal ramus with three or four teeth situated in distal half; and two or three terminal denticles.

TRIUNGUIS FEMALE: Shell length 450-490 μ (average, 470 μ); height, 290-310 μ (average, 300 μ). Shell more ovate than that of male and slightly larger. Dorsal margin of the valve slightly arched; anterior half higher than that of the male. Extreme anterior end slightly pointed, anteroventral margin bearing conspicuous concavity. Prominent amiculum situated in posterodorsal portion of shell; deeply staining mass in this area often obliterates details of the genital appartaus.

Type-locality. — Twenty-five Mile Creek, 4.5 mi E of Blythewood, Richland Co., South Carolina, U.S.A. (types, USNM: holotype, 103567; allotype, 103568; paratype, 103569). Hosts, *Cambarus latimanus* and *C. acuminatus.*

Distribution (in addition to type-locality). —

United States

GEORGIA — BANKS CO. — 3.7 mi N of Commerce on rt 441 at jct of rt 85 (host, *Cambarus* (*Depressicambarus*) sp.). MORGAN CO. — 7.6 mi N of Jasper on rt 73 (hosts, *Procambarus spiculifer* and *C. latimanus*). MURRAY CO. — 1 mi E of Chatsworth on rt 76 (hosts, *C.* (*Depressicambarus*) sp., *C.* (*Puncticambarus*) sp., and *Orconectes spinosus*).

NORTH CAROLINA — CHATHAM CO. — 9.8 mi N of Sanford on rt 501 (hosts, *Cambarus reduncus, C. uhleri,* and *C. acuminatus*). DURHAM CO. — Cypress swamp in Duke Forest near Durham (hosts, *Fallicambarus uhleri, C. b. bartonii,* and *C. d. diogenes*). EDGECOMBE CO. — 9 mi NE of Saratoga at jct of 4 roads (host, *C. d. diogenes*); Otter Creek, trib of Tar R, 9 mi NE of Saratoga (host, *Procambarus a. acutus*). ORANGE CO. — trib of New Hope Ck E of Piney Mountain (host, *C. b. bartonii*). RANDOLPH CO. — 2.8 mi NE of Randolph Co. line on rt 49 (hosts, *P. a. acutus* and *C. acuminatus*).

PENNSYLVANIA — YORK CO. — trib of Codom's Creek 4 mi S of York on rt 616 (host, *Orconectes limosus*).

SOUTH CAROLINA — RICHLAND CO. — 3 localities in Broad R drainage (hosts, *C. spicatus, C. latimanus, C. b. bartonii,* and *Procambarus acutus acutus*) (Crawford, 1959: 166); 1 locality at jct of Broad and Saluda R (hosts, *C. spicatus, C. latimanus,* and *C. bartonii*) (Crawford, 1959: 166); 1 locality in Wateree River drainage (hosts, *C. latimanus* and *C. acuminatus*) (Crawford, 1959: 166).

TENNESSEE — MCMINN CO. — 7.5 mi SW of Athens on US 11 (host, *Orconectes erichsonianus*).

VIRGINIA — HALIFAX CO. — Grassy Ck 4 mi E of South Boston on rt 58 (host, *Cambarus acuminatus*); Brandon Ck 5 mi E of Pittsylvania on rt 58 (host, *C. acuminatus*). MECKLENBURG CO. — about 2 mi S of Brodnax on VA 659 (hosts, *C. acuminatus* and *Procambarus acutus acutus*); trib of Roanoke R 3.2 mi NE of Ebony (host, *C. acuminatus*); St. John's Ck, trib of Middle Meherrin R, 7 mi N of Chase City on VA 49 (host, *C. acuminatus*); S. Meherrin R 4.5 mi N of Chase City on VA 49 (host, *C. acuminatus*); N Meherrin R 11 mi N of Chase City on VA 49 (host, *C. acuminauts*). NOTTOWAY CO. — Small stream flowing E 10 mi N of Victoria (hosts, *P. a. acutus* and *C. acuminatus*). PATRICK CO. — 1.6 mi S of Franklin-Patrick Co. line on VA 40 (hosts, *C. longulus* and *C. acuminatus*).

Dactylocythere ungulata (Hart and Hobbs, 1961)

Plate XXII, figures 1-2
Plate XLIX

Entocythere ungulata Hart and Hobbs, 1961b: 174 (K), 177-178 (M, H, R, A, D), figs. 12-14.

Dactylocythere ungulata. — Hart, 1962: 131 (C, D, S). — Hart and Hart, 1966a: 8, 9 (A). — Hobbs, 1967: 5 (K). — Ferguson, 1968: 501 (C, D). — Hart and Hart, 1969a: 187 (B). — Hobbs and Hobbs, 1970: 5-7 (K). — Hobbs III, 1971: 144 (M). — Hobbs and Barr, 1972: 81 (D, H).

Diagnosis. — MALE: Shell length 450-510 μ (average, 480 μ); height, 240-320 μ (average, 280 μ).

Shell subovate; posterior portion higher than anterior; setae sparsely situated on anterior, anteroventral, and posterior margins. Pigmented eyespot, situated approximately ⅕ the shell length from the anterior end.

Copulatory complex with peniferum extending ventrally to level of internal border of horizontal ramus of clasping apparatus; distal extremity truncate. Accessory groove extending dorsally to dorsal level of loop of spermatic duct. Finger guard heavy, short, and with a rounded protuberance posterodistally. Dorsal finger with a single terminal seta. Clasping apparatus strongly curved with extensions of the horizontal and vertical rami forming an acute angle; horizontal ramus conspicuously thickened dorsoventrally and distinctly longer than vertical ramus, three teeth on internal border — the distal one contiguous with a terminal group of three teeth; external border of horizontal ramus entire. The proximal tooth on the internal border may show considerable variation in size between specimens.

TRIUNGUIS FEMALE: Shell length 460-530 μ (average, 490 μ); height, 260-320 μ (average, 280 μ). Shell subovate with margins entire; submarginal setae present in limited numbers anteriorly, ventrally, and posteriorly, absent dorsally; genital complex consisting of amorphous mass surrounding subcircularly curved rod and supported by U-shaped ribs; rami of ribs radiating from rod; anterodorsalmost pair of supporting ribs with one ramus extending dorsally, other anteriorly; pair of accessory ribs parallel and posterior to dorsal arms of anterodorsal supporting ribs; upper portion of J-shaped rod bifid, with genital papilla positioned between rami. Eyes pigmented, situated one-fifth shell length from anterior margin.

Relationships. — *D. ungulata* has its closest affinities with *D. arcuata, D. steevesi,* and *D. prionata.* It may be distinguished from them by the thickened horizontal ramus of its clasping apparatus.

Type-locality. — Upper Parrot Cave, 10 mi NW of Livingston, Overton Co., Tennessee, U.S.A. (types, ANSP). Host, *Cambarus tenebrosus.*

Distribution (in addition to type-locality). —

United States

KENTUCKY — CARROLL CO. — Ellis Ck, 4.4 mi SW of Kentucky R bridge at Carrollton on rt 42 (hosts, *Cambarus ornatus* and *Orconectes rusticus rusticus*). FAYETTE CO. — 100 yards downstream from Russel's Cave, 6.5 mi N of Lexington (host, *C. b. bartonii*); stream several hundred feet below Bryan Station Spring (hosts, *O. rusticus* and *C. laevis*); Waveland Farm Spring (host, *C. tenebrosus*). HARDIN CO. — Ck from cave at Crystal Lake 2 mi N of Muldraugh off rt 31 (host, *C.* sp.). HART CO. — Buckner Hollow Cave (hosts, *C. tenebrosus* and *O. pellucidus*). JESSAMINE CO. — Spring on Keene Road just S of Fayette-Jessamine Co. line (host, *C. laevis*). ROCKCASTLE CO. — Pine Hill Cave at Pine Hill on US 25 (host, *O. australis packardi*) (Hobbs III, 1971: 144). WARREN CO. — Pruitts Saltpeter Cave, N of Bowling Green, S of Anna (host, *C. tenebrosus*). WAYNE CO. — Horse Hollow Cave (hosts, *C. distans* and *O. a. packardi*).

TENNESSEE — CLAY CO. — Oak Grove W of Celina (host, *Orconectes putnami*). DAVIDSON CO. — Richland Ck at Bellemeade W of Nashville (host, *Cambarus tenebrosus*). DICKSON CO. — Montgomery Bell State Park (host, *O.* species). FRANKLIN CO. — Sinking Cove Cave (host, *C. tenebrosus*). GRUNDY CO. — Big Mouth Cave, 5 mi NE of Pelham (host, *O. a. australis*). LINCOLN CO. — Branch of Swan Ck 6.7 mi E of McBurg on rt 64 (hosts, *O. r. mirus* and *O. erichsonianus*). OVERTON CO. — 1.4 mi N of Putnam Co. line on TN 43 (hosts, *O.* sp. and *O. placidus*). PICKETT CO. — Small stream along Co. rt 42, 0.5 mi S of Kentucky-Tennessee state line in NE corner (host, *C. tenebrosus*) (Hobbs III, 1971: 144). PUTNAM CO. — Blind Fish Cave, Long Hollow (host, *O. a. australis*). SCOTT CO. —White Oak Ck 100 yards above mouth at Clear Fork R NE of Rugby (host, *C.* (*Jugocambarus*) sp.).

Dactylocythere xystroides (Hobbs and Walton, 1963)

Plate XXII, figures 3-7
Plate XLVIII

Dactylocythere xystroides Hobbs and Walton, 1963a: 460-461 (M, H, R, A, D), figs. 7-10. — Hobbs, 1967: 5 (K). — Ferguson, 1968: 501 (C, D). — Hobbs and Walton, 1970: 851-863 (M). — Hobbs and Hobbs, 1970: 5-7 (K).

Diagnosis. — MALE: Shell length 490 μ; height, 280 μ. Shell ovate with a slight long shallow emargination along ventral margin anterior to midlength. Marginal setae evenly dispersed around entire shell. Eyes present.

Copulatory complex with peniferum bearing a distinct hump on caudal margin and with distal extremity directed cephalically. Penis small, situated distal to base of clasping apparatus and just distal to midpoint between dorsal margin of spermatic loop and ventral extremity of peniferum. Accessory groove prominent and extending proximad to level of dorsal margin of spermatic loop. Finger guard comparatively slender with a slight undulating posterior margin and a shallowly cleft tip. Dorsal and ventral fingers slender, the latter gently curved cephalically just proximal to midlength.

Clasping apparatus not clearly divisible into vertical and horizontal rami because of thickened juncture between them. Proximal end of vertical ramus expanded and this followed by a conspicuously slendered shaft directed posteroventrally almost to midlength of posterior margin; distal portion of ramus flared to fuse imperceptibly with proximal broadened horizontal ramus. External border of horizontal ramus entire; internal border with four step-like serrations along distal half; no terminal denticles discernible.

TRIUNGUIS FEMALE: Shell length 420 μ; height, 230-250 μ (average, 240 μ). Shell somewhat smaller than that of male and a little more suddenly vaulted anteriorly.

Relationships. — *Dactylocythere xystroides* has its closest affinities with *D. chalaza, D. daphnioides, D. runki,* and *D. steevesi.* They share in common an accessory groove of the peniferum which reaches proximally to about the level of the dorsal margin of the loop of the spermatic duct, a somewhat broad junction between the vertical and horizontal rami of the clasping apparatus, and, except in *D. steevesi,* distinct serrations rather than teeth along the internal border of the clasping apparatus. *D. steevesi* may be distinguished from *D. xystroides* by the subparallel axes on the proximal and distal ends of the clasping apparatus. In *D. chalaza,* the distal three serrations of the horizontal ramus are crowded distally. In *D. runki* and *D. daphnioides,* there are more serrations and the junction between the vertical and horizontal rami is not nearly so broad as in *D. xystroides.*

Type-locality. — Hurricane and Tumbling Creeks, SE of Waverly, Humphreys Co., Tennessee, U.S.A. (types, USNM: holotype, 108029; allotype, 108030; paratypes, 108031 and 108032). Hosts, *Orconectes compressus, O. placidus, O.* species, and *Cambarus striatus.*

Distribution (in addition to type-locality). —

United States

TENNESSEE — CHEATHAM CO. — 1 mi W of Davidson Co. line, W of Belleview (host, *Orconectes* species). DAVIDSON CO. — Richland Ck at Bellemeade W of Nashville (host, *O.* species and *O. placidus*); Buffalo Ck at jct of rts 70S and 70N, S of Nashville (host, *O.* species). DICKSON CO. — Montgomery Bell State Park (host, *O. rusticus* subsp.). FRANKLIN CO. — trib to Elk River 11.4 mi W of Winchester on rt 64 (hosts, *Cambarus girardianus* and *C.* (*Depressicambarus*) sp.). GILES CO. — ck 4.5 mi E of Pulaski on rt 64 (hosts, *C. striatus* and *C. tenebrosus*). LINCOLN CO. — Swan Ck about 11 mi W of Fayetteville on rt 64 (hosts, *O. erichsonianus* and *O.* species); branch of Swan Ck 6.7 mi E of McBurg on rt 64 (hosts, *O.* sp., *C. striatus,* and *C. tenebrosus*). MAURY CO. — Little Bigbe Ck at Mt. Pleasant (hosts, *O. placidus* and *O.* species). WAYNE CO. — Forty-eight Ck, trib to Buffalo R, 7.0 mi E of Waynesboro on rt 64 (hosts, *O. alabamensis, O. compressus, O. spinosus,* and *O. forceps*).

DONNALDSONCYTHERE (HART, 1962)

Subgenus *Donnaldsoncythere* Rioja, 1942b: 688 (M). — Rioja, 1943a: 568, 584 (R). — Hoff, 1944b: 330 (R). — Hartman, 1963: 145 (C).

Donnaldsonensis Group. — Hart, 1960: 1 (M). — Hobbs and Walton, 1961: 384 (R).

Humesi Group. — Hart and Hobbs, 1961a: 25 (D, H).

Donnaldsoncythere Hart, 1962: 125 (K), 131 (M, S, C), type-species, *Entocythere donnaldsonensis* Klie, 1931. — Hart, 1963: 51 (R, D). — Crawford, 1965: 151 (R). — Hobbs III, 1965: 161 (K). — Hart and Hart, 1966a: 3 (A). — Hobbs, Holt, and Walton, 1967: 36, 37 (M), 67 (E). — Hart and Hart, 1969a: 188 (D, E, R). — Hobbs and Hobbs, 1970: 2 (K).

Diagnosis. — Terminal tooth of mandible with cusps. No finger guard on copulatory complex. External border of clasping apparatus entire; internal border with several teeth, or entire; apex with several small denticles. Horizontal and vertical rami of clasping apparatus merging imperceptibly. Peniferum large, but extending vertically beyond clasping apparatus only slightly, or not at all. Penis curved at almost right angle near midlength, with proximal extremity level, or nearly so, with base of clasping apparatus; penis length greater than that of clasping apparatus; peniferum without accessory groove; amiculum lacking in female.

Relationships. — Hobbs and Walton (1963b: 369-370) have discussed the intrageneric relationships of this poorly understood genus, and point out that although anatomic details of the type-species, *Donnaldsoncythere donnaldsonensis,* are not adequately defined, there seems to be no question that the genus is composed of a very closely allied group of species, the combined ranges of which extend from New York to northern Georgia, as far east as Philadelphia and west to the Cumberland Plateau in Tennessee.

It seems likely, they say, that *D. humesi* is the most generalized species of the group — possessing a combination of (1) a clasping apparatus which exhibits a near-angular bend corresponding to the junction of the vertical and horizontal rami of most entocytherids, (2) an angular distal extension of the peniferum and (3) a massive penis which may be extended to the posteroventral margin of the peniferum posterior to the angular distal extension. From a stock possessing these characteristics, it may be postulated that two main lines of evolution developed. One culminated in *D. truncata,* in which the angular clasping apparatus was retained; the angular distal extremity of the peniferum was broadened and shortened but remained angular, and the broadening (in the longitudinal axis of the body) resulted in the penis-passage lying above the angular extension rather than posteroventral to it.

In the other line, the clasping apparatus tended to lose its angular bend, the penis remained unchanged in position; however, marked changes occurred in the angular extensions of the peniferum. Examples of the postulated early stages of alteration may be observed in *D. tuberosa* and *D. hiwasseensis,* in which the primitive posterodistal angle of the distal portion of the peniferum became thickened mesially with an accompanying rounding of the posterior angle; the unaltered anterodistal angle remained essentially in its primitive state. A more advanced stage of this same tendency may be seen in *D. ileata* and *D. pennsylvanica,* in which the thickening of the posterodistal angle of the distal portion of the peniferum increased to such an extent that the anterodistal angle became rounded and the thickening affected almost the entire distal border, causing it to extend obliquely posterodorsally-anteroventrally. In *D. pennsylvanica,* the posterodistal angle became so thickened that a subcircular prominence resulted. The extreme of this general trend was reached in *D. ardis,* in which the mesial thickening extended over much of the terminal extension and, in mesial aspect, reduced the distal extension into an acuminate projection resembling a plow point. At about the same stage of development as occurs in *D. tuberosa* and *D. hiwasseensis,* an innovation of directing the anterior margin of the extension mesially resulted in the scoop-like structure found in *D. scalis.*

Type-species. — *Entocythere donnaldsonensis* Klie, 1931: 334 (by original designation, Hart, 1962: 131).

Distribution. — United States: Northern Georgia northwestward to Indiana and northeastward to Maine.

Although there are no records, the genus probably occurs in southern Ontario and Quebec, Canada.

Host. — Freshwater crayfishes of the family Astracidae.

Figure 24. — Distribution of the genus *Donnaldsoncythere.*

Key to the Species of the Genus Donnaldsoncythere

1	Distal extremity of peniferum acute or bilobed	9
1′	Distal extremity of peniferum truncate, rounded, or scoop-like	2
2(1′)	Distal extremity of peniferum scoop-like	*D. scalis*
2′	Distal extremity of peniferum truncate or rounded	3
3(2′)	Distal extremity of peniferum rounded	4
3′	Distal extremity of peniferum angular	8
4(3)	Distal extremity of peniferum distinctly attenuated	5
4′	Distal extremity of peniferum not conspicuously attenuated	6
5(4)	Distal portion of peniferum with subcircular disc on posterolateral surface	*D. pennsylvanica*
5′	Distal portion of peniferum with heavily sclerotized thickening on posterolateral surface	*D. ileata*
6(4′)	Tip of penis may be extended to a level little beyond that of mid-length of clasping apparatus	*D. donnaldsonensis*
6′	Tip of penis may be extended much beyond mid-length of clasping apparatus	7
7(6′)	Eyes reduced or absent; terminal extension of peniferum distinctly delimited at base by posterior emargination	*D. tuberosa*
7′	Eyes well-developed; terminal extension of peniferum not distinctly delimited at base by posterior emargination	*D. hiwasseensis*
8(3′)	Penis guide situated dorsal to terminal extension of peniferum	*D. truncata*
8′	Penis guide situated posterodorsal to terminal extension of peniferum	*D. humesi*
9(1)	Distal extremity of peniferum acute	*D. ardis*
9′	Distal extremity of peniferum bilobed	*D. cayugaensis*

Donnaldsoncythere ardis (Hobbs and Walton, 1963)

Plate XXII, figures 8-12
Plate L

Donnaldsoncythere ardis Hobbs and Walton, 1963b: 364-368, 370 (M, H, R, A, D), figs. 6-10, 26. — Hobbs, Holt, and Walton, 1967: 35 (K), 37-38 (M, D, H, A, S), 67-69 (D, E), 18, 20, 25, 44, 45, 50, 51, 54, 57, 58, 61, 62 (A, H). 72, 73, fig. 8b. — Ferguson, 1968: 501 (C, D).

Diagnosis. — MALE: Shell length 393-422 μ (average, 496 μ); height, 200-230 μ (average, 211 μ). Eyes present; shell ovate with ventral margin entire; marginal setae evenly spaced anteriorly, ventrally, and posteriorly. Copulatory complex with peniferum terminating distally in a conspicuously slender, pointed extension; dorsal and ventral fingers slender and subparallel proximally; distal portion of ventral finger gently curved anteriorly with base of terminal setae lying in longitudinal axis of body; clasping apparatus only slightly curved along shaft, not divisible into vertical and horizontal rami; distal portion slightly enlarged and directed anteroventrally; external border entire; internal border with two teeth, and apical border with four serrations.

TRIUNGUIS FEMALE: Shell length 393-415 μ (average, 404 μ); height, 200-259 μ (average, 239 μ). Shell much higher posteriorly than anteriorly, with a distinct elongate emargination ventrally.

Type-locality. — Little Oregon Creek, Near Maggie, 0.7 mi NW of jct of rts 636 and 658, Craig Co., Virginia, U.S.A. (types, USNM: holotype, 108018; allotype, 108019; paratype, 108020, 122494, and 122495). Hosts, *Cambarus b. bartonii* and *C. longulus.*

Distribution (in addition to type-locality). —

United States

VIRGINIA — CRAIG CO. and MONTGOMERY CO. — 15 unspecified localities (hosts, *C. b. bartonii, C. longulus,* and *C. acuminatus*). This ostracod is known only from the James River Drainage in tributaries of Craig and Johns Cks at elevations below 2000 feet.

Donnaldsoncythere cayugaensis (Hobbs and Walton, 1966)

Plate XXIII, figures 1-5
Plate L

Donnaldsoncythere cayugaensis Hobbs and Walton, 1966: 4-6 (M, H, R, A, D), figs. 1h-k.

Diagnosis. — MALE: Shell length 435-460 μ (average, 450 μ); height, 240-260 μ (average, 248 μ). Eye present. Shell with highest portion posterior to midlength; dorsal margin tapering more gently anteriorly than posteriorly, and ventral margin entire; submarginal setae somewhat evenly, if widely, spaced around entire perimeter of shell.

Copulatory complex with peniferum terminating distally in a bilobed rounded prominence; anteroventral margin of posterior lobe thickened. Dorsal and ventral fingers slender and almost straight with terminal seta of dorsal finger reaching level of penis guides on peniferum. Clasping appartaus with distal portion directed at about a 45° angle to basal portion and not clearly divisible into vertical and horizontal rami; margins entire except distal portion of internal border with two teeth and distal extremity with four denticles.

TRIUNGUIS FEMALE: Shell length of allotype 360 μ; height, 240 μ. Shell of triunguis female much higher than that of male and higher posteriorly than anteriorly; ventral margin with a distinct concavity just anterior to midlength.

Relationship. — This ostracod is most closely related to *D. ileata,* but differs from it in the bilobed ventral extension of the peniferum and the slightly more angular clasping apparatus.

Type-locality. — Tributary to Fall Creek, 1 mi E of Summer Hill Village, on rt 90, Cayuga Co., New York, U.S.A. (types, USNM: holotype and allotype, 111255; paratype, 111256). Host, *Cambarus robustus.*

Distribution (in addition to type-locality). —

United States

NEW YORK — OSWEGO CO. — Salmon R 8 mi E of Pulaski on rt 13 in Lake Ontario drainage (host, *C. robustus*) (Hobbs and Walton, 1966: 6). TOMPKINS CO. — Fall Creek, Cayuga Lake drainage (host, *C. robustus*) (Hobbs and Walton, 1966: 6); Salmon Ck in Cayuga Lake drainage (host, *C. robustus*) (Hobbs and Walton, 1966: 6).

Donnaldsoncythere donnaldsonensis (Klie, 1931)

Plate XXIII, figure 6
Plate L

Entocythere donnaldsonensis Klie, 1931: 334-341 (M, H, R, D), figs. 1-9. — Wolf, 1934-1938: vol. 3: 43 (C). — Rioja, 1941a: 177 (C), 182 (R), 189 (R). — Rioja, 1941b: 193 (C, D, H), 195-197 (R). — Hoff, 1944b: 327, 328 (M), 330 (C), 331 (K). — Tressler, 1959: 733 (K), fig. 28.196. — Hart, 1960: 1, 4 (R), fig. 8. — Hart and Hobbs, 1961b: 173 (B, C), 174 (K), 184 (R). — Hart, 1962: 121 (C, H), 123 (C).

Entocythere donnaldsoni. — Wolf, 1934-1938: vol. 2: 489 (C).

Entocythere (Cytherites) donnaldsonensis. — Hoff, 1942a: 66 (K), 69 (R). — Hoff, 1943: 280 (R).

Entocythere (Donnaldsoncythere) donnaldsonensis. — Rioja, 1942b: 688 (K), 691 (R, E), 683, 694 (R). — Howe, 1961: 300 (C).

Entocythere donnalsonensis. — Dobbin, 1941: 185 (R). — Rioja, 1941b: fig. 1A. — Rioja, 1951: 170 (B).

Entocythere donalsonensis. — Rioja, 1943b: 568 (R).

Entocythere donaldsonensis. — Tressler, 1947: 705 (C). — Rioja, 1949: 328 (R). — Rioja, 1953: 292 (C).

Donnaldsoncythere donnaldsonensis. — Hart, 1962: 131-132 (C, D, S). — Hobbs and Walton, 1963b: 363 (R), 364 (K), 369 (R). — Hobbs and Barr, 1972: 81 (D, H).

Diagnosis. — MALE: Shell length 460 μ (see Klie, 1931: 340). Copulatory complex with conspicuous penis situated in distal $\frac{1}{3}$ of peniferum; peniferum terminating bluntly. Clasping apparatus with vertical ramus longer than horizontal ramus; horizontal ramus indistinctly merging with vertical ramus, bearing two teeth on internal border and three terminal denticles; external border of both rami entire.

Type-locality. — Donaldson's Cave, Lawrence Co., Indiana, U.S.A. (types, unknown). Hosts, probably *Cambarus pellucidus* (= *Orconectes inermis inermis*) and *C. bartonii tenebrosus* (= *C. tenebrosus*). See Fage, 1931: 361-374.

Distribution. — Known only from the type-locality.

Donnaldsoncythere hiwasseensis (Hobbs and Walton, 1971)

Plate XXIII, figures 7-10
Plate L

Entocythere hiwasseensis Hobbs and Walton, 1961: 281-384 (M, H, R, D), figs. 1, 4, 5, 6. — Hart, 1962: 123, 132 (C).

Donnaldsoncythere hiwasseensis. — Hart, 1962: 132 (C, D, S). — Hobbs and Walton, 1963b: 364 (K), 370 (E, R), fig. 22. — Hart and Hart, 1966a: 388: 8 (A). — Ferguson, 1968: 501 (C, D). — Hobbs and Barr, 1972: 81 (D, H).

Diagnosis. — MALE: Shell length 375-405 μ (average, 391 μ); height, 200-230 μ (average, 213 μ). Shell subelliptical with maximum height at anterior portion of posterior quarter; marginal setae relatively evenly dispersed dorsally and ventrally.

Copulatory complex with peniferum sclerotized and anterodistal extremity produced into a prominence which in lateral aspect appears angular posteriorly and evenly rounded anteriorly, extreme distal surface with a subcircular heavily sclerotized area; dorsal finger stout and directed ventrally, ventral finger slender and strongly curved at distal portion of proximal third, its tip directed anteroventrally; penis prominent, large, strongly curved, with tip extending ventrally immediately posterior to posterodistal portion of angle of peniferum; clasping apparatus extending ventrally to level of distal extremity of peniferum, not divisible into vertical and horizontal rami; margins entire except for two teeth on internal border near distal end, its distal extremity bearing four teeth.

FEMALE: Shell length of triunguis female 380-415 μ (average, 396 μ); height, 205-240 μ (average, 225 μ). Shell somewhat asymmetrical, with a nearly straight ventral margin; right valve with posterior margin extended beyond that of left; a small, corneous, apically acute, cylindrical prominence present in anterodorsal portion of posteroventral quarter.

Smaller biunguis females without a corneous prominence in the posterodorsal region of body. Shell length 355-360 μ (average, 357 μ); height 195-205 μ (average, 201 μ).

Relationships. — *Donnaldsoncythere hiwasseensis* is closely related to *D. scalis, D. ileata, D. humesi,* and *D. tuberosa* — and it is likely that further study of these ostracods will show that they are but one species and should be synonymized.

Type-locality. — Stream 2.5 mi E of Brasstown Bald, Towns Co., Georgia, U.S.A. (types, USNM: holotype, 105963; allotype, 105964; paratype, 105963). Hosts, *Cambarus bartonii bartonii* and an undescribed species of *Cambarus* (*Puncticambarus*).

Distribution (in addition to type-locality). —

United States

GEORGIA — DADE CO. — Spring from cave on NW slope of Lookout Mt., 3.5 mi NE of Trenton (host, *Cambarus b. bartonii*). DAWSON CO. — Shoal Ck (host, *C.* (*Puncticambarus*) sp.). LUMPKIN CO. — Cane Ck at Dahlonega (host, *C. b. bartonii*). BANKS CO. — 0.9 mi SW of Homer on rt 98 (hosts, *Procambarus raneyi, C. b. bartonii, C. latimanus,* and *C.* (*Depressicambarus*) sp.).

KENTUCKY — MEADE CO. — Doe Run (host, *Cambarus tenebrosus*); Rockhaven Cave in Otter Creek State Park (host, *C. tenebrosus*).

Horton H. Hobbs, Jr., has indicated (personal communication) that this species is also found in northwestern South Carolina and southwestern North Carolina.

Donnaldsoncythere humesi (Hoff, 1943)

Plate XXIII, figures 11-14
Plate L

Entocythere (*Cytherites*) *humesi* Hoff, 1943: 282-285 (M, H, R, D), figs. 2a-h. — Rioja, 1943a: 565 (R).

Entocythere humesi. — Hoff, 1944b: 331 (K). — Tressler, 1947: 705 (C). — Rioja, 1949: 328 (R). — Hobbs, 1955: 329 (A), 332, 333 (D). — Ferguson, 1958: 199 (C). — Tressler, 1959: 733 (K), fig. 28.197. — Hart, 1960: 1, 4 (R), fig. 9. — Hobbs and Walton, 1960: 19 (R). — Crawford, 1961: 241 (R), 242 (R), 244 (A). — Hart and Hobbs, 1961b: 184 (R). — Hobbs and Walton, 1961: 384 (R). — Hart, 1962: 122 (C).

Donnaldsoncythere humesi. — Hart, 1962: 132 (C, D, S). — Hobbs and Walton, 1963b: 363 (D), 364 (K), 369 (R), fig. 21. — Hart, 1964: 243, 245 (R). — Crawford, 1965: 150 (A).

Diagnosis. — MALE: Shell length 450 μ; height, 240 μ. Shell reniform, with greatest height in posterior one-third.

Copulatory complex with peniferum terminating in heavily chitinized truncated lobe; penis large, and directed ventrally. Clasping apparatus with indistinct separation between vertical and horizontal rami; horizontal ramus with two teeth on internal border and three terminal denticles.

Type-locality. — Stream near Boston, Erie Co., New York, U.S.A. (types, USNM: holotype, allotype, and paratype, 81293). Most, *Cambarus robustus.*

Distribution. — Known only from type-locality, not from localities in North Carolina, Virginia, and West Virginia listed in Hobbs 1955: 332-333.

Donnaldsoncythere ileata Hobbs and Walton, 1963

Plate XXIV, figures 1-5
Plate L

Donnaldsoncythere ileata Hobbs and Walton, 1963b: 364 (K), 368, 369 (M, H, R, A, D), 370 (E, R), figs. 11-15, 23. — Hobbs and Walton, 1966: 7 (D, A). — Hobbs, Holt, and Walton, 1967: 35 (K), 38 (H, A), 38 (M, D, H, A, S), 67-69 (D, E), 18, 20, 25, 27, 30, 39, 41, 42, 44, 45, 47, 49, 51, 54, 55, 56-59, 61-64 (A, H), 72, 73, 74, fig. 8c. — Ferguson, 1968: 501 (C, D). — Walton and Hobbs, 1971: 87-103 (B). — Peters, 1971: 100.

Diagnosis. — MALE: Shell length 415-459 μ (average, 445 μ); height, 215-259 μ (average, 242 μ). Eyes present; shell oblong-ovate with posterior portion distinctly higher than anterior; ventral margin entire; marginal setae sparse but present anteriorly, ventrally, and posteriorly.

Copulatory complex with peniferum terminating distally in a rounded extension, with a posterior thickened border; anterior border always rounded; dorsal and ventral fingers slender; ventral finger gently curved cephalically; clasping apparatus with proximal shaft portion almost straight and distal third directed at an angle of 45 to 90 degrees to shaft; external border entire; internal border of distal fourth with two teeth and terminal margin with four or five serrations.

TRIUNGUIS FEMALE: Shell length 437-474 μ (average, 460 μ); height, 237-282 μ (average, 269 μ). Shell more highly vaulted than in male and ventral margin with a distinct emargination immediately anterior to midlength.

Type-locality. — Greenbriar Branch at Newport, Giles Co., Virginia, U.S.A. (types, USNM: holotype, 108027; allotype, 108028; paratypes, 108035, 122497). Host, *Cambarus b. bartonii.*

Distribution (in addition to type-locality). —

United States

VIRGINIA — CRAIG CO. — Localities in New and James Rivers (hosts, *Cambarus b. bartonii, C. sciotensis, C. longulus, C. acuminatus,* and *Orconectes juvenilis*) (Hobbs and Walton, 1963: 368). GILES CO. — Greenbriar Branch, trib of Sinking Ck, at Newport on rt 42 (host, *C. b. bartonii*); Rapids in New R at Pembroke (host, *C. sciotensis*). MONTGOMERY CO. — Localities in New and James R (hosts, *C. b. bartonii, C. sciotensis, C. longulus, C. acuminatus,* and *O. juvenilis*) (Hobbs and Walton, 1963: 368).

WEST VIRGINIA — MONROE CO. — Localities in trib of New R (hosts, *Cambarus b. bartonii, C. sciotensis, C. longulus, C. acuminatus,* and *Orconectes juvenilis*) (Hobbs and Walton, 1963: 368).

This ostracod has also been found from central Tennessee northeastward into central New York state (Horton H. Hobbs, Jr., personal communication).

Donnaldsoncythere pennsylvanica (Hart, 1960)

Plate XXIV, figures 6-7
Plate L

Entocythere pennsylvanica Hart, 1960: 1-4 (M, R, D, H), figs. 1-7. — Hart and Hobbs, 1961b: 184 (R). — Hart, 1962: 123 (C).

Donnaldsoncythere pennsylvanica. — Hart, 1962: 132 (C, D, S). — Hobbs and Walton, 1963b: 364 (K), 370 (E, R), fig. 23. — Ferguson, 1968: 501 (C, D).

Diagnosis. — MALE: Shell length 440-450 μ (average, 445 μ); height, 230-270 μ (average, 254 μ). Shell subelliptical in outline; slightly higher posteriorly than anteriorly; margin entire; lacking setae. Eye located approximately ⅕ the shell length from the anterior margin.

Copulatory apparatus with distal portion of peniferum not extending significantly distad of clasping apparatus; terminal portion of peniferum characterized by corneous disc-like truncate structure. Penis heavy and directed ventrally. Vertical and horizontal rami of clasping apparatus merging imperceptibly; internal border of horizontal ramus with two or three teeth; teminating in three denticles.

TRIUNGUIS FEMALE: Shell similar to male but slightly larger. Shell length 440-460 μ (average, 446 μ); height, 240-300 μ (average, 278 μ).

Type-locality. — Ridley Creek, ¾ mi SW of Goshenville, Delaware Co., Pennsylvania, U.S.A. (types, ANSP). Host, *Cambarus b. bartonii.*

Distribution (in addition to type-locality). —

United States

MARYLAND — Northern Maryland (personal communication from Horton H. Hobbs, Jr.).

MAINE — AROOSTOOK CO. — Beau Lake on St. Francis R (host, *Cambarus bartonii*).

NEW YORK — CAYUGA CO. — Trib to Fall Ck 1 mi E of Summer Hill Village (host, unknown) (Hobbs and Walton, 1966: 6).

OHIO — JACKSON CO. — 6.8 mi S of Rocky Hill on rt 35 (hosts, *Orconectes s. sanborni* and *Cambarus laevis*).

VIRGINIA — GILES CO. — Tribs of New and James R on Salt Pond, Butt, and Doe Mts (hosts, *C. bartonii* and *C. sciotensis*). MONTGOMERY CO. — Small stream 21.3 mi S of New Castle on Craig Creek Road (hosts, *C. b. bartonii* and *C. longulus*).

WEST VIRGINIA — CLAY CO. — 3.1 mi S of Hartland on rt 16 (host, *Cambarus robustus*).

Donnaldsoncythere scalis Hobbs and Walton, 1963

Plate XXIV, figures 8-12
Plate L

Donnaldsoncythere scalis Hobbs and Walton, 1963: 364-366 (M, H, R, A, D), 370 (E, R), figs. 1-5, 25. — Hobbs and Walton, 1966: 6, 7 (D, A). — Hobbs, Holt, and Walton, 1967: 7 (B), 12 (B, H), 35 (K), 39, 40 (M, R, H, A, S), 67-69 (D, E), 27, 29, 30, 38, 45, 47, 52, 58, 61, 64 (A, H), 72, 73, figs. 7d, 9a. — Ferguson, 1968: 502 (C, D). — Hobbs and Walton, 1970: 860 (A).

Diagnosis. — MALE: Shell length 437-452 μ (average, 441 μ); height, 222-245 μ (average, 233 μ). Eyes present; shell ovate with ventral margin entire; marginal setae sparse, but evenly spaced anteriorly, ventrally, and posteriorly.

Copulatory complex with peniferum terminating distally in a scoop-like expansion; dorsal and ventral fingers slender; clasping apparatus forming a gentle arc with external border entire; internal border with two teeth near distal end and four terminal serrations.

TRIUNGUIS FEMALE: Shell length 422-452 μ (average, 439 μ); height, 245-274 μ (average, 256 μ). Shell more highly vaulted than in male and ventral margin with a distinct emargination anterior to mid-length.

Type-locality. — Little Stony Creek, 2.8 mi N of Mountain Lake Biological Station on rt 613, Giles Co., Virginia, U.S.A. (types, USNM: holotype, 108024; allotype, 108025; paratypes, 108026, 122498, and 122499). Host, *Cambarus b. bartonii.*

Distribution (in addition to type-locality). —

United States

KENTUCKY — LEWIS CO. — 2.5 mi E of Garrison on rt 10 (host, *Orconectes juvenilis, Cambarus robustus,* and *C. b. bartonii*).

Donnaldsoncythere truncata Hobbs and Walton, 1963

Plate XXIV, figures 13-17
Plate L

Donnaldsoncythere truncata Hobbs and Walton, 1963b: 364 (K), 369 (E, M, H, R, A, D), figs. 16-20, 24. — Hobbs, Holt, and Walton, 1967: 35 (K, D, A), 38 (A, H), 40 (M, D, H, A, S), 67-69 (D, E), 18, 20, 25, 27, 43, 49, 52, 55, 59, 61, 63, 64 (A, H). — Ferguson, 1968: 502 (C, D).

Diagnosis. — MALE: Shell length 430-445 μ (average, 438 μ); height, 222-259 μ (average, 242 μ). Eyes present; shell oblong-ovate with posterior portion higher than anterior (although contrast not so great as in *D. ileata*); ventral margin entire; marginal setae sparse but present anteriorly, ventrally, and posteriorly.

Copulatory complex with peniferum terminating distally in a broad short terminal extension with subangular distal corners, neither corner conspicuously thickened; distal end of penis situated above terminal extension rather than posterodorsal to it as in other species of the genus; dorsal and ventral fingers slender; ventral finger gently curved throughout length; clasping apparatus with proximal shaft portion only slightly curved and distal third directed at an angle of about 90 degrees to proximal third; external border entire; internal border of distal fifth with six serrations, the proximal three tooth-like.

TRIUNGUIS FEMALE: Shell length 445-482 μ (average, 462 μ); height, 267-282 μ (average, 277 μ). Shell more highly vaulted than in male and ventral margin with a distinct emargination anterior to midlength; most specimens with one valve longer than the other and a few with a distinct posterior extension.

Type-locality. — Whitleys Branch (tributary to Walker Creek and New River), at jct of rts 730 and 622, Giles Co., Virginia, U.S.A. (types, USNM: holotype, 108021; allotype, 108022; paratypes, 108023 and 122500). Hosts, *Cambarus b. bartonii.*

Distribution (in addition to type-locality). —

United States

VIRGINIA — MONTGOMERY AND GILES COS. — Trib of Spruce Run and Walker Ck in New R drainage and in trib of the North Fork of the Roanoke R (hosts, *C. bartonii, C. sciotensis, C. longulus,* and *C. acuminatus*) (Hobbs and Walton, 1963: 369).

Donnaldsoncythere tuberosa (Hart and Hobbs, 1961)

Plate XXV, figures 1-2
Plate L

Entocythere tuberosa Hart and Hobbs, 1961b: 182-184 (M, H, R, A, D), 178 (A), 180 (A), figs. 29-32.

Entocythere tuberculata (*lapsus calami*). — Hart and Hobbs, 1961b: 174 (K).

Donnaldsoncythere tuberosa. — Hart, 1962: 132 (C, D, S). — Hobbs and Walton, 1963b: 364 (K), 370 (E, R), fig. 22. — Hart and Hart, 1966a: 9 (A). — Ferguson, 1968: 502 (C, D). — Hobbs and Walton, 1968: 328 (A, H), 242 (A). — Hobbs and Walton, 1970: 855, 857, 862 (A). — Hobbs and Barr, 1972: 81 (D, H).

Diagnosis. — MALE: Shell length 500-520 μ (average, 510 μ); height, 300-330 μ (average, 310 μ). Eyes reduced or absent. Shell subovate with the greatest height in posterior fifth. Setae scattered on anterior, dorsal, and ventral margins.

Copulatory complex with ventral extremity of peniferum extending ventrally to about the same level as the clasping apparatus; terminal portion with a distinct constriction, thus forming a prominent tubercle on posteroventral margin; a second tubercle present just ventral to base of penis. Clasping apparatus forming a gentle arc, not divisible into the usual vertical and horizontal rami; distal internal border with three teeth and an apical row of four acute ones.

TRIUNGUIS FEMALE: Shell length 530 μ; height, 320 μ. Shell similar to that of male except somewhat higher in posterior half. A corneous papilla directed posteroventrally present near posterodorsal extremity.

Type-locality. — Upper Parrott Cave, 10 mi NW of Livingston, Overton Co., Tennessee, U.S.A. (types, ANSP). Host, *Cambarus tenebrosus.*

Distribution (in addition to type-locality). —

United States

KENTUCKY — WAYNE CO. — Horse Hollow Cave (host, *Orconectes australis packardi*) (Hobbs and Barr, 1972: 83).

TENNESSEE — OVERTON CO. — Upper Parrott Cave 10 mi NW of Livingston (host, *Cambarus tenebrosus*); Roaring R 4.7 mi N of Rickman (host, *C. tenebrosus*). PUTNAM CO. — Blind Fish Cave, Long Hollow (host, *Orconectes australis australis*).

WEST VIRGINIA — POCAHONTAS CO. — Durbin, Greenbriar R (hosts, *Cambarus robustus* and *C. chasmodactylus*).

This ostracod also occurs in southwestern Virginia.

ENTOCYTHERE MARSHALL, 1903

Entocythere Marshall, 1903 (in part): 118-121, 125, 128 (M, H). — Hart, 1962: 125 (K), 132 (M, S, C) (restricted), type-species, *Entocythere cambaria* Marshall, 1903. — Hart, 1963: 51 (R, D). — Hartmann, 1963: 145 (C). — Hobbs III, 1965: 160 (K). — Hobbs, 1966: 67 (S). — Hobbs, Holt, and Walton, 1967: 44 (M). — Hart and Hart, 1969a: 185 (D, E, R). — Hobbs and Hobbs, 1970: 2 (K), 11 (M). —Hobbs, 1971: 40 (M, D). — Hart, 1971: 1-11 (M).

Cambaria Group. — Rioja, 1943a: 560 (C, R). — Hoff, 1944b: 330 (by implication) (R). — Crawford, 1959: 149, 152 (R). — Crawford, 1961: 244 (A).

Illinoisensis Group. — Rioja, 1943a: 560 (C, R). — Rioja, 1949: 327 (R). — Rioja, 1955: 196 (R). — Hobbs, 1956 (1957): 431 (R). — Hart, 1959: 201 (M).

Subgenus *Entocythere.* — Rioja, 1944: 686 (M). — Hartmann, 1963: 145 (C).

Diagnosis. — Terminal mandibular tooth not pectinate (except in *E. ruibali, E. tyttha,* and *E. lepta*), but occasionally with two or three cusps proximal to apex. Copulatory complex of male without finger guard or accessory groove; ventral portion of peniferum usually rounded and partially sclerotized, frequently with anteroventral sclerotized spine; penis directed anteroventrally or obliquely mesially, simple, with spermatic and prostatic elements contiguous throughout their lengths. Clasping apparatus well-developed, extending well beyond peniferum, with rami seldom forming angle greater than 90° (except in *E. dorsorotunda* Hoff, 1944); external border of horizontal ramus entire, internal border with two or more teeth, and distal extremity of ramus with three to six denticles. Triunguis female with pectinate process on distal podomere of second antenna, genital area lacking J-shaped rod and amiculum.

Type-species. — *Entocythere cambaria* Marshall, 1903.

Distribution. — Mexico (Campeche and Chiapas) to Wisconsin and eastward to western Pennsylvania, North Carolina and Florida. Also present on Cuba and the Isla de Pinos.

Hosts. — Freshwater crayfishes of the family Astacidae.

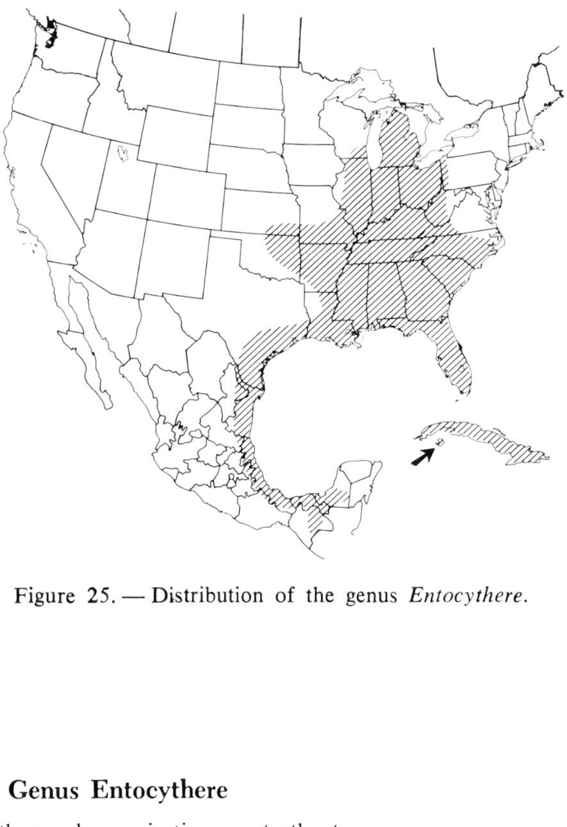

Figure 25. — Distribution of the genus *Entocythere.*

Key to the Species of the Genus Entocythere

1	Mesial surface of clasping apparatus with angular projections or teeth at or near junction of horizontal and vertical rami	2
1′	Mesial surface of clasping apparatus without angular projections or teeth at or near junction of horizontal and vertical rami	3
2(1)	Mesial surface of clasping apparatus with two teeth, neither extending above level of upper margin of horizontal ramus	*E. internotalus*
2′	Mesial surface of clasping apparatus with two angular projections extending above level of upper margin of horizontal ramus	*E. reddelli*

3(1′) External border of clasping apparatus at junction of horizontal and vertical rami produced posteriorly in subacute angle .... *E. illinoisensis*
3′ External border of clasping apparatus at junction of horizontal and vertical rami never produced posteriorly in angle 4
4(3′) Distal tooth of mandible pectinate 5
4′ Distal tooth of mandible terminating in large median denticle with smaller lateral ones, or tapering without denticles 6
5(4) Major axes of clasping apparatus directed at angle of 90° or less 12
5′ Major axes of clasping apparatus directed at angle of more than 90° .... *E. ruibali*
6(4′) Internal borders of vertical and horizontal rami of clasping apparatus disposed at angle less than 90° .... *E. elliptica*
6′ Internal borders of vertical and horizontal rami of clasping apparatus disposed at angle of at least 90° 7
7(6′) Internal borders of vertical and horizontal rami of clasping apparatus disposed at angle distinctly greater than 90° 8
7′ Internal borders of vertical and horizontal rami of clasping apparatus disposed at angle of approximately 90° 9
8(7) Internal borders of horizontal ramus of clasping apparatus with five teeth subequally spaced; three apical denticles .... *E. dorsorotunda*
8′ Internal border of horizontal ramus of clasping apparatus with three teeth subequally spaced; six apical denticles .... *E. kanawhaensis*
9(7′) Teeth on internal border of horizontal ramus of clasping apparatus blade-like .... *E. claytonhoffi*
9′ Teeth on internal border of horizontal ramus of clasping apparatus acute 10
10(9′) External border of clasping apparatus subangular at junction of vertical and horizontal rami .... *E. cambaria*
10′ External border of clasping apparatus rounded at junction of vertical and horizontal rami 11
11(10′) Shells of both sexes oblong-ovate .... *E. dentata*
11′ Shells of both sexes strongly convex dorsally and concave ventrally .... *E. mexicana*
12(5) Horizontal ramus of clasping apparatus with two teeth .... *E. lepta*
12′ Horizontal ramus of clasping apparatus with five teeth .... *E. tyttha*

Entocythere cambaria Marshall, 1903

Frontispiece
Plate XXV, figures 3-4
Plate LI

Entocythere cambaria Marshall, 1903: 117-144 (M, H, D, B), Pls. X-XIII, figs. 1-30. — Muller, 1912: 317 (M, D, C). — Klie, 1926: 54 (B, H). — Klie, 1930: 272 (H, R). — Klie, 1931: 334, 340, 341 (R). — *not E. cambaria* of Allen, 1933: 119, 120. — Wesenberg-Lund, 1939: 485 (H, C). — Rioja, 1940b: 593, 596-598, 602, 604, 607-609 (R). — Rioja, 1940c: 57, 58 (C, M, R, A, H). — *not* Hobbs, 1941: 14. — Dobbin, 1941: 185 (R). — Rioja, 1941a: 177 (C), 182, 189 (R). — Rioja, 1941b: 195-197 (R), fig. 1B. — Ferguson, 1944: 32(3): 715 (C, R). — Hoff, 1944: 330, 331 (R, K). — Tressler, 1947: 705 (C). — Pratt, 1948: 413 (C). — Rioja, 1949: 327 (R). — Rioja, 1951: 169, 170 (C). — Howe, 1955: 66 (C). — Tressler, 1959: 728 (K), fig. 28, 184. — Crawford, 1959: 152 (R). — Howe, 1961, *in* Moore and Pitrat, 1961: 300 (C). — Howe, 1962: 80 (C). —Hart, 1962: 121 (C), 133 (C, D, S). — Hart, 1963: 51 (R, D). — Hart, Nair, and Hart, 1967: 5 (B, M). — Hobbs, Holt, and Walton, 1967: 6, 14 (B, C). — Hart and Hart, 1971: 115 (C); the reference to *E. cambaria* in line 4, p. 115, should be placed following the diagnosis of the genus *Entocythere* on p. 114 — *lapsus calami.*

Entocythere (*Entocythere*) *cambaria.* — Hoff, 1942a: 63-67 (C, M, R). — Rioja, 1942a: 203 (R) (by implication). — Rioja, 1942b: 686 (K), 695 (R). — Hoff, 1943: 280 (R). — Rioja, 1943b: 577 (R). — Howe, *in* Moore and Petrat, 1961: 300 (C).

When the entocytherid ostracods were revised (Hart, 1962), no specimens of *Entocythere cambaria* Marshall (1903) were available for study — nor were any known to be extant. Furthermore, Marshall made no mention of depositing specimens in any permanent collection, and inquiries at the University of Wisconsin and elsewhere were fruitless.

Marshall's treatment of *E. cambaria* was detailed, and many of his observations on their habits and their relationships with free-living ostracods are of considerable value. Unfortunately — but understandably — Marshall was not aware of the systematic importance of the copulatory apparatus of these ostracods, and although his drawing of the copulatory apparatus (1903: 141, fig. 10) is detailed enough to allow positive identification of specimens, it does not include sufficient detail of certain structures now recognized as having taxonomic importance.

Since, according to Article 68 of the International Code of Zoological Nomenclature (1964), *Entocythere cambaria* Marshall must be considered the type-species of the genus *Entocythere*, both by original designation and by monotypy, it is important that a specimen assignable to the species be made available for reference and study.

Recently several specimens have been identified that can be assigned to the species described by Marshall. Those specimens, collected from localities in Missouri, Oklahoma, and Wisconsin, are listed below. The specimen collected on 1 July 1958 by R. E. Hampton from the Yahara River on Route 19 near Sun Prairie, Dane Co., Wisconsin, was found on the crayfish *Orconectes propinquus* (Girard), and is in keeping with Marshall's implication that *E. cambaria* was found on "different species of *Cambarus* . . . from the vicinity of Madison" (Dane Co., Wisconsin). This specimen is now deposited in the U.S. National Museum in Washington, D.C., and is designated the neotype.

Diagnosis. — MALE: Shell length 555-590 μ (average, 573 μ); height, 290-320 μ (average, 304 μ). Ventral portion of peniferum terminating bluntly; extreme distal margin corneous; with ventrally directed tooth-like process on anterodistal face. Penis small, vertically situated, approximately 25 microns from distal end of peniferum. Clasping apparatus distinctly L-shaped; horizontal and vertical rami subequal in length; external margin of vertical ramus concave, internal margin sinuous; external margin of horizontal ramus sinuous, internal margin bearing five well-developed teeth; horizontal ramus terminating in three denticles. Dorsal finger approximately ⅔ length of clasping apparatus vertical ramus; ventral finger subequal in length to entire clasping apparatus.

FEMALE: Unknown.

Type-locality. — Streams near Madison, Dane Co., Wisconsin, U.S.A. (original material unknown). Host, *Cambarus* sp. Neotype designated above from the Yahara River near Sun Prairie, where river crosses rt 19 near Madison, Dane Co., Wisconsin. Host, *Cambarus propinquus.*

Distribution (in addition to type-locality). —

United States

MISSOURI — BUTLER CO. — 10 mi W of Poplar Bluff on US 60 (host, *Orconectes ozarkae*). CARTER CO. — Van Buren, 8 mi NW of Butler-Carter Co. line on US 60 (host, *Procambarus acutus acutus*). IRON CO. — Middlebrook Ck 1 mi SW of Middlebrook (hosts, *O. hylas* and *O. quadruncus*). WASHINGTON CO. — Mill Ck at Mineral Point (hosts, *O. punctimanus* and *O. medius*).

OKLAHOMA — CHEROKEE CO. — Stream about 10 mi E of Tahlequah (host, *Orconectes neglectus neglectus*). LATIMER CO. — Bondy Ck 1 mi S of Wilburton (host, *O. difficilis*).

Entocythere claytonhoffi Rioja, 1942

Plate XXV, figures 5-7
Plate LII

Entocythere cambaria. — Hobbs, 1941: 4 (H, A, B) (in part).

Entocythere (Entocythere) claytonhoffi. — Rioja, 1942b: 687 (K), 689 (R), 690-695 (M, D, R, H, A), figs. 4-13, 15-17. — Hoff, 1943: 280 (R). — Rioja, 1943a: 558, 560 (R). — Rioja, 1943b: 567 (C), 577, 579, 580 (R), 582 (M), figs. 5, 6, 27.

Entocythere claytonhoffi Rioja, 1942a: 201-203 (M, H, R, A, D), figs. 1-3. — Hoff, 1944b: 330 (R), 331 (K), 345 (R), 349 (R). — Tressler, 1947: 705 (C). — Rioja, 1949: 321 (A), 327 (D, H, R). — Rioja, 1953: 287, 291 (C), 292 (R, E). — Tressler, 1954: 138 (D). — Rioja, 1955: 194, 196 (R), fig. 9. — Hobbs, 1956 (1957): 431, 433 (R). — Tressler, 1959: 727 (K), fig. 28.181. — Crawford, 1959: 152 (R). — Hart, 1962: 133 (C, D, S). — Hobbs, 1971: 41-42 (M, H, R, A, D), fig. 26. — Reddell and Mitchell, 1971: 142 (C).

Diagnosis. — MALE: Shell length 460-480 μ (average, 470 μ); height, 250-260 μ (average, 255 μ). Shell with pronounced concavity ventrally, almost angular anteriorly, and with anteroventral margin almost straight. Distal tooth of mandible simple. Clasping apparatus with external junction of horizontal and vertical rami broadly rounded and junction not conspicuously thickened; external border of horizontal ramus curved, internal border with four teeth, proximal three broad and blade-like; apex with four denticles.

TRIUNGUIS FEMALE: Shell length 550-590 μ (average, 560 μ); height, 320-390 μ (average, 350 μ).

Type-locality. — La Cueva Chica, 3 km SE of Ciudad Valles and 609 m SE of Pujal, San Luis Potosi, Mexico (original types no longer extant but topotypes in USNM collection). Host, *Procambarus acutus cuevachicae.*

Distribution (in addition to type-locality). —

Mexico

CAMPECHE — Palizada (host, *Procambarus llamasi*); Santa Rita, 50 km E of Escarega (host, *P. llamasi*); Estacion de Haro, Escarcega (host, *P. llamasi*).

CHIAPAS — Villa Margaritas (host, *Procambarus pilosimanus*); Gruta de Zapaluta, 30 km S of Comitan (host, *P. pilosimanus*).

SAN LUIS POTOSI — Sotano de Huitzmolotitla, 8 km N of Xilitla (host, *P. toltecae*); 3 km N of Xilitla (host, *P. toltecae*).

TABASCO — 37 km N of Villa Hermosa (host, *Procambarus llamasi*).

TAMAULIPAS — La Cueva de San Nicolas, about 2 km SE of El Pachon (host, *Procambarus toltecae*).

VERACRUZ — El Mirador de Zacuapan, about 10 km NE of Huatusco (host, *Procambarus mexicanus*) (Rioja, 1949: 327); Hacienda de Jalapilla, about 5 km SE of Orizaba (host, *P. mexicanus*) (Rioja, 1949: 327); 3 km S of Coscomatepec (host, *P. aztecus*) (Rioja, 1949: 327); Rincon de la Doncella, Ciudad Mendoza (host, *P. mexicanus*); Zapoapan de Cabana, Catemaco (host, *P. ruthveni zapoapensis*); El nacimiento de la Laguna Encantada, San Andres Tuxtla (host, *P. vazquezae*); Tres Puentes, Col. Emiliana Zapala, Jalapa (host, *P. mexicanus*); El Tulat San Simon, N of San Andres Tuxtla (host, *P. ruthveni zapoapensis*); Tomatlan (host, *P. aztecus*); Cuitlahuac (host, *P. aztecus*).

Unless otherwise indicated, all of the above localities are from Hobbs, 1971: 41.

Entocythere dentata Crawford, 1965

Plate XXV, figures 8-12
Plate LI

Entocythere dentata Crawford, 1965: 151-154 (M, R, A, D), figs. 10-16. — Hobbs and Perkins, 1967: 146 (H, A, D). — Ferguson, 1968: 502 (C, D). — Hobbs and Hobbs, 1970: 12 (R).

Diagnosis. — MALE: Shell length 460 μ; height, 310 μ. Shell subovate in lateral view, with highest point just posterior to midlength. Setae sparse along free margin, with few or none on lateral surface of the valves. Eye situated one-fifth of valve length from anterior end. Margins smoothly rounded.

Peniferum wide, without conspicuous irregularities on edges but with strongly sclerotized inferior margin. Clasping apparatus clearly divisible into vertical and horizontal rami, with two forming rounded angle of about 90° at juncture. Vertical ramus subequal in length to horizontal ramus, and bearing prominence on superior posterior margin. Exterior margin of horizontal ramus smooth, flexor surface bearing five prominent denticles with three terminal teeth.

TRIUNGUIS FEMALE: Shell length 495-690 μ (average, 593 μ); height, 345-420 μ (average, 383 μ). Genitalia of triunguis female composed of system of fine, tortuous, sclerotized tubules surrounded by heterogenous, granular mass.

Type-locality. — City of Greenboro, Guilford Co., North Carolina, U.S.A. (types, USNM: holotype, 113505; allotype, 113506; morphotype, 113507). Host, *Cambarus catagius.*

Distribution (in addition to type-locality). —

United States

GEORGIA — COOK CO. — 5.6 mi N of Hahira on rt 41 (host, *Procambarus advena*). WARE CO. — 12 mi N of Waycross on Alma Road (host, *P. pubischelae*).

NORTH CAROLINA — GUILFORD CO. — Dug from burrows in city limits of Greensboro (host, *C. catagius*); burrows in lawn at E Whittington St., Greensboro (host, *C. catagius*) (Hobbs and Perkins, 1967: 145). ORANGE CO. — 5 mi E of Chapel Hill (host, *C. reduncus*).

PENNSYLVANIA — CRAWFORD CO. — Conneautville (*Orconectes obscurus*?).

Entocythere dorsorotunda Hoff, 1944

Plate XXV, figures 13-15
Plate LI

Entocythere dorsorotunda Hoff, 1944b: 332 (K), 341-345 (M, H, R, D), figs. 11-14. — Tressler, 1947: 705 (C). — Tressler, 1959: 733 (K), fig. 28.195. — Rioja, 1949: 328 (R). — Crawford, 1959: 152 (R). — Hart, 1959: 194 (K),

201 (D, H), fig. 12. — Hart, 1962: 133-134 (C, D, S). — Hobbs and Walton, 1966: 7 (R).

Diagnosis. — MALE: Shell length 420-460 μ (average, 436 μ); height, 230-260 μ (average, 243 μ). Shell elliptical with all margins, including ventral one, convex. Posterior part of shell much broader than anterior. Very slight break in arc of anterior margin near midpoint, but this interruption not marked. Eye located slightly more than one-fifth length of shell from anterior end.

Peniferum terminating in blunt, somewhat chitinous, lobe. Clasping apparatus with nearly straight vertical ramus joined at obtuse angle with horizontal ramus. Horizontal ramus proximally straight, distally curved, external border entire, internal border with five teeth, terminal portion with three denticles.

Type-locality. — Stream near Fitzgerald, Ben Hill Co., Georgia, U.S.A. (types, USNM: holotype, allotype, and paratype, 81559). Host, *Procambarus advena.*

Distribution (in addition to type-locality). —

United States

ALABAMA — MOBILE CO. — Spring Ck about 15 mi SW of Mobile on US 90 (host, *Procambarus lecontei*).

FLORIDA — ALACHUA CO. — Hogtown Ck about 2 mi from Gainesville city limits on 9th Street (host, *Procambarus advena*); 3 mi N, ½ mi W of Gainesville on rt 441 (host, *P. advena*); 4 mi N of Gainesville (host, *P. advena*); burrows 4 mi N of Gainesville in seepage area (host, *P. advena*). BAY CO. — Swamp adjoining Econfina R W of Bennet (host, *P. rogersi* (intergrades)). CHARLOTTE CO. — Englewood (host, *P. alleni*) (Hoff, 1944b: 345). CLAY CO. — Farm of H. C. Smith (host, *P. advena*) (Hoff, 1944b: 345). DADE CO. — Burrows in a marsh in Miami (host, *P. alleni*) (Hoff, 1944b: 345). DUVAL CO. — Burrows 1 mi N of jct of Edgewood Avenue of Waycross Road (hosts, *P. advena* and *P. paeninsulanus*).

GEORGIA — BEN HILL CO. — Burrows near Fitzgerald (host, *Procambarus advena*). CLINCH CO. — Near Fargo (host, *P. seminolae*) (Hoff, 1944b: 345). ECHOLS CO. — 3.6 mi NE of state line on rt 89 (host, *P. advena*). PIERCE CO. — Near Blackshear (host, *P. advena*) (Hoff, 1944b: 345).

Entocythere elliptica Hoff, 1944

Plate XXVI, figures 1-2
Plate LI

Entocythere elliptica Hoff, 1944b: 328 (R), 330 (R), 331 (K), 345-349 (M, H, R, D), figs. 15-21. — Tressler, 1947: 705 (C) (*sic eliptica*). — Rioja, 1949: 327 (R) (*sic eliptica*). — Crawford, 1959: 157 (R). — Hart, 1959: 193 (D), 94 (K), 201 (D, H, M), figs. 15, 16. — Walton and Hobbs, 1959: 115 (K). — Tressler, 1959: 728 (K), fig. 28.183. — Hart, 1962: 122 (C), 134 (C, D, S). — Hobbs and Walton, 1966: 246 (R). — Hart and Hart, 1969a: 185 (D).

Diagnosis. — MALE: Shell length 510-560 μ (average, 528 μ); height, 250-280 μ (average, 272 μ). Shell regularly elliptical except for slight angle at juncture of dorsal and posterior margins. Anterior end rounded, but little narrower than posterior. Ventral margin almost straight, a slight sinuation sometimes near anterior one-third. Eye large, usually oval in shape, and located about one-fifth of shell length from anterior end.

Distal portion of peniferum truncate, usually flat or slightly concave. Penis situated in distal one-fourth of peniferum. Clasping apparatus with distinct vertical and horizontal rami, forming acute angle with one another; external border entire, slightly convex; internal border with five or six small teeth; terminal portion with three denticles.

Type-locality. — North of Fargo, Clinch Co., Georgia, U.S.A. (types, USNM: holotype, allotype, and paratype, 81560). Host, *Procambarus seminolae.*

Distribution (in addition to type-locality). —

United States

ALABAMA — BULLOCK CO. — 5.5 mi SE of Midway on rt 82 (host, *Procambarus verrucosus*). BUTLER CO. — 2.6 mi W of Greenville on rt 10 (hosts, *P. a. acutus* and *Cambarus diogenes* subsp.). COVINGTON CO. — 7 mi SE of Andulusia on rt 55 (hosts, *P. oklaoosae* and *P. versutus*). HENRY CO. — 2.7 mi N of Columbia on rt 95 (host, *P. spiculifer*). HOUSTON CO. — 0.8 mi on side rd off rt 52, 1 mi W of Columbia (host, *P. spiculifer*). LAUDERDALE CO. — 18¾ mi W of Athens (host, *Orconectes spinosus*). MACON CO. — 7.2 mi W of Tuskegee on rt 80 (hosts, *P. acutissimus, P. verrucosus,* and *C.* (*Depressicambarus*) sp.). MOBILE CO. — Moore's Branch, about 8 mi SW of Mobile on US 90 (host, *P. lecontei*). ST. CLAIR CO. — 4.1 mi E of Blount-St. Clair line on rt 231 (hosts, *O. spinosus, C. diogenes* subsp., *C. jordani,* and *C.* (*Puncticambarus*) sp.). TUSCALOOSA CO. — N shore of Black Warrior R at Lock II (host, *P. a. acutus*).

FLORIDA — ALACHUA CO. — Ck 1.5 mi W of LaCrosse on Gainesville Road (host, *Procambarus paeninsulanus*). BAY CO. — 6.3 mi W of Beacon Hill on rt 98 (host, *P. rogersi* intergrades). CALHOUN CO. — 1.7 mi S of Chipola R on rt 6 (host, *P. spiculifer*); 6 mi N of Blountstown, 27.3 mi SE of Marianna (host, *P. r. rogersi*). ESCAMBIA CO. — 2 mi E of Bluff Springs on Escambia R (hosts, *Fallicambarus* sp., *C. diogenes* subsp., *P. a. acutus, P. bivittatus,* and *P. shermani*); sloughs along Escambia R on rt 62 near Jay and Century (hosts, *P. a. acutus, P. okaloosae,* and *P. bivittatus*). FLAGLER CO. — 10.8 mi E of San Mateo, 3.9 mi N of rt 28 (host, *P. paeninsulanus*). GADSDEN CO. — Kilby Ck 3.25 mi S of Quincy (host, *P. spiculifer*); 12.6 mi N of Bristol on rt 12 (hosts, *P. spiculifer* and *P. paeninsulanus*); 6.3 mi N of Hosford on rt 65 (host, *P. spiculifer*). GULF CO. — 5.8 mi W of Weewahitchka on rt 52 (hosts, *P. pycnogonopodus, P. latipleurum,* and *P. kilbyi*). JEFFERSON CO. — 3 mi S of Georgia line on FL 19 (host, *P. paeninsulanus*). LEVY CO. — Ck under rts 19 and 98, 6 mi SE of Gulf Hammock (host, unknown). LIBERTY CO. — 1.6 mi N of Hosford on rt 65 (host, *P. spiculifer*); 11.1 mi N of Bristol on rt 12 (host, *P. spiculifer*). ORANGE CO. — Near Orlando (host, *P. alleni*) (Hoff, 1944b: 345). SANTA ROSA CO. — Flooded McCaskill's Mill Ck (hosts, *P. shermani, P. a. acutus,* and *Fallicambarus* sp.). TAYLOR CO. — 7.3 mi NW of Perry on rt 19 (hosts, *C. leonensis* and *P. paeninsulanus*). WASHINGTON CO. — Eoro (host, *P. pycnogonopodus*).

GEORGIA — BURKE CO. — Silty ck 5 mi N of Waynesboro on Augusta Rd (host, *Procambarus pubescens*); Boggy Gut Ck 11.4 mi NE of Waynesboro (hosts, *P. pubescens* and *Cambarus latimanus*). CLARKE CO. — Middle Oconee R at Princeton (host, *P. spiculifer*). DECATUR CO. — 3 mi W of Stone Mt on rt 78 (hosts, *P. spiculifer* and *C. latimanus*); Mosquito Ck, 0.9 mi off rt 97 turnoff SW of Faceville (host, *P. spiculifer*); 4.6 mi S of Bainbridge on rt 27 (host, *P. spiculifer*); Mosquito Ck 1.1 mi SE of Faceville (hosts, *P. spiculifer* and *P. paeninsulanus*); Butler Ck 5 mi from home of C. G. Duke on IPC property (host, *P. spiculifer*); trib of Mosquito Ck 100 yards N of Florida-Georgia line on rt 97 (host, *P. spiculifer*). EMANUEL CO. — Jacks Ck at Lexsy (host, *P. enoplosternum*). EVANS CO. — 5.5 mi N of Claxton on US 301 (host, *P. litosternum*). FANNIN CO. — Near Flat Top Mountain (host, *C. bartonii*). GWINNETT CO. — 4.6 mi W of Loganville on rt 78 (host, *P. spiculifer*); 5.5 mi W of Loganville on rt 78 (hosts, *P. spiculifer* and *C. latimanus*). HEARD CO. — 5.3 mi W of Newman on rt 34 (hosts, *C. latimanus* and *P. spiculifer*). LIBERTY CO. — 7.6 mi SW of Midway on rt 38 (hosts, *P. advena, P. troglodytes,* and *P. barbatus*). MADISON CO. — 5.4 mi SE of Jackson Co. line on GA 98 (hosts, *C.* (*Depressicambarus*) sp., *C. sp.,* and *P. raneyi*). McINTOSH CO. — 5.4 mi S of Eulonia (host, *P. troglodytes*); 16.9 mi N of Darien on rt 17 (hosts, *P. advena* and *P. barbatus*). MURRAY CO. — 1 mi E of Chatsworth on rt 76 (hosts, *C.* (*Depressicambarus*) sp., *C.* (*Puncticambarus*) sp., and *Orconectes spinosus*); 3 mi W of Chatsworth on rt 52 (host, *C.* (*Puncticambarus*) sp.). SCREVEN CO. — Beaver Dam Ck 3 mi N of Sylvania (host, *P. pubescens*); 6.8 mi S of Sylvania on rt 301 (host, *P. epicyrtus*). WHITE CO. — 7.4 mi N of Cleveland on rt 75 (host, *P. spiculifer*). WHITFIELD CO. — 9.3 mi S of Dalton (hosts, *C. spinosus, C.* (*Puncticambarus*) sp., and *C.* (*Depressicambarus*) sp.).

LOUISIANA — ALLEN PARISH — Ten Mile Ck (hosts, *Procambarus acutus acutus, P. simulans simulans, P. dupratzi, Orconectes difficilis,* and *Cambarellus puer*). LINCOLN PARISH — 2 mi N of Ruston (host, *P. vioscai*). ST. TAMMANY PARISH — Locality unknown (host, *P. a. acutus*).

MISSISSIPPI — LEAKE CO. — 11.9 mi S of Carthage on rt 35 (host, *Procambarus acutus acutus*). LOWNDES CO. — 2.3 mi W of Columbus on rt 82 (hosts, *P. a. acutus, P. viaeviridis, P. hybus,* and *P. Fallicambarus fodiens*). OKTIBBEHA CO. — 6.4 mi SW of Mississippi State University campus on rt 12 (host, *P. acutissimus*); 15.2 mi SW of Mississippi State University campus on rt 12 (host, *P. acutissimus*). SCOTT CO. — 8.6 mi S of Leake-Scott Co. line on rt 35 (host, *P. a. acutus*).

SOUTH CAROLINA — HAMPTON CO. — Locality unknown (host, unknown) (Hoff, 1944b: 349).

TEXAS — LIBERTY CO. — 7 mi W of Dayton on rt 90 (hosts, *Procambarus hinei, P. acutus acutus,* and *Cambarellus puer*).

Entocythere illinoisensis Hoff, 1942

Plate XXVI, figures 3-4
Plate LI

Entocythere (*Entocythere*) *illinoisensis* Hoff, 1942a: 66-69 (M, H, R, D, K), figs. 1-8. — Hoff, 1942b: 166 (D). —

Rioja, 1942a: 202, 203 (R), fig. 4. — Rioja, 1942b: 687 (K), 689-691, 693-695 (R), figs. 1, 18, 19. — Hoff, 1943: 286 (H, A, D). — Rioja, 1943a: 558, 560 (R). — Rioja, 1943b: 569, 572 (R), 573 (R) (*sic ilinoisensis*), 577, 579, 580 (R), figs. 29, 30.

Entocÿthere illinoisensis. — Hoff, 1944b: 330 (R), 331 (K), 349 (R). — Tressler, 1947: 705 (C). — Rioja, 1949: 327 (R). — Stamper, 1957: 50 (B). — Crawford, 1959: 152 (*sic illinoisensus*), 157 (R). — Tressler, 1959: 727 (K), fig. 28.180. — Hobbs and Walton, 1960: 22 (R). — Hart, 1962: 122, 134 (C, D, S). — Hart and Hart, 1966a: 3 (A). — Hobbs, Holt, and Walton, 1967: 7 (B, C). — Hart and Hart, 1969a: 185 (D).

Diagnosis — MALE: Shell length 540-610 μ (average, 575 μ); height, 280-290 μ (average, 285 μ). Shell subelliptical in outline, with greatest height in posterior half; eye pigmented.

Peniferum rounded distally, with flange-like protuberance on anterodistal face; penis small, located in distal $\frac{1}{4}$ of peniferum. External border of horizontal ramus of clasping apparatus convex and entire; internal border with four irregular teeth; four terminal denticles.

TRIUNGUIS FEMALE: Shell similar to male but larger — averaging 630 μ in length, 370 μ in height.

Type-locality. — "East central Illinois," U.S.A. Probably Vermillion Co., Illinois (types, USNM: holotype, 81089, allotype and paratype, 81090). Hosts, *Orconectes propinquus, O. virilis,* and *O. immunis.*

Distribution (in addition to type-locality). —

United States

ALABAMA — LIMESTONE CO. — 6 mi E of Athens on US 72 (hosts, *Orconectes compressus* and *O. spinosus*).

ARKANSAS — PERRY CO. — Stream near Casa (host, *Orconectes palmeri longimanus*) (Hoff, 1943: 286). PULASKI CO. — Stream near Ivesville (hosts, *O. meeki* and *Procambarus acutus acutus*) (Hoff, 1943: 286). SCOTT CO. — 6.0 mi E of Cauthron on rt 28 (host, *O. p. longimanus*). SHARPE CO. — About 9 mi W of Hardy, 0.7 mi E of jct of US 62 and AR 11 (host, *O. ozarkae*). WASHINGTON CO. — East Fork of White River near rt 74 bridge, $\frac{1}{4}$ mi off rt 16 (host, *O. longidigitus*).

ILLINOIS — CHAMPAIGN CO. — Locality unknown (hosts, one or more from among *Orconectes propinquus, O. virilis,* and *O. immunis*) (Hoff, 1942a: 69). CRAWFORD CO. — Locality unknown (hosts, one or more from among *O. propinquus, O. virilis,* and *O. immunis*) (Hoff, 1942a: 69).

INDIANA — DEKALB CO. — 6.6 mi W of Butler on US 6 (host, *Orconectes propinquus*). HENDRICKS CO. — Big White Lick Ck, 2 mi N of Brownsburg at railroad bridge (host, *O. rusticus*).

KENTUCKY — BATH CO. — 4.2 mi E of Owingsville on rt 60 (hosts, *Orconectes rusticus, O. juvenilis,* and *Cambarus* (*Depressicambarus*) sp.). BREATHITT CO. — Buckhorn Creek, 5.2 mi NE of rt 15 on dirt road, at Camp Robertson, near Noble (hosts, *C.* (*Jugicambarus*) sp. and *O. juvenilis*). FRANKLIN CO. — creek at jct of rt 60 and 151 (hosts, *O. rusticus* and *C. ornatus*). JEFFERSON CO. — Long Run, 1.9 mi N of rt 60 and 5.6 mi E of Middletown (hosts, *O. rusticus* and *C. laevis*). LETCHER CO. —— Paine Gap Creek and Patter's Fork near Secon on rt 119 (host, *O. juvenilis*). MORGAN CO. — 4 mi E of Grassy Creek on US 460 (hosts, *C.* sp. and *O. juvenilis*).

MICHIGAN — MONTMORENCY CO. — 5.4 mi W of jct of rts 32 and 33 (host, *Orconectes propinquus*). NEWAYGO CO. — White River at jct of rt 3 in White Cloud (host, *O. virilis*).

MISSISSIPPI — RANKIN CO. — 0.5 mi from Rankin-Scott Co. line on rt 80 (host, *Procambarus acutus acutus*).

OHIO — CRAWFORD CO. — Locality unknown (host, *Orconectes rusticus*). FRANKLIN CO. — Banks of Big Walnut, $\frac{1}{2}$ mi above Lockbourne Road (host, *Cambarus b. bartonii*). SCIOTO CO. — Scioto Brush Creek near Rushtown on rt 104 (*O. r.* rusticus). WYANDOT CO. — Locality unknown (host, *O. r. rusticus*).

TENNESSEE — CLAY CO. — Oak Grove, W of Celina (host, *Orconectes putnami*). DAVIDSON CO. — Seven Mile Creek N of Oglesby (host, *O. rusticus* ssp.). DEKALB CO. — Smithville (host, *O. r.* ssp. and *O. placidus*). GILES CO. — 10.6 mi N of Pulaski on rt 31A (host, *O. spinosus*). HICKMAN CO. — Locality unknown (hosts, *O. compressus, O. r.* ssp., and *O. placidus*). HUMPHREYS CO. — Hurricane Creek S of Waverly (hosts, *O. compressus, O.* sp., and *O. placidus*). LAWRENCE CO. — Shoal Creek 1.4 mi E of Lawrenceburg on rt 64 (host, *Cambarus longirostris*). LINCOLN CO. — Branch of Swan Creek 6.7 mi E of McBurg on rt 64 (hosts, *O.* sp., *C. striatus,* and *C. tenebrosus*). MEIGS CO. — 1.9 mi N of Georgetown (hosts, *C. longirostris* and *O.* sp.) (Hobbs and Walton, 1960: 22). PUTNAM CO. — Falling Water River below Burgess Falls Lake Dam, Coney Fork of Cumberland River (host, *O. placidus*); Southern Baxter on TN 56 (host, *O. placidus*). RUTHERFORD CO. — Murfreesboro (host, *O. placidus*). WAYNE CO. — Forty-eight Mile Creek, trib to Buffalo River, 7 mi E of Waynesboro on rt 64 (hosts, *O. alabamensis, O. compressus, O. spinosus,* and *O. forceps*). WILLIAMSON CO. — Franklin (host, *O. rusticus* ssp.).

VIRGINIA — BUCHANAN CO. — Garden Creek at Skegge (host, *Orconectes juvenilis*).

Entocythere internotalus Crawford, 1959

Plate XXVI, figures 3-7
Plate LII

Entocythere internotalus Crawford, 1959: 152-157 (M, H, R, D), 150 (B), 151 (H, K), 178, 180, 181 (A), figs. 1-9. — Crawford, 1961: 238 (A). — Hart, 1962: 123 (C), 134 (C, D, S). — Hobbs, Holt, and Walton, 1967: 34 (K), 38, 43, 44-45, 72, 73, 77, 78 (M, D, H, A, S), 69 (D, E), 18, 20, 25, 41, 44, 50, 54, 55, 58, 61, 63, (A, H), figs. 7b, 11a. — Ferguson, 1968: 502 (C, D). — Hobbs and Walton, 1968: 246 (R). — Hart and Hart, 1969a: 185 (D). — Hart and Hart, 1971: 107, 111.

Entocythere internotalis, Hobbs and Walton, 1967: 35 (D, A) (erroneous spelling).

Diagnosis. — MALE: Shell length 510-615 μ (average, 536 μ); height, 255-330 μ (average, 276 μ). Shell elliptical in lateral aspect, dorsal margin slightly more rounded than ventral. Anterior and posterior margins evenly rounded. Ventral margin with slight concavity approximately one-third of length of valve from anterior end. Setae conspicuous near free margins, being more numerous at posterior end than at anterior and sparsely scattered along ventral margin. Small eye located one-sixth of valve length from anterior end.

Copulatory complex with terminal portion of peniferum truncate; small depression on posteroventral margin; penis near level of base of clasping apparatus. Vertical and horizontal rami of clasping apparatus at right-angles to one another; juncture arched, thickened, and bearing heavy mesial talon; horizontal ramus tapering distally, internal border with distinct proximal tooth, and three or four relatively indistinct teeth more distally; three terminal denticles.

TRIUNGUIS FEMALE: Shell length 630-680 μ (average, 654 μ); height, 370-430 μ (average, 397 μ). Shell similar in shape to that of male.

Type-locality. — Colonel's Creek, on route 262, 14.6 mi E of city limits of Columbia, Richland Co., South Carolina, U.S.A. (types, USNM: holotype, 103559; allotype, 103560; morphotype, 103561; paratype, 103562). Hosts, *Procambarus enoplosternum* and *Cambarus acuminatus.*

Distribution (in addition to type-locality). —

United States

ALABAMA — SUMPTER CO. — locality unknown (host, *Procambarus acutissimus*).

FLORIDA — LEVY CO. — small creek under US 19 and 98, 6 mi SE of Gulf Hammock on rt 19 (host, *Procambarus paeninsulanus*). TAYLOR CO. — Scanlan (host, *P. paeninsulanus*).

GEORGIA — ATKINSON CO. — 3.1 mi S of Douglas on US 441 (hosts, *Procambarus advena* and *P. seminolae*); N side of Pearson on US 441 (hosts, *P. pubischelae* and *P. seminolae*). BRANTLEY CO. — city limits of Hoboken on rt 121 (host, *P. seminolae*). CHATTAHOOCHEE CO. — 2.8 mi S of Cusseta on US 27 (host, *P. spiculifer*). DOOLY CO. — 5.2 mi NW of Vienna on GA 90 (hosts, *P. spiculifer, P. howellae,* and *Cambarus (Depressicambarus)* sp.). ECHOLS CO. — 3.6 mi NE of Georgia state line on rt 89 (host, *P. advena*). FLOYD CO. — 23 mi NW of Rome (host, *C. (Puncticambarus)* sp.). GLYNN CO. — 6 mi N of Brunswick on rt 17 (hosts, *P. advena* and *P. pubischelae*). LOWNDES CO. — 14.1 mi E of Valdosta (host, *P. pubischelae*). MONTGOMERY CO. — 1.3 mi S of Truetlen Co. line on US 221 (hosts, *P. a. acutus* and *P. enoplosternum*). PAULDING CO. — rt 101 at Tallapoosa River (hosts, *C. latimanus* and *C. halli*). RANDOLPH CO. — 11.8 mi S of Cuthbert on US 27 (host, *P. spiculifer*). TRUETLEN CO. — 16.0 mi SE of East Dublin on rt 29 (hosts, *P. enoplosternum* and *P. a. acutus*); 19.5 mi SE of East Dublin on rt 29 (hosts, *P. truculentus* and *P. enoplosternum*). WORTH CO. — 2.5 mi E of Poulon on US 83 (hosts, *P. advena* and *P. seminolae*).

KENTUCKY — HICKMAN CO. — Obion Creek in area of Murphy's Pond (host, *Procambarus acutus acutus*).

LOUISIANA — VERNON PARISH — Billy Goat Bayou (hosts, *Procambarus natchitochae, P. vioscai, P. clarkii, Cambarellus shufeldtii,* and *Orconectes lancifer*).

MISSISSIPPI — MONROE CO. — Muldon (host, *Cambarus diogenes* ssp). NEWTON CO. — 10.7 mi N of Chunky on rt 80 (host, *Procambarus acutus acutus*). SMITH CO. — in city limits of Mize on rt 35 (hosts, *P. a. acutus* and *C.* sp.).

NORTH CAROLINA — BEAUFORT CO. — roadside ditch on rt 264, 0.5 mi E of jct with rt 1714,

4 mi W of Pantego (host, *Procambarus acutus acutus*). BLADEN CO. — 3.7 mi S of Dublin on rt 41 (host, *P. pearsei pearsei*). RANDOLPH CO. — 2.8 mi NE of Randolph Co. line on rt 49 (hosts, *P. a. acutus* and *Cambarus acuminatus*).

SOUTH CAROLINA — ABBEVILLE CO. — small intermittent stream on rt 20, 1.4 mi N of Abbeville (host, *Cambarus latimanus*) (Crawford, 1961: 238). DORCHESTER CO. — ditch 1.4 mi NW of Harleyville on rt 178 (hosts, *Procambarus troglodytes* and *C. uhleri*). RICHLAND CO. — 26 localities in Congaree River drainage (hosts, *C. latimanus, C. acuminatus, C. uhleri, C. reduncus, P. a. acutus, P. enoplosternum, P. troglodytes*, and *P.* sp.) (Crawford, 1959: 156); 13 localities in the Wateree River drainage (hosts, *C. latimanus, C. acuminatus, C. reduncus, P. a. acutus, P. enoplosternum, P. troglodytes,* and *P.* sp.) (Crawford, 1959: 156); 3 localities in the Broad River drainage (hosts, *C. spicatus, C. latimanus, C. bartonii, P. a. acutus, P. enoplosternum,* and *P. troglodytes*) (Crawford, 1959: 156).

TEXAS — BLANCO CO. — locality unknown (host, *Procambarus clarkii*).

Entocythere kanawhaensis Hobbs and Walton, 1966

Plate XXVI, figures 8-12
Plate LI

Entocythere kanawhaensis Hobbs and Walton, 1966: 6-7 (M, H, R, A, D), figs. 11-16. — Hobbs, Holt, and Walton, 1967: 34 (K), 45 (M, D, H, A, S), 69 (D, E), 20, 29, 38, 39, 47, 52, 54, 55 (*sic kanawaensis*), 57, 61, 64 (A, H). — Hart and Hart, 1969a: 185 (D).

Diagnosis. — MALE: Shell length 600-615 μ (average, 607 μ); height, 300 μ. Eye present. Shell distinctly elongate, highest just posterior to midlength, and with dorsal margin evenly contoured, tapering gently both anteriorly and posteriorly; ventral margin entire; submarginal setae present except dorsally between level of eye and anus; setae mostly regularly spaced but somewhat closer together posteriorly.

Copulatory complex with posterior and ventral margins of peniferum entire, latter rounded; entire peniferum of almost uniform thickness and without ornamentation. Penis small, its length approximately one-fourth anterior-posterior dimension of distal portion of peniferum, and situated dorsal to ventral extremity of peniferum. Dorsal finger slender with terminal seta reaching level of penis; ventral finger, also slender, and sinuous. Clasping apparatus with distinct vertical and horizontal rami meeting at angle of approximately 100°. Vertical ramus slightly convex anteriorly with both margins entire; horizontal ramus slightly convex ventrally with external border entire, internal border with three distinct teeth and six apical denticles.

TRIUNGUIS FEMALE: Shell length 698 μ; height, 398 μ (one specimen). Shell proportionately higher than that of male and larger; posterior declivity of dorsal margin more sudden than in that of male; ventral margin entire. Submarginal setae as in male.

Relationships. — *Entocythere kanawhaensis* seems to have its closest affinities with *E. dorsorotunda* Hoff (1944) and *E. elliptica* Hoff (1944) but differs from both species by possessing only three teeth along the internal border of the horizontal ramus of the clasping apparatus of the male.

Type-locality. — Rapids in New River at Pembroke, Giles Co., Virginia, U.S.A. (types, USNM: holotype and allotype, 111257). Host, *Cambarus sciotensis*.

Distribution (in addition to type-locality). —

United States

VIRGINIA — GILES CO. — Cascades of Little Stony Creek (hosts, *C. sciotensis* and *C. b. bartonii*) (Hobbs and Walton, 1966: 7).

Entocythere lepta Hart and Hart, 1971

Plate XXVII, figures 1-5
Plate LII

Entocythere lepta Hart and Hart, 1971: 114-115 (M, H, R, A, D), figs. 8a-e.

Diagnosis. — MALE: Shell length 320 μ height, 190 μ (single specimen). Shell subovoid, posterior and blunt; greatest height $\frac{1}{3}$ shell length from posterior end. Eye pigmented, located $\frac{1}{4}$ shell length from anterior end. Distal tooth of mandible pectinate.

Copulatory complex with peniferum bluntly terminated ventrally. Peniferum short, extending only to midlength of vertical ramus of clasping apparatus. Spermatic loop short; penis horizontal, located in distal $\frac{1}{5}$ of peniferum, and extending $\frac{3}{4}$ width of peniferum.

Clasping apparatus with approximately 60° angle between vertical and longer horizontal rami. External border of vertical ramus evenly convex; internal border evenly concave. External border of horizontal ramus convex with apex directed anterodorsally. Internal border of horizontal ramus with two teeth and four terminal denticles.

TRIUNGUIS FEMALE: Shell length 270-330 μ (average, 300 μ); height, 170-210 μ (average, 190 μ). Shell subovoid, greatest height near midlength. Eye pigmented, situated ⅕ shell length from anterior end. Papilla situated at posterior of shell.

Relationships. — Its small size in relation to other members of this genus distinguishes this from any other *Entocythere* except *E. tyttha,* from which it differs in the shape of the peniferum and the number of teeth on the clasping apparatus. The pectinate tooth on the mandible and the shape of the clasping apparatus are characteristic of *Uncinocythere;* the blunt peniferum is typical of *Entocythere.*

Type-locality. — Kenzel Springs, 25 mi NW of Gatlinburg, Blount Co., Tennessee, U.S.A. (types, USNM: holotype, 138461; allotype and paratype, 138462). Host, *Cambarus bartonii cavatus.*

Distribution. — Known only from type-locality.

border with five or more simple teeth, and apex with four denticles.

Remarks. — Should the La Magdalena record cited below prove to have been for *E. claytonhoffi,* as indicated by Rioja, then this is the only Puebla record for the species and would place it within range of *E. mexicana,* a close relative, and one which presumably vicariates for *E. claytonhoffi* in the area immediately north of the Cordillera Volcanica Transversal.

Type-locality. — Villa Juarez, Puebla, Mexico (types no longer extant). Hosts, *Procambarus hoffmanni* and *P. caballeroi.*

Distribution (in addition to type-locality). —

Mexico

PUEBLA — Huauchinango (host, *P. riojai*) (Rioja, 1943a: 560); La Magdalena, 3 km N of Zihuateutla (host, *P. erichsoni*) (Rioja, 1949: 327); Zihuateutla (host, *P. hortonhobbsi*) (Hobbs, 1971: 43).

VERACRUZ — Sabana de El Poste, Miahuapan, 8-10 km N of Coyutla (host, *Procambarus hoffmanni*) (Rioja, 1949: 327).

Entocythere mexicana Rioja, 1943

Plate XXVII, figures 6-10
Plate LII

Entocythere (Entocythere) mexicana Rioja, 1943a: 553-560 (M, H, R, A, D), figs. 1-10. — Rioja, 1943b: 572, 573, 576, 577 (R), 579 (M, R), 582 (R), figs. 26, 28, 31-39.

Entocythere mexicana. — Hoff, 1944b: 330, 345, 349 (R), 331 (K). — Tressler, 1947: 95 (C). — Rioja, 1949: 32 (D, H). — Tressler, 1954: 138 (D). — Hobbs, 1956 (1957): 431 (D, R). — Tressler, 1959: 729 (K), fig. 28.185. — Crawford, 1959: 152 (R). — Hart, 1962: 122 (C), 134 (C, D, S). — Crawford, 1965: 153 (R). — Hobbs and Walton, 1968: 246 (R). — Hobbs, 1971: 42-43 (M, H, R, A, D), fig. 28.

Entocythere claytonhoffi. — Rioja, 1949: 327 (in part).

Diagnosis. — MALE: Shell length 490-520 μ (average, 500 μ); height, 270-280 μ (average, 275 μ). Shell with or without broad shallow concavity ventrally and with anteroventral margin gently curved. Distal tooth of mandible simple. Clasping apparatus with external junction of horizontal and vertical rami not produced into angular "heel" but rounded and not conspicuously thickened; external border of horizontal ramus not strongly curved, occasionally concave or straight from base almost to apex, internal

Entocythere reddelli Hobbs and Walton, 1968

Plate XXVII, figures 11-15
Plate LII

Entocythere reddelli Hobbs and Walton, 1968: 243-246 (M, H, R, A, D), figs. 2a-d.

Diagnosis. — MALE: Shell length 560-590 μ (average, 570 μ); height, 270-300 μ (average, 290 μ). Eye pigmented. Shell oblong-ovate with shallow concavity ventrally anterior to midlength. Submarginal setae subtending entire valve margin, most numerous anteriorly and posteriorly and least abundant dorsally.

Copulatory complex: Peniferum with ventral extremity broadly truncate or rounded with sclerotized areas, anteroventral portion with apparent spine-like plate; penis U-shaped and situated diagonally anterodorsally-posteroventrally with apex directed ventrally and slightly posteriorly. Dorsal finger of moderate proportions with bifid apical projection; ventral finger moderately heavy, straight along basal three-fifths followed by gentle curve in following fifth, straight in distal fifth and terminating in simple apical seta. Clasping apparatus with horizontal and vertical rami situated at angle of approximately 80 degrees with gradual, but strong, thickening in area of juncture of two rami; external border of both rami entire; internal border with prominent biangular plate mesial to junction of rami and with four teeth on horizontal ramus; apex of apparatus with three teeth.

TRIUNGUIS FEMALE: Shell length 640-700 μ (average, 680 μ); height, 330-400 μ (average, 370 μ). Eye pigmented. Shell subreniform with distinct concavity on ventral margin slightly anterior to midlength. Submarginal setae as in holotype. Genital area with hyaline J-shaped rod immediately anterior to posterior sclerotized prominence.

Relationships. — *Entocythere reddelli* has its closest affinities with those members of the genus having a "thickened" junction of the horizontal and vertical rami of the clasping apparatus — *E. elliptica, E. internotalus,* and *E. mexicana.* It differs from all other entocytherids, however, in possessing the prominent biangular plate at the junction of the horizontal and vertical rami of the clasping apparatus.

Type-locality. — Golden Fawn Cave, 8 mi NNE of Boerne, Kendall Co., Texas, U.S.A. (types, USNM: holotype and allotype, 123330). Host, *Procambarus clarkii.*

Distribution (*in addition to type-locality*). —

United States

ARKANSAS — GREENE CO. — trib of St. Francis River 4.9 mi NE of Paragould on rt 1 (hosts, *Cambarellus puer, Fallicambarus hedgepethi, Faxonella clypeata, Procambarus acutus acutus, P. viaeviridis*).

KANSAS — SUMNER CO. — 5.6 mi N of jct of rts 166 and 81 on 81 (host, *Procambarus simulans simulans*).

Entocythere ruibali Rioja, 1955

Plate XXVII, figures 16-18
Plate LI

Entocythere ruibali Rioja, 1955: 194-196 (M, H, R, D), figs. 2-8. — Hobbs, 1956 (1957): 431-432 (D, M), 433 (R). — Hart, 1959: 201. — Hart, 1962: 122 (C), 134 (C, D, S). — Ferguson, 1968: 502 (C, D). — Straskraba, 1969: 9 (D, H, S). — Hobbs and Hobbs, 1970: 11 (R). — Hobbs, 1971: 44 (M, H, R, A, D), fig. 29.

Diagnosis. — MALE: Shell length 360-430 μ (average, 390 μ); height, 240-250 μ (average, 245 μ). Shell with ventral margin straight and with anteroventral margin gently curved. Distal tooth of mandible with cusps. Clasping apparatus with external junction of horizontal and vertical rami broadly rounded and junction of rami only slightly thickened; external border of horizontal ramus curved; internal border with proximal two teeth sometimes truncate (blade-like) but never with third and fourth truncate; apex with four denticles.

Type-locality. — Fourteen km W of Camagüey, Camagüey Province, Cuba (types not extant). Host, *Procambarus cubensis cubensis.*

Distribution (*in addition to type locality*). —

Cuba

ISLA DE PINOS — La Ceiba (host, *Procambarus atkinsoni*) (Hobbs, 1956: 432); Nueva Gerona (host, *P. cubensis cubensis*) (Hobbs, 1956: 432).

MATANZAS — Union de Reyes (host, *Procambarus cubensis cubensis*) (Hobbs, 1956: 432).

PINA DEL RIO — Guane (host, *Procambarus cubensis cubensis*) (Hobbs, 1956: 432).

Entocythere tyttha Hobbs and Hobbs, 1970

Plate XXVIII, figures 1-4
Plate LI

Entocythere tyttha Hobbs and Hobbs, 1970: 11-13 (M, H, R, A, D), figs. 5a-e.

Diagnosis. — MALE: Shell length 410-460 μ (average, 430 μ); height, 220-260 μ (average, 240 μ). Eye pigmented, located one-fourth shell length from anterior margin. Shell subovate with margins entire. Marginal setae moderate in number and evenly spaced along anterior, posterior, and ventral margins. Terminal tooth of mandible pectinate.

Copulatory complex with peniferum sclerotized and truncate ventrally, possessing acute projection at anteroventral extremity and low, rounded protuberance posteroventrally; penis comparatively large, located in distal third of peniferum; clasping apparatus with horizontal and vertical rami joined by gentle arc; vertical ramus convex anteriorly and with longitudinal groove on proximal external border; internal border of horizontal ramus entire; internal border of horizontal ramus bearing five teeth; distal extremity terminating in three denticles; dorsal and ventral fingers slender; dorsal finger bifurcate distally; ventral finger anteriorly convex proximally and with subangular bend of approximately 90 degrees near mid-length, thus paralleling contour of clasping apparatus.

TRIUNGUIS FEMALE: Shell length 480 μ; height, 280-290 μ (average, 295 μ). Eye pigmented, located as in male; shell slightly larger than that of male but with submarginal setae similarly dispersed. Genital complex consisting of prominent tubercle, partially surrounded by coiled amorphous mass, and long, slender anteroventrally projecting process.

Relationships. — *Entocythere tyttha* has a very close relative in *E. dentata* Crawford, 1965: 151. Not only does the latter have a similar clasping apparatus but also has three lateral cusps on the distal tooth of the mandible, thus approaching the pectinate condition found in *E. tyttha.* In addition, the female genital complex in the two is remarkably similar. The two species differ, however, in that the clasping apparatus of *E. dentata* is more angular and lacks a longitudinal groove on the external border of the vertical ramus, and, as pointed out immediately above, the mandible is not pectinate.

Type-locality. — Tributary of the Obion River, 6 mi SW of Wingo on rt 45, Graves Co., Kentucky, U.S.A. (types, USNM: holotype and allotype, 126258). Host, *Fallicambarus hedgepethi.*

Distribution. — Known only from type-locality.

GEOCYTHERE HART, 1962

Geocythere Hart, 1962: 125 (K), 134-135 (M, S, C), type-species, *Entocythere geophila* Hart, 1959. — Hart, 1963: 51 (R, D). — Hart, 1965b: 161 (K). — Hobbs, 1967: 3 (R). — Hart and Hart, 1969a: 187 (D, E, R). — Hart and Hart, 1969b: 155, 160, 163, 164 (M, R). — Hobbs III, 1969a: 170 (R). — Hobbs III, 1970: 175 (R). — Hobbs and Hobbs, 1970: 2 (K).

Geophila Group. — Hart, 1959: 195 (M). — Hobbs and Walton, 1962: 44 (R).

Diagnosis. — Terminal tooth of mandible pectinate. Male copulatory complex lacking finger guard and accessory groove; peniferum with ventral portion large, extending ventrally considerably beyond clasping apparatus, and with conspicuous anteriorly or anterodorsally directed projection; projection obliquely truncate and extending from anterior surface rather than being continuous with ventral margin; penis large, U-shaped, extending anteriorly from base but curved with apex directed posteroventrally and with spermatic and prostatic elements contiguous throughout their lengths. Clasping apparatus somewhat L-shaped, but only in *G. nessoides* clearly divisible into vertical and horizontal rami; internal border entire except near distal extremity where annulate or emarginate; external border with one to four prominences variously situated along distal half of apparatus. Triunguis female with second antenna lacking accessory pectinate process on distal podomere; genital complex without J-shaped rod or amiculum but possessing heavily sclerotized genital papilla posterodorsally.

Type-species. — *Geocythere geophila* Hart, 1959: 159; by original designation, Hart, 1962: 135.

Distribution. — A possibly disjunct distribution encompassing Louisiana, Mississippi, the panhandle of Florida and southwestern Georgia on the south and Indiana, Illinois, western Tennessee, and western Kentucky on the north.

Hosts. — Freshwater crayfishes of the family Astacidae.

Figure 26. — Distribution of the genus *Geocythere.*

Key to the Species of the Genus Geocythere

1	External border of clasping apparatus with single protuberance	*G. gyralea*
1′	External border of clasping apparatus with more than one protuberance	2
2(1′)	Clasping apparatus with prominences along external border restricted to small area distal to mid-length of horizontal ramus, and distalmost portion bearing denticles	*G. geophila*
2′	Clasping apparatus with prominences along external border at apparent junction of vertical and horizontal rami, and distalmost portion annulate	3
3(2)	Distal portion of peniferum acute	*G. acuta*
3′	Distal portion of peniferum truncate	*G. nessoides*

Geocythere acuta Hart and Hart, 1971

Plate XXVIII, figures 5-7
Plate LII

Geocythere acuta Hart and Hart, 1971: 115-116 (M, H, R, A, D), figs. 9a-c.

Diagnosis. — Male: Shell length 450-470 μ (average, 460 μ); height, 270-290 μ (average, 280 μ). Shell ovoid, venter and dorsum equally rounded, greatest height at midlength. Eye pigmented, situated ⅕ shell length from anterior end.

Copulatory complex with peniferum slender at midlength, widening dorsally and ventrally. Terminal portion of peniferum with anteriorly-directed acute apex bearing posteriorly directed barb proximoventral to tip. Penis situated in distal ⅙ of peniferum, on level with horizontal ramus of clasping apparatus.

Clasping apparatus with horizontal and vertical rami not well differentiated. Internal border of clasping apparatus concave from near base to tip; distal portion of horizontal ramus with six annulations, resulting in resemblance to stacked cones. External border of clasping apparatus evenly convex with two protrusions proximal to proximalmost annulus.

Biunguis female: Shell length 380-440 μ (average, 410 μ); height, 210-260 μ (average, 240 μ). Shell ovoid, with greatest height near midlength. Eye pigmented.

Relationships. — *Geocythere acuta* is easily recognized by the posteriorly directed barb on the distal portion of the peniferum. The protrusion on the distal portion of the peniferum of *G. nessoides* is directed ventrally. The clasping apparatus of *G. nessoides* is most similar to *G. acuta* but that of the former is approximately ½ of the length of that of *G. acuta*. *G. gyralea* has a much longer, curved penis, extending beyond the border of the peniferum.

Type-locality. — Two miles NW of Bellflower, McLean Co., Illinois, U.S.A. (types, USNM: holotype, 138463; allotype, 138464). Host, *Cambarus d. diogenes.*

Distribution. — Known only from type-locality.

Geocythere geophila (Hart, 1959)

Cover Drawing
Plate XXVIII, figures 8-11
Plate LII

Entocythere geophila Hart, 1959: 194 (K), 195-198 (M, H, R, D), 200 (R), figs. 1-6. — Walton and Hobbs, 1959: 115 (K). — Hart, 1960: 2 (R). — Hart, 1962: 123 (C). — Levinson, 1962: 94 (C).

Geocythere geophila. — Hart, 1962: 135 (C, D, S). — Hart, 1965: 257, 258 (R). — Ferguson, 1968: 502 (C, D). — Hobbs and Hobbs, 1970: 13 (K), 14, 15 (R).

Diagnosis. — Male: Shell length 400-430 μ (average, 410 μ); height, 220-250 μ (average, 237 μ). Shell subelliptical in outline and somewhat vaulted posterior to midlength. Ventral margin straight or slightly convex, and continuous with well rounded arcs that form anterior and posterior margins. Setae scarce, but occur singly or in pairs near anterior and posterior margins. Eye located approximately ⅕ shell length from anterior margin.

Copulatory complex with distal portion of peniferum extending beyond clasping apparatus for distance subequal to length of horizontal ramus of clasping apparatus. Horizontal and vertical rami clearly distinguishable; vertical ramus about 1.5 times as long as horizontal ramus and S-shaped; horizontal ramus not curved, internal border straight (or nearly so), and bears two or three distal denticles, external border convex and bearing 2- or 3-lobed protuberance at about mid-length.

TRIUNGUIS FEMALE: Shell length 370-400 μ (average, 388 μ); height, 220-250 μ (average, 233 μ). Shell similar in shape to that of male.

Relationships. — See *G. nessoides.*

Type-locality. — Stream and burrows on rt 71, 11.4 mi N of Altha, Jackson Co., Florida, U.S.A. (types, USNM: holotype and allotype, 101464; paratype, 101465). Hosts, *Procambarus spiculifer, P. paeninsulanus,* and *Cambarus d. diogenes.*

Distribution (in addition to type-locality). —

United States

FLORIDA — JACKSON CO. — Dry Creek, 5.5 mi S of Marianna on rt 84 (hosts, *C. d. diogenes* and *P. spiculifer*).

GEORGIA — DECATUR CO. — Spring Creek, 4.8 mi E of Iron City on rt 84 (hosts, *Cambarus d. diogenes* and *Procambarus spiculifer*).

TENNESSEE — OBION CO. — burrows in bank of bayou near Reelfoot Lake Biological Station (host, *Cambarus diogenes* subsp.).

Geocythere gyralea Hart, 1965

Plate XXVIII, figures 12-14
Plate LII

Geocythere gyralea Hart, 1965: 257-258 (M, R, D), figs. 3, 4. — Hart and Hart, 1966a: 5 (A). — Hart and Hart, 1967a: 3 (M). — Ferguson, 1968: 502 (C, D). — Hart and Hart, 1969b: 157 (M). — Hobbs and Hobbs, 1970: 14, 15 (R).

Diagnosis. — MALE: Shell length 460-500 μ (average, 470 μ); height, 250-280 μ (average, 260 μ). Eyes present, situated one-fourth shell length from anterior end; shell subelliptical in outline with greatest height at midpoint; setae scarce, situated on anterior end.

Copulatory complex with ventral extremity of peniferum extending considerably beyond clasping apparatus; distal portion of peniferum cornified and terminating in two points directed anterodorsally; ventral point larger than more dorsal one; falcate corneous penis conspicuous, situated distally on peniferum, and extending slightly beyond posteroventral margin; clasping apparatus consisting of easily distinguishable horizontal and vertical rami; vertical ramus about 2.5 times longer than horizontal ramus; external border of vertical ramus convex; internal border concave; external border of horizontal ramus straight (or nearly so), internal border slightly concave and bearing two teeth; distal extremity bearing three denticles; external border of angle between horizontal and vertical rami with posteroventrally directed protuberance.

TRIUNGUIS FEMALE: Shell similar to that of male, but slightly larger. Shell length 470-520 μ (average, 490 μ); height, 300-310 μ (average, 305 μ).

Relationships. — See *G. nessoides.*

Type-locality. — Bird Haven, near Olney, Richland Co., Illinois, U.S.A. (types, USNM: holotype, 110943; paratypes, 110944 and 110945). Host, *Cambarus d. diogenes.*

Distribution (in addition to type-locality). —

United States

ILLINOIS — EDGAR CO. — Twin Lakes, 1 mi N of Paris (host, *Orconectes virilis*). HARDIN CO. — Peters Creek, 2¼ mi E of Elizabethtown (host, *O. kentuckiensis*).

INDIANA — POSEY CO. — trib of Big Creek, 1 mi SE of Wadeville (host, *Cambarus diogenes* subsp.).

KENTUCKY — CHRISTIAN CO. — Sinking Fork of Little River, 7.1 mi W of Hopkinsville on rts 68 and 80 (host, *Orconectes tricuspis*).

Geocythere nessoides Hobbs and Hobbs, 1970

Plate XXIV, figures 1-5
Plate LII

Geocythere nessoides Hobbs and Hobbs, 1970: 13 (K), 13-15 (M, H, R, A, D), figs. 6a-d.

Diagnosis. — MALE: Shell length 420-470 μ (average, 450 μ); height, 250-270 μ (average, 260 μ). Eye pigmented, located one-fourth shell length from anterior margin. Shell subelliptical with greatest height slightly posterior to midlength. Submarginal setae scarce but somewhat evenly distributed anteriorly, posteriorly, and ventrally.

Copulatory complex with peniferum extending ventrally beyond clasping apparatus for distance greater than length of vertical ramus of apparatus; portion ventral to level of dorsal margin of loop of spermatic duct 1.5 times longer than portion dorsal to loop; ventral swollen portion with heavy, obliquely truncate, and horizontally grooved projection extending anteriorly, and much smaller subacute tuberculiform prominence directed posteroventrally from ventral surface. Clasping apparatus with distinct vertical and horizontal rami forming angle of approximately

90 degrees, former third longer than latter with margins entire (although internal border strongly convex) and slender proximal fourth bent posterodorsally; horizontal ramus with external border bearing four rounded prominences along proximal half, distal half of ramus with three annulations and single apical denticle. Dorsal finger with single apical seta; ventral finger slender and sinuous, resembling contour of anterior margin of peniferum.

TRIUNGUIS FEMALE: Shell length 350-500 μ (average, 450 μ); height, 230-290 μ (average, 260 μ). Eye pigmented and situated as in male. Shell similar to that of male but with concavity on ventral margin anterior to midlength. Genital complex consisting of small posterodorsally situated conical papilla, surrounded by amorphous mass, directed anteroventrally.

Relationships. — *Geocythere nessoides* is perhaps more closely allied to *G. geophila* than to the only other member of the genus, *G. gyralea;* however, the latter two are distinctly more similar to each other than is either to *G. nessoides*. The most obvious dissimilarities between *G. geophila* and *G. gyralea* are seen in the much longer clasping apparatus and the longer peniferum in which the portion ventral to the dorsal level of the spermatic loop is approximately 2.5 times longer than that dorsal to the loop. The external border of the clasping apparatus with several prominences, instead of only one, and the shorter penis, together with the occasional occurrence of a very small ventral tuberculiform prominence on the peniferum suggest a closer affinity of *G. nessoides* to *G. geophila* than to *G. gyralea.* The distinguishing characters of the three are summarized in the preceding key.

Type-locality. — Burrows in roadside ditch along rt 49, 0.7 mi N of Harrison, Stone Co., Mississippi, U.S.A. (types, USNM: holotype, 126259; allotype, 126260; paratype, 126261). Hosts, *Cambarus d. diogenes, C. d. ludovicianus,* and *Fallicambarus hedgepethi.*

Distribution (in addition to type-locality). —

United States

LOUISIANA — EAST FELICIANA, ST. HELENA, and TANGIPAHOA PARISHES — localities unknown (hosts, *Cambarus diogenes diogenes* and/or *C. d. ludovicianus* and/or *Fallicambarus hedgepethi*) (Hobbs and Hobbs, 1970: 14).

MISSISSIPPI — RANKIN CO. — 5.7 mi SW of jct of rts 468 and 469 on rt 469 (host, *C. d. ludovicianus*).

HARPAGOCYTHERE HOBBS III, 1965

Harpagocythere Hobbs III, 1965: 160 (K), 162 (M), type-species, *Harpagocythere georgiae* Hobbs III, 1965. — Hobbs and Walton, 1968: 247 (K). — Hart and Hart, 1969a: 185-187 (D, E, R). — Hart and Hart, 1969b: 158 (M, R).— Hobbs and Hobbs, 1970: 2 (K).

Diagnosis. — Terminal tooth of mandible with cusps. Copulatory complex of male with finger guard; ventral portion of peniferum slender and terminating in a small recurved projection, its ventral margin extending ventrally almost to same level as clasping apparatus; simple penis less than one-fifth as long as clasping apparatus and situated opposite ventral portion of finger guard; accessory groove absent; external border of clasping apparatus entire, internal border of horizontal ramus with two prominences, apex with four small denticles, upper one of which not acute; extensions of horizontal and vertical rami forming angle of approximately 70°.

Type-species. — *Harpagocythere georgiae* Hobbs III, 1965.

Distribution. — Apparently limited to northwestern South Carolina and southwestern North Carolina.

Hosts. — Freshwater crayfishes of the family *Astacidae.*

Figure 27. — Distribution of the genus *Harpagocythere.*

Harpagocythere georgiae Hobbs III, 1965

Plate XXIX, figures 6-8
Plate LII

Harpagocythere georgiae Hobbs III, 1965: 163 (M, H, R, D), figs. 4-6. — Ferguson, 1968: 502 (C, D). — Hobbs and Walton, 1968: 247, 250 (R).

Diagnosis. — MALE: Shell length 399 μ; height, 210 μ. Eyes present. Shell ovate, ventral margin with shallow emargination slightly anterior to midlength, greatest height slightly posterior to midlength. Submarginal setae evenly and widely spaced ventrally and few posteriorly, absent dorsally and anteriorly. Terminal tooth of mandible with cusps.

Copulatory complex described in generic diagnosis. Dorsal and ventral fingers slender; ventral finger strongly curved at distal end of proximal third. Finger guard less than half length of clasping apparatus with acute terminal projection, latter sometimes with small posterior tooth at base.

FEMALE: Unknown.

Relationships. — *Harpagocythere georgiae* has its closest affinities with those entocytherids possessing a small penis located high in the peniferum and a finger guard on the copulatory complex (members of the genera *Dactylocythere, Sagittocythere,* and *Cymocythere*). In the absence of an accessory groove, it resembles *C. cyma* and *C. phyma.* The clasping apparatus is more similar to that of *S. barri* than to that of any species. It differs from all entocytherids in possessing a comparatively slender peniferum with an anteroventral recurved projection. The superficial resemblance of the latter to that of *D. leptophylax* was pointed out above.

Type-locality. — Crayfish burrows in ravine 1 mi E of Saluda River and 0.5 mi N on rt 183 on road near Farrs Bridge, Greenville Co., South Carolina, U.S.A. (types, USNM: holotype, 111262; paratype, 111263). Hosts, *Cambarus carolinus* and *C. latimanus.*

Distribution. — Known only from type-locality.

Harpagocythere tertius Hobbs and Walton, 1968

Plate XXIV, figures 9-10
Plate LII

Harpagocythere tertius Hobbs and Walton, 1968: 246-247 (M, H, R, A, D), 250 (R), figs. 3a-c.

Diagnosis. — MALE: Average shell length 380 μ; height, 210-220 μ (average, 215 μ). Eye pigmented. Shell ovate with posterior half of shell distinctly higher than anterior. Submarginal setae present except dorsally but less abundant ventrally than anteriorly and posteriorly.

Copulatory complex: Peniferum with ventral portion strongly curved anteriorly and tapering to short, but conspicuous hook-like protuberance at anteroventral angle. Penis L-shaped, simple, almost horizontally situated at level slightly below tip of finger guard. Finger guard with broad tapering basal portion and short, slender, subacute distal portion. Dorsal finger large with basal portion reaching tip of finger guard and terminating in long, simple seta; ventral finger slender and assuming contour of internal border of clasping apparatus — broadly U-shaped. Clasping apparatus heavy, gently curved without distinct horizontal and vertical rami; external border entire and regularly curved; internal border entire to base of distal third, there interrupted by tuberculiform prominence followed distally by second one, and apical curvature marked by six low lobes.

FEMALE: Unknown.

Relationships. — *Harpagocythere tertius* may be distinguished from *H. georgiae* Hobbs III, 1965: 163 by the clasping apparatus, which in *H. tertius* is broader distally than at midlength, and by the much more strongly bent ventral portion of the peniferum.

Type-locality. — Tributary to Saluda River (Santee River drainage) along rt 25 S of jct with rt 11, Greenville Co., South Carolina, U.S.A. (types, USNM: holotype, 123329). Host, *Cambarus* sp.

Distribution (in addition to type-locality). —

United States

NORTH CAROLINA — JACKSON CO. — Whitewater River at NC 107 (host, *Cambarus* sp.) (Hobbs and Walton, 1968: 247).

HARTOCYTHERE HOBBS III, 1970

Hartocythere Hobbs III, 1970: 180-181 (M), type-species, *Entocythere torreya* Hart, 1959.

Diagnosis. — Terminal tooth of mandible pectinate. Male copulatory complex lacking finger guard and accessory groove; ventral portion of peniferum heavily sclerotized, angled anteroventrally, deeply cleft, and extending ventrally much beyond clasping apparatus; corneous penis large and curved anteroventrally, with prostatic and spermatic elements contiguous throughout their lengths. Clasping apparatus curved approximately at right angle but not distinctly divisible into vertical and horizontal rami, mesiodistal surface bearing linear series of small tubercles and apex of apparatus with four denticles. Triunguis female without J-shaped rod or amiculum but possessing heavy, sclerotized genital papilla.

Types-species. — *Entocythere torreya* Hart, 1959. The genus *Hartocythere* was proposed by Hobbs III (1970) to receive this single somewhat disjunct species described by Hart, which in his generic revision (1962) he assigned to the genus *Geocythere.* With the discovery of additional species of the latter and the erection of the closely allied genus *Ornithocythere* Hobbs, 1967, of which five species are now known, uncertainty arose as to the limits of variation in the two genera. The removal of Hart's *G. torreya* from the genus makes *Geocythere* a more unified assemblage, and permits rather concise definitions of the three genera.

Distribution. — United States: Alabama — Escambia, Coctawhatchee, and Appalachicola drainage systems in Alabama (Bullock, Clark, Conecuh, and Geneva countries). Florida — Escambia and Liberty countries. Georgia — Randolph Co.

Host. — Freshwater crayfishes of the family Astacidae. Insofar as is known, this ostracod is restricted to *Cambarus d. diogenes.*

Figure 28. — Distribution of the genus *Hartocythere.*

Hartocythere torreya (Hart, 1959)

Plate XXIX, figures 11-13
Plate LIII

Entocythere torreya Hart, 1959: 198-200 (M, H, R, A, D), 194 (K), figs. 7-11. — Walton and Hobbs, 1959: 115 (K). — Levinson, 1962: 94 (C). — Hart, 1962: 123 (C).

Geocythere torreya. — Hart, 1962: 135 (C, D, S). — Ferguson, 1968: 502 (C, D).

Hartocythere torreya. — Hobbs III, 1970: 181-182 (M, H, R, A, D, S), figs. 2L-O.

Diagnosis. — Male: Shell length 430-480 μ (average, 460 μ); height, 250-340 μ (average, 270 μ). Eyes pigmented, situated one-fourth shell length from anterior margin. Shell subelliptical with greatest height slightly posterior to midlength; margins entire.

Submarginal setae few in number but evenly spaced anteriorly, posteriorly, and ventrally.

Copulatory complex described in diagnosis of genus.

TRIUNGUIS FEMALE: Shell length 480-510 μ (average, 500 μ); height, 260-280 μ (average, 270 μ). Eyes as in male. Shell similarly shaped but slightly larger than that of male and with shallow ventral concavity slightly anterior to midlength. Submarginal setae, although sparse, evenly spaced around entire margin.

Genital complex consisting of long, slender, sclerotized papilla directed slightly anteroventrally; two posteroventrally curved alate processes situated dorsal to papilla; all of these structures surrounded by amorphous mass with adhering detritus.

Relationships. — This species has its closest affinities with the members of the genera *Geocythere* and *Ornithocythere* but may be readily distinguished from all of them by the highly sclerotized, deeply bifid ventral portion of the peniferum.

Type-locality. — Burrows near Blue Springs, Torreya State Park, Liberty Co., Florida, U.S.A. (types, USNM: holotype, 101462; paratype, 101463). Host, *Cambarus d. diogenes.*

Distribution (in addition to type-locality). —

United States

ALABAMA — BULLOCK, CLARKE, CONECUH, and GENEVA COS. — localities unknown (host, *Cambarus diogenes diogenes*) (Hobbs III, 1970: 182).

FLORIDA — ESCAMBIA CO. — locality unknown (host, *Cambarus diogenes diogenes*) (Hobbs III, 1970: 182).

GEORGIA — RANDOLPH CO. — locality unknown (host, *Cambarus diogenes diogenes*) (Hobbs III, 1970: 182).

LITOCYTHERE HOBBS AND WALTON, 1968

Litocythere Hobbs and Walton, 1968: 247 (K, M), type-species, *Litocythere lucileae* Hobbs and Walton, 1968. — Hobbs and Hobbs, 1970: 2 (K).

Diagnosis. — Terminal tooth of mandible with cusps. Copulatory complex with finger guard; peniferum extending ventrally to approximately same level as clasping apparatus, strongly compressed laterally and thus appearing plate-like, the anteroposterior plane broader at ventral extremity than at level of penis, and ventral margin broadly concave; penis complex with contiguous spermatic and prostatic elements; accessory groove lacking. Clasping apparatus large, C-shaped, without distinct demarcation between horizontal and vertical rami; horizontal portion thickened, short, bearing single prominence near terminal group of 4 to 6 small, rounded teeth.

Type-species. — *Litocythere lucileae* Hobbs and Walton, 1968.

Distribution. — Known only from type-locality.

Host. — Freshwater crayfishes of the family Astacidae. Limited to *Cambarus asperimanus.*

Figure 29. — Distribution of the genus *Litocythere.*

Litocythere lucileae Hobbs and Walton, 1968

Plate XXV, figures 1-3
Plate LIII

Litocythere lucileae Hobbs and Walton, 1968: 247-250 (M, H, R, A, D), figs. 3d-f.

Diagnosis. — MALE: Shell length 360-410 μ (average, 390 μ); height, 200-230 μ (average, 220 μ). Eye pigmented. Shell ovate but with posterior portion distinctly higher than anterior. Submarginal setae abundant anteriorly and posteriorly, less so ventrally, and not evident dorsally.

Copulatory complex: as described in generic diagnosis. Finger guard broad basally but with distal portion tapering to narrow rounded tip; dorsal finger comparatively large, tapering from base to tip of setiform terminal; ventral finger slender and broadly U-shaped with proximal bend occurring at distal end of proximal third of finger and distal bend at base of distal sixth of finger. Clasping apparatus much thicker in vicinity of prominence on internal border than elsewhere.

FEMALE: — Unknown.

Relationships. — *Litocythere lucileae* is most closely allied to *Harpagocythere georgiae* and *H. tertius.* It shares so many features with the latter that but for the fact that the peniferum of *L. lucileae* is so completely different from that of both, it might have been assigned to *Harpagocythere.* The most striking similarities occur in the finger guard and in the clasping apparatus. As indicated above, *L. lucileae* is readily distinguishable from all other entocytherids by the broad concave ventral margin of the peniferum.

Type-locality. — Tributary to Linville River (Santee R. drainage) near rt 221, Linville, Avery Co., North Carolina, U.S.A. (types, USNM: holotype, 123327; paratype, 123328). Host, *Cambarus asperimanus.*

Distribution. — Known only from type-locality.

LORDOCYTHERE HOBBS AND HOBBS, 1970

Lordocythere Hobbs and Hobbs, 1970: 2 (K), 15 (M), type-species, *Lordocythere petersi* Hobbs and Hobbs, 1970.

Diagnosis. — Terminal tooth of mandible pectinate. Copulatory complex of male without finger guard; peniferum extending ventrally slightly beyond clasping apparatus with portion ventral to base of latter slender, its distal fifth tapering to subacute apex, and posterior margin with distinct acute eminence at level of penis. Penis C-shaped, consisting of contiguous prostatic and spermatic elements, and directed posteroventrally. Clasping apparatus slightly curved and tapering from base with one small subapical tooth on internal border and four apical denticles; external border entire. Dorsal finger heavy and two-thirds as long as clasping apparatus; ventral finger slender and disposed subparallel to clasping apparatus. Triunguis female with second antenna lacking accessory pectinate process on distal podomere; genital complex consisting of large vertically disposed acute projection suspended in posterodorsal portion of body.

Type-species. — *Lordocythere petersi* Hobbs and Hobbs, 1970.

Distribution. — Kentucky and Tennessee, U.S.A.

Hosts. — Freshwater crayfishes of the family Astacidae. Limited to *Cambarus d. diogenes* and *C. sphenoides.*

Figure 30. — Distribution of the genus *Lordocythere.*

Lordocythere petersi Hobbs and Hobbs, 1970

Plate XXX, figures 4-7
Plate LIII

Lordocythere petersi Hobbs and Hobbs, 1970: 16 (M, H, R, A, D), figs. 9a-d.

Diagnosis. — MALE: Shell length 430-470 μ (average, 450 μ); height, 260-380 μ (average, 270 μ). Eye pigmented, located one-fifth shell length from anterior margin. Shell highly arched dorsally with posterior portion noticeably higher than anterior, and posterior margin sloping rather sharply anteroventrally, thus shortening flattened portion of ventral margin; greatest height almost at midlength. Submarginal setae present anteriorly, ventrally, and posteriorly. Copulatory complex as described in generic diagnosis.

TRIUNGUIS FEMALE: Shell length 430-490 μ (average, 460 μ); height, 260-290 μ (average, 280 μ). Eye pigmented, located approximately one-sixth shell length from anterior margin. Shell more highly vaulted than that of male with posterior margin partially vertically disposed and bearing slight concavity; ventral margin also with shallow concavity anterior to midlength; greatest height posterior to midlength; submarginal setae as in male. Genital complex described in generic diagnosis.

Relationships. — This ostracod has no close affinities with any other, and while the arclike clasping apparatus of the male somewhat resembles that of certain members of the genera *Ascetocythere, Plectocythere,* and *Donnaldsoncythere,* in the former two, not only is the penis directed anteriorly rather than posteriorly, but also it consists of widely separated prostatic and spermatic elements. Whereas the members of the genus *Donnaldsoncythere* have a penis in which the two elements are combined as they are in *L. petersi,* it is much more massive, highly sclerotized, and directed anteriorly or anteroventrally (at least basally); furthermore, only in the monotypic *Rhadinocythere serrata* is the peniferum nearly so attenuated ventrally. This, to us, seems to be a very superficial resemblance, for the serrate condition of the internal border of the clasping apparatus, the highly sclerotized ventral portion of the peniferum of the latter, together with the anteriorly directed penis pose a combination of characters which together are as unique as are those of *L. petersi.* As has been indicated above, *L. petersi* is the only entocytherid known in which the penis is, from its base, directed posteriorly.

Type-locality. — Creek and burrows in seepage area, 4.8 mi SW of jct of routes 25W and 75 on route 25W SW of Corbin, Whitley Co., Kentucky, U.S.A. (types, USNM: holotype, 126262; allotype, 126263). Hosts, *Cambarus sphenoides* and *C. diogenes* subsp.

Distribution (*in addition to type-locality*). —

United States

TENNESSEE — SCOTT CO. — trib of Big South Fork of Cumberland River, 8.6 mi N of Oneida on US 27 (hosts, *Cambarus diogenes* subsp. and *C.* sp.) (Hobbs and Hobbs, 1970: 16).

OKRIOCYTHERE HART, 1964

Okriocythere Hart, 1964: 243 (M), type-species, *Okriocythere cheia* Hart, 1964. — Hart and Hart, 1969a: 188 (D, E, R). — Hobbs III, 1970: 175 (R). — Hobbs and Hobbs, 1970: 2 (K).

Diagnosis. — Terminal tooth of mandible with cusps. No finger guard on copulatory complex. Clasping apparatus heavy, club-shaped, with horizontal ramus reduced and possessing teeth on internal and external borders. Peniferum extending only slightly beyond clasping apparatus; penis large, curved, and situated distally on peniferum.

Type-species. — *Okriocythere cheia* Hart, 1964.

Distribution. — Washington, D.C.; Maryland; eastern Virginia; and eastern North Carolina, U.S.A.

Hosts. — Freshwater crayfishes of the family Astacidae. Limited to *Cambarus d. diogenes.*

Figure 31. — Distribution of the genus *Okriocythere.*

Okriocythere cheia Hart, 1964

Plate XXX, figures 8-10
Plate LIII

Okriocythere cheia Hart, 1964: 243-245 (M, R, A, D), figs. 1, 2. — Hobbs, 1967: 3, 8 (R, A). — Hobbs III, 1969a: 168, 170 (R). — Peters, 1971: 100.

Diagnosis. — MALE: Shell length 470-480 μ (average, 473 μ); height, 280 μ. Eyes reduced or absent; when present, situated $\frac{1}{3}$ shell length from anterior end. Shell subelliptical in outline with greatest height about midpoint. Setae scarce; situated on anterior and posterior ends.

Copulatory complex with ventral extremity of peniferum extending ventrally slightly beyond clasping apparatus; distal prominence not heavily cornified. Penis situated in ventral portion of peniferum, and extending slightly beyond base of ventral prominence. Clasping apparatus heavy, club-shaped, with acute caudoventral angle at junction of horizontal and vertical rami; horizontal ramus much reduced, when compared with other entocytherid ostracods (except members of the genus *Donnaldsoncythere*). External border of horizontal ramus with five teeth, from which conspicuous elevations extend dorsally on mesial and lateral surfaces as ridges, and distal two apparently encircle the ramus; internal border of horizontal ramus with two teeth.

FEMALE: Shell of one biunguis female identified, similar to male but slightly smaller (length, 0.43; height, 0.25).

Variations. — Slight variations were noted among the four male specimens, the most conspicuous being that of the external border of the clasping apparatus in which four or five teeth may be present; the internal border possesses one or two teeth.

Type-locality. — Prince Georges Co., Maryland, U.S.A. (types, USNM: holotype, 110940; paratype, 110941). Host, *Cambarus d. diogenes* Girard.

Distribution (in addition to type-locality). —

United States

DISTRICT OF COLUMBIA — WASHINGTON — Zoological Park (host, *Cambarus d. diogenes*).

MARYLAND — MONTGOMERY CO. — Plummer Island (host, *C. d. diogenes*). PRINCE GEORGES CO. — Laurel (host, *C. d. diogenes*).

NORTH CAROLINA — EDGECOMB CO. — 9 mi E of Saratoga at jct of four roads (host, *C. d. diogenes*).

VIRGINIA — PRINCESS ANNE CO. — Landstown Rd off Princess Anne Rd, Virginia Beach (host, *C. d. diogenes*).

ORNITHOCYTHERE HOBBS, 1967

Ornithocythere Hobbs, 1967: 2 (M), type-species, *Ornithocythere waltonae* Hobbs, 1967. — Hobbs III, 1968: 40-41 (D, H). — Hart and Hart, 1969a: 187 (D, E, R). — Hobbs III, 1970: 171, 172 (M). — Hobbs and Hobbs, 1970: 2 (K).

Diagnosis. — Terminal tooth of mandible pectinate. Male copulatory complex lacking finger guard and accessory groove; peniferum extending ventrally beyond clasping apparatus; acute, sclerotized, beak-like prominence (resembling inverted head of a bird) directed anterodorsally, its ventral base arising from ventral border of peniferum rather than from anterior surface. Corneous penis large, directed anterodorsally with prostatic and spermatic elements propinquant throughout length; clasping apparatus well-developed, sometimes divisible into horizontal and vertical rami; vertical ramus with margins entire; horizontal ramus with or without entire borders, ramus terminating in three to five apical denticles. Triunguis females without J-shaped rod or amiculum, but possessing sclerified genital papilla; campanula present or absent; no pectinate process on distal podomere of second antenna.

Type-species. — *Ornithocythere waltonae* Hobbs, 1967.

Distribution. — Eastern Virginia and Maryland, Mississippi, Alabama, Kentucky, U.S.A.

Host. — Freshwater crayfishes of the family Astacidae. Limited to *Cambarus d. diogenes.*

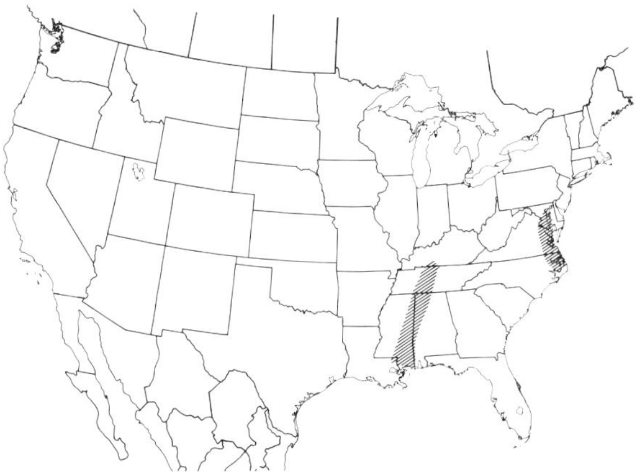

Figure 32. — Distribution of the genus *Ornithocythere.*

Key to the Species of the Genus Ornithocythere

(From Hobbs III, 1970)

1	External border of clasping apparatus entire	*O. waltonae*
1′	External border of clasping apparatus annulate or with rounded prominences	2
2(1′)	Clasping apparatus L-shaped and divisible into vertical and horizontal rami	*O. rhea*
2′	Clasping apparatus with sinuous or simple curve, not divisible into vertical and horizontal rami	3
3(2′)	Clasping apparatus S-shaped, with three rounded prominences on external border, terminating in four apical denticles	*O. aetodes*
3′	Clasping apparatus with simple curve, mesiodistal surface bearing linear series of four small tubercles, terminating in three to five apical denticles	4
4(3′)	Beak-like projection on anteroventral portion of peniferum subacute; two tubercles situated on proximoventral angle of beak-like prominence, third one on ventral margin near apex; clasping apparatus with four mesial prominences proximal to five apical denticles	*O. gypodes*
4′	Beak-like projection on anteroventral portion of peniferum acute, without tubercles; clasping apparatus possessing four annulations proximal to apical denticles	*O. popi*

Ornithocythere aetodes Hobbs III, 1970

Plate XXX, figures 11-13
Plate XXXI, figure 1
Plate LIII

Diagnosis. — MALE: Shell length 520-550 μ (average, 540 μ); height, 290-320 μ (average, 300 μ). Eyes pigmented, located approximately one-fourth shell length from anterior margin. Shell subovate, highest posterior to midlength; margins entire; submarginal setae scarce, absent dorsally, present anteriorly, posteriorly, and ventrally.

Copulatory complex with peniferum extending ventrally beyond clasping apparatus, its ventral portion highly sclerotized, beak-like prominence directed anterodorsally with swelling on proximoventral angle; penis large, U-shaped, situated in distal one-fifth of peniferum, and supported by sclerotized penis guide, latter extending dorsally beyond level of penis. Clasping apparatus S-shaped, not clearly divisible into horizontal and vertical rami; external and internal borders entire except external border bearing three smooth prominences at distal one-fifth and distalmost portion terminating in four apical denticles; ventral finger four times length of dorsal finger — former slender, L-shaped, paralleling clasping apparatus, latter slender and directed ventrally.

TRIUNGUIS FEMALE: Shell length 530-580 μ (average, 540 μ); height, 300-340 μ (average, 320 μ). Eyes pigmented; shell subovate with concavity slightly anterior to midlength; margins entire; submarginal setae like those of male; genital complex consisting of sclerified conical, acute papilla directed anteroventrally, bordered by two alate processes, anteriormost directed ventrally, posterior ala directed posteroventrally and terminating distally in recurved spine; both papilla and alae enclosed in amorphous mass.

Relationships. — *Ornithocythere aetodes* has its closest affinities with *O. waltonae.* The penifera of both are quite similar in shape and structure, particularly in the ventral portion. *O. aetodes,* however, possesses an S-shaped clasping apparatus, the external border of which bears three smooth prominences, whereas *O. waltonae* supports a C-shaped clasping apparatus in which the external border is entire. *O. gypodes* and *O. popi* are similar in that they share with *O. aetodes* the "bird-head" peniferum. *O. aetodes* demonstrates the close relationship that exists between the genera *Ornithocythere* and *Geocythere.*

The structure of the male copulatory complex of *O. aetodes* provides some insight into a probable common origin of the general *Geocythere, Hartocythere, Okriocythere,* and *Ornithocythere.* Like the members of the genus *Geocythere, O. aetodes* possesses excrescences on the external border of the clasping apparatus as does the monotypic *Okriocythere;* also, the length of the peniferum from the dorsalmost level of the clasping apparatus to its ventral extremity approaches that of the members of the genus *Geocythere.* Its affinities to *Hartocythere* are most obviously indicated in the alate process of the female genital complex.

Type-locality. — Burrow in ditch on rt 28, 0.5 mi W of jct of rts 28 and 35, Smith Co., Mississippi, U.S.A. (Leaf River drainage) (types, USNM: holotype, 126264; paratype, 126265). Host, *Cambarus d. diogenes.*

Distribution. — Known only from type-locality.

Ornithocythere gypodes Hobbs III, 1969

Plate XXXI, figures 2-3
Plate LIII

Ornithocythere gypodes Hobbs III, 1969a: 169-170 (M, H, R, D), figs. 2, 3. — Hobbs III, 1970: 173 (K), 175 (R), 176 (M), 177 (D, H, M).

Diagnosis. — MALE: Shell length 450-500 μ (average, 460 μ); height, 230-290 μ (average, 260 μ). Eye pigmented, located approximately one-fourth shell length from anterior margin. Shell subelliptical, highest anterior to midlength with dorsal margin tapering to small eminence at posterior end of shell; ventral margin entire.

Copulatory complex without finger guard; peniferum extending ventrally beyond clasping apparatus with ventral portion highly sclerotized and produced into beak-like prominence directed anteriorly; accessory groove absent; penis complex with spermatic and prostatic elements propinquant throughout their length, not separated; penis complex slightly longer than anterior-posterior dimension of peniferum at level of base of penis. Two tubercles located on proximoventral angle of beak-like prominence, third one on ventral margin near apex. Clasping apparatus not divisible into horizontal and vertical rami; internal and external borders entire, except at distal portion where mesiodistal surface bears linear series of four small tubercles; clasping apparatus terminating in three to five apical denticles. Dorsal finger moderately slender, bearing one terminal spine; ventral finger slender and subparallel to clasping apparatus, bearing one terminal spine.

TRIUNGUIS FEMALE: Shell length 480-500 μ (average, 485 μ); height, 250-290 μ (average, 270 μ). Eyes pigmented; shell subovate with greatest height slightly posterior to midlength; margins entire; submarginal setae present except dorsally; genital complex consisting of sclerified, conical papilla directed posteroventrally and surrounded by amorphous mass.

Relationships. — *Ornithocythere gypodes* has its closest affinities with *O. waltonae.* The penifera are distinctly similar in that both emphatically resemble the head of a bird. The penis complexes are markedly similar, but the penis of *O. gypodes* is longer than that of *O. waltonae.* Likewise, *O. gypodes* is similar to *Okriocythere cheia,* to the members of the genus *Geocythere,* and less so to members of the genera *Ascetocythere* and *Plectocythere.* The combination of the more ventrally situated penis, the three spines located on the beak-like prominence of the peniferum, and the quite distinct clasping apparatus are unique to *O. gypodes.*

Type-locality. — Burrow in flood plain of Tombigbee River ¼ mi E of river, 1 mi W of Alabama-Mississippi state line, Lowndes Co., Mississippi, U.S.A. (types, USNM: holotype, 123879). Host, *Cambarus d. diogenes.*

Distribution (in addition to type-locality). —

United States

ALABAMA — PICKENS CO. — "two localities" (host, *Cambarus d. diogenes*) (Hobbs III, 1970: 177).

MISSISSIPPI — NOXUBEE CO. — "two localities" (host, *C. d. diogenes*) (Hobbs III, 1970: 177).

Ornithocythere popi Hobbs III, 1970

Plate XXXI, figures 4-7
Plate LIII

Ornithocythere popi Hobbs III, 1970: 177-179 (M, H, R, A, D), figs. 2g-k.

Diagnosis. — MALE: Shell length 370-450 μ (average, 430 μ); height, 220-250 μ (average, 230 μ). Eyes pigmented, located approximately one-fourth shell length from anterior margin; shell subovate with posterior region higher than anterior; greatest height slightly posterior to mid-length; margins entire; submarginal setae scarce, but occurring ventrally and along anterior and posterior margins.

Copulatory complex with peniferum extending ventrally beyond clasping apparatus, ventral portion sclerified and terminating anteroventrally in beak-like prominence; penis large, U-shaped, and situated approximately in distal one-fourth of peniferum; clasping apparatus not clearly divisible into horizontal and vertical rami, but gently curved; external border of disal half with four annulations; distalmost portion terminating in three apical denticles; dorsal and ventral fingers slender, latter three times length of former; ventral finger curved parallel to clasping apparatus; both terminating in single spines.

TRIUNGUIS FEMALE: Shell length 400-440 μ (average, 420 μ); height, 220-250 μ (average, 240 μ). Eyes pigmented, located one-fourth shell length from anterior margin; shell subovate with greatest height slightly posterior to midlength; margins entire; genital complex situated posterodorsally, with sclerotized genital papilla directed ventrally, enclosed in ventrally tapering campanula with orifice at ventral extremity.

Relationships. — *Ornithocythere popi* is most closely related to *O. gypodes.* The penifera of both species resemble the head of a bird and the penes are U-shaped. The clasping apparatus are markedly similar in that both possess annulations and terminate in three apical denticles; the annulations of *O. popi* are larger and extend farther proximally. This species is also closely related to *O. waltonae* and *O. aetodes,* having the characteristic beak-like prominence on the anteroventral portions of the peniferum. Like *O. waltonae,* the genital complex of the triunguis female of *O. popi* is composed of a papilla which is enclosed in a campanula.

Type-locality. — Burrow in ditch on rt 49, 0.7 mi N of Stone-Harrison Co. line, Stone Co., Mississippi, U.S.A. (types, USNM: holotype, 126266; allotype, 126267; paratype, 126268). Hosts, *Cambarus d. diogenes* and *C. l. ludovicianus.*

Distribution (in addition to type-locality). —

United States

MISSISSIPPI — PEARL RIVER CO. — Burrows in bank of small creek on rt 11, 8.8 mi SW of jct of rts 11 and 26.

Ornithocythere rhea Hobbs III, 1970

Plate XXXII, figures 1-2
Plate LIII

Ornithocythere rhea Hobbs III, 1970: 179-180 (M, H, R, A, D), figs. 2a-f.

Diagnosis. — MALE: Shell length 460-500 μ (average, 480 μ); height, 250-280 μ (average, 270 μ). Eyes pigmented, located approximately one-fourth shell length from anterior margin. Shell subovate with greatest height slightly posterior to midlength; margins entire; submarginal setae evenly spaced anteriorly, ventrally, and posteriorly, but absent dorsally.

Copulatory complex with peniferum elongate, slender, and extending well beyond clasping apparatus; projection on anteroventral portion of peniferum beak-like, subacute, and extending anterodorsally; penis large, U-shaped, and extending slightly beyond ventral margin of peniferum; penis resting in groove bordered by sclerotized penis guide; clasping apparatus variable but always clearly divisible into horizontal and vertical rami; vertical ramus 2.5 times length of horizontal ramus; two rami forming angle less than 90°; external and internal borders of vertical ramus entire; mesial internal border with or without single tooth; mesial external border with row of three to five rounded prominences; apical portion with four denticles; dorsal and ventral fingers slender each terminating in single spine; dorsal finger straight, directed anteroventrally; ventral finger gently curved anteriorly.

TRIUNGUIS FEMALE: Shell length 480-490 μ (average, 480 μ); height, 290-300 μ (average, 300 μ). Eyes pigmented, located approximately one-fourth shell length from anterior margin; shell subovate with truncate posterior margin, slightly larger than that of male; margins entire; submarginal setae as in male; genital complex consisting of small sclerotized papilla directed anteroventrally and surrounded by amorphous mass; sclerotized prominence with sac-like extension directed posterodorsally from base of papilla.

Relationships. — The male of *Ornithocythere rhea,* like all the members of the genus *Ornithocythere,* possesses a peniferum which resembles the inverted head of a bird, however, it is most closely allied to *O. aetodes.* Both species possess a relatively long clasping apparatus; the mesial external border of which bear a row of rounded prominences. The clasping apparatus of *O. rhea,* however, has an angular bend resulting in the horizontal and vertical rami being disposed at less than a right angle whereas in all other members of the genus, the clasping apparatus is never clearly divisible into two rami. *O. rhea* is also distinct in that the penis is very long and slender in comparison with the other members of the genus.

The triunguis female also shows unique characters which suggest that this species is the most remote member of the genus. Among them is the possession of a sac-like extension directed posterodorsally from the base of the papilla of the genital complex.

Type-locality. — Trammel Creek, 1.4 mi from jct of rts 221 and 31W at Alvaton, Allen Co., Kentucky, U.S.A. (Barren River drainage) (types, USNM: holotype and allotype, 126269). Host, *Cambarus diogenes* ssp.

Distribution. — Known only from type-locality.

Ornithocythere waltonae Hobbs, 1967

Plate XXXII, figures 3-6
Plate LIII

Ornithocythere waltonae Hobbs, 1967: 2-4 (M, H, R, A, D), 8 (A), figs. 1a-c. — Hobbs III, 1968: 41 (H, D). — Hobbs III, 1969a: 167, 170 (R). — Hobbs III, 1970: 171, 176 (M), 172-173 (K), 175-179 (R), figs. 1f-g. — Peters, 1971: 100.

Diagnosis. — MALE: Shell length 470-500 μ (average, 490 μ); height, 250-270 μ (average, 270 μ). Eye present. Shell highest posterior to midlength with dorsal margin tapering ventrally only slightly less suddenly anteriorly than posteriorly; ventral margin of shell entire; submarginal setae rather evenly spaced anteriorly, ventrally, and posteriorly, very few dorsally.

Copulatory complex as described in generic diagnosis. In addition, dorsal finger moderately slender with apex reaching level of base of penis complex; ventral finger slender and subparallel to clasping apparatus, latter extending ventrally only slightly beyond ventral margin of peniferum.

TRIUNGUIS FEMALE: Shell length 490-520 μ (average, 510 μ); height, 280-310 μ (average, 300 μ). Eye present. Shell of triunguis female distinctly larger than that of male with maximum height posterior to midlength; ventral margin of shell with distinct emargination just anterior to midlength; submarginal setae as in male although somewhat more abundant anteriorly and posteriorly. Amiculum and

J-shaped rod absent but posterodorsal area with papilla projecting into cup-like cavity containing material resembling that composing J-shaped rod of members of the genus *Dactylocythere*.

Type-locality. — Roadside ditch at Acerdale, 0.4 mi from jct of Great Bridge and Indian River, Princess Anne Co., Virginia, U.S.A. (types, USNM: holotype, 113472; allotype, 113473; paratype, 113474). Hosts, *Procambarus a. acutus* and *Cambarus d. diogenes*.

Distribution (in addition to type-locality). —

United States

MARYLAND — ANNE ARUNDEL CO. — 2.5 mi N of Davidsonville, Kings Branch at Sands Rd (host, *C. d. diogenes*).

VIRGINIA — KING AND QUEEN CO. — 6.4 mi E of West Point on rt 14 (hosts, *C. acuminatus* and *C. d. diogenes*). NORFOLK CO. — Foot of Little John Drive (host, *C. d. diogenes*); roadside ditch near Great Bridge (host, *C. d. diogenes*). NORTHUMBERLAND CO. — Lyells, on rt 202 (host, *C. d. diogenes*); 9.3 mi E of Lyells on rt 202 (host, *C. d. diogenes*). PRINCESS ANNE CO. — Roadside ditch from Stumpy Lake on rt 605, Virginia Beach (host, *C. d. diogenes*); Landstown Rd off Princess Anne Rd, Virginia Beach (host, *C. d. diogenes*). SURRY CO. — 9.5 mi N of Surry Court House (host, *C. d. diogenes*).

PHYMOCYTHERE HOBBS AND HART, 1966

Cymocythere Hart, 1962: 125, 129 (M) (in part). — Crawford, 1965: 148, 150, 151 (R). — Hobbs III, 1965: 160 (K).

Phymocythere. — Hobbs and Hart, 1966: 36 (K), 48 (M, S), type-species, *Entocythere phyma* Hobbs and Walton, 1962. — Hobbs, Holt, and Walton, 1967: 45-46 (M), 68 (R, E). — Hart and Hart, 1969a: 188, 189 (D, E, R). — Hart and Hart, 1969b: 158 (M, R). — Hobbs and Hobbs, 1970: 2 (K).

Asceta Group. — Hobbs and Walton, 1962: 42 (in part).

Diagnosis. — Terminal tooth of mandible with cusps. Copulatory complex with finger guard; peniferum extending ventrally much beyond clasping apparatus, elongate, comparatively slender, entirely corneous with terminal bulbous enlargement lacking superficial extensions; penis complex consisting of separate dorsal spermatic and ventral prostatic elements; spermatic element longer than clasping apparatus; accessory groove lacking. Clasping apparatus reduced, not clearly divisible into vertical and horizontal rami; external border of horizontal ramus entire, internal border also entire except at apex, latter may be expanded and bear three to six denticles.

Type-species. — *Entocythere phyma* Hobbs and Walton, 1962.

Distribution. — United States: southwestern Virginia and eastern West Virginia.

Hosts. — Freshwater crayfishes of the family Astacidae.

Figure 33. — Distribution of the genus *Phymocythere*.

Phymocythere phyma (Hobbs and Walton, 1962)

Plate XXXII, figures 7-9
Plate LIII

Entocythere phyma Hobbs and Walton, 1962: 42-46 (M, H, R, D, K), figs. 10-13. — Hart, 1962: 123 (C). — Hobbs and Hart, 1966: 35, 36.

Cymocythere phyma. — Hart, 1962: 129 (C, S). — Crawford, 1965: 150 (R). — Hobbs III, 1965: 159 (M, R), 163 (R).

Phymocythere phyma. — Hobbs and Hart, 1966: 35, 48-49, 53, figs. 30, 31 (H, D, R). — Hobbs and Walton, 1966: 7. — Hobbs, Holt, and Walton, 1967: 34 (K), 46 (M, D, H, A, S), 67-69 (D, E), 20, 29, 30, 38, 39, 45, 52, 54, 55, 57-59, 61, 62, 64, 72, 73, fig. 11c (A, H). — Ferguson, 1968: (C, D). — Walton and Hobbs, 1971: 88, 91, 93-94, 97.

Diagnosis. — MALE: Shell length 375-410 μ (average, 389 μ); height, 210-230 μ (average, 220 μ).

Shell subovate with margins entire; maximum height slightly posterior to midlength. Marginal setae very sparse anteriorly, ventrally, and posteriorly; absent dorsally. For other diagnostic characters, see generic diagnosis above.

Relationships. — *Phymocythere phyma* has its closest affinities with the members of the genera *Ascetocythere* and *Plectocythere*. It shares with them the separate spermatic and prostatic elements of the penis complex. It differs from both, however, in the unadorned terminal bulbous enlargement of the peniferum and in the reduced clasping apparatus.

Type-locality. — White Rock Branch, 6.7 mi N of Mountain Lake Biological Station on rt 613, Giles Co., Virginia, U.S.A. (types, USNM: holotype, 108178). Hosts, *Cambarus bartonii* and *C. sciotensis*.

Distribution (in addition to type-locality). —

United States

VIRGINIA — GILES CO. — Laurel Branch, trib of Big Stony Creek, 0.6 mi E of rt 635 on rt 628; Cascades on Little Stony Creek. CRAIG CO. — Dicks Creek, 2.2 mi from jct of rts 636 and 658 (Hobbs and Walton, 1962: 46). ROCKINGHAM CO. — Dry River at Rawley Springs, 13 mi W of Harrisonburg on rt 33 (Hobbs and Hart, 1966: 49).

WEST VIRGINIA — GREENBRIER CO. — Matt's Black Cave, 2.0 mi S of Renick; Crossbones Cave, NW of Renick; Buckeye Creek, 3.4 mi E of White Sulphur Springs on rt 60. PENDLETON CO. — Mystic Cave, near Onego. RANDOLPH CO. — Griffeth's Creek, 15.5 mi W of Durbin on rt 250; Brooks Run, 17.2 mi W of Elkins on rt 33; 6.2 mi E of Elkins on rt 33. SUMMERS CO. — Bargers Springs, 12 mi from Hinton; 3.5 mi W of Talcott on rt 63 (Hobbs and Hart, 1966: 49).

Hosts. — In the Virginia localities, *P. phyma* was found on *Cambarus b. bartonii* (Fabricius) or *Cambarus sciotensis* Rhoades; in the West Virginia localities, it was associated with *Cambarus robustus* Girard, *Cambarus bartonii carinirostris* Hay, or *Cambarus nerterius* Hobbs.

PLECTOCYTHERE HOBBS III, 1965

Plectocythere Hobbs III, 1965: 160 (K), 161-162 (M), type-species, *Plectocythere crotaphis* Hobbs III, 1965. — Hobbs and Hart, 1966: 36 (R, K), 47 (M). — Hobbs, 1967: 3 (R). — Hobbs, Holt, and Walton, 1967: 8 (H). — Hart and Hart, 1969a: 188-189 (D, E, R). — Hart and Hart, 1969b: 158 (M, R). — Hobbs III, 1969a: 16, 168, 170 (R). — Hobbs and Hobbs, 1970: 2 (K).

Diagnosis. — Terminal tooth of mandible with cusps. Copulatory complex of male without finger guard; peniferum extending ventrally much beyond clasping apparatus, with bulbous area ventral to level of base of clasping apparatus and terminating ventrally in long, tapering projection; penis consisting of separate spermatic and prostatic elements not contiguous along most of their courses, each element as long as, or longer than, clasping apparatus and reaching (or almost reaching) ventral extremity of peniferum; accessory groove absent; external border of clasping apparatus entire, internal border with two prominences, apex with four small rounded scallops.

Type-species. — *Plectocythere crotaphis* Hobbs III, 1965.

Distribution. — United States: eastern Kentucky.

Hosts. — Freshwater crayfishes of the family Astacidae.

Figure 34. — Distribution of the genus *Plectocythere*.

Plectocythere crotaphis Hobbs III, 1965

Plate XXXII, figures 10-12
Plate LIII

Plectocythere crotaphis Hobbs III, 1965: 162 (M, H, R, D), figs. 1-3. — Hobbs and Hart, 1966: 47 (M), figs. 26, 27. — Ferguson, 1968: 502 (C, D).

Diagnosis. — Shell length 420 μ; height, 259 μ. Eye lacking. Shell ovate with ventral margin entire; greatest height slightly anterior to midlength. Submarginal setae widely spaced anteriorly, posteriorly, and ventrally, absent dorsally. Terminal tooth of mandible with cusps.

Copulatory complex described in generic diagnosis. Dorsal finger broad at base and two-thirds as long as slender ventral finger, latter slightly curved at distal end of basal third. Clasping apparatus not clearly divisible into horizontal and vertical rami; short basal portion proximal to bend probably corresponds to vertical ramus.

FEMALE: Unknown.

Relationships. — See *Plectocythere johnsonae* (below).

Type-locality. — Crayfish burrows near spring, 11 mi NE of jct of rts 25E and 119, Bell Co., Kentucky, U.S.A. (types, USNM: holotype, 111264).

Distribution. — Known only from type-locality.

Plectocythere johnsonae Hobbs and Hart, 1966

Plate XXXIII, figures 1-3
Plate LIII

Plectocythere johnsonae Hobbs and Hart, 1966: 47-48 (M, H, R, D), figs. 28, 29. — Ferguson, 1968: 502 (C, D).

Diagnosis. — MALE: Shell length 430 μ; height, 270 μ. Eye without pigment. Shell subovate with ventral margin slightly concave anterior to midlength; greatest height slightly posterior to midlength. Submarginal setae widely spaced except dorsally where absent.

Ventral portion of peniferum slightly swollen, but produced in tapering, slender, slightly curved projection; penis consisting of separate spermatic and prostatic elements not contiguous along entire length; both elements longer than clasping apparatus and reaching ventral extremity of peniferum. Clasping apparatus not clearly divisible into horizontal and vertical rami; external border with rounded eminence at base of distal fifth; internal border entire; mesial border with rounded eminence slightly distal to midlength; apex with four small denticles.

Dorsal finger approximately two-thirds as long as ventral finger, latter curved at end of basal fourth.

FEMALE: Unkonwn.

Relationships. — This species appears to have its closest affinities with *P. crotaphis,* the only other species of the genus, from which it can be readily distinguished by the slightly curved, as opposed to undulating, penes and ventral projection of the peniferum.

Type-locality. — Crayfish burrows in Jenkins, Letcher Co., Kentucky, U.S.A. (types, USNM: holotype, 113454). Host, *Cambarus carolinus.*

Distribution. — Known only from type-locality.

RHADINOCYTHERE HART, 1962

Rhadinocythere Hart, 1962: 125 (K), 135 (M, S, C) (Note that diagnosis should read "*No* finger guard on copulatory complex."), type-species, *Entocythere serrata* Hoff, 1944. — Hart, 1965: 255 (H). — Hobbs III, 1965: 160 (K). — Hobbs, Holt, and Walton, 1967: 8 (H). — Hart and Hart, 1969a: 187 (D, E, R). — Hobbs and Hobbs, 1970: 2 (K).

Diagnosis. — Terminal tooth of mandible with cusps. No finger guard on copulatory complex. Horizontal and vertical rami of clasping apparatus merging imperceptibly. Distal portion of peniferum simple, slender, and unadorned. Penis situated at level of articulation of clasping apparatus. Female without amiculum.

Distribution. — United States: northern Indiana to southwestern Missouri and southeastern Minnesota.

Hosts. — Freshwater burrowing crayfishes of the family Astacidae.

Figure 35. — Distribution of the genus *Rhadinocythere.*

Rhadinocythere serrata (Hoff, 1944)

Plate XXXIII, figures 4-7
Plate LIII

Entocythere serrata Hoff, 1944b: 332-337 (M, H, R, D), 331 (K), figs. 1-5. — Tressler, 1947: 705 (C). — Rioja, 1949: 328 (R). — Tressler, 1959: 726 (K), fig. 28, 179. — Hart, 1962: 122 (C).

Rhadinocythere serrata. — Hart, 1962: 135 (C, D, S). — Hart, 1965b: 258, 259 (D). — Hobbs and Hobbs, 1970: 16 (K).

Diagnosis. — MALE: Shell length 480-500 μ (average, 486 μ); height, 250-300 μ (average, 278 μ). Shell subelliptical in outline with anterior end less high than posterior. Shell nearly void of setae. Eye well developed.

Copulatory complex with peniferum narrowed and extended to form long, gently curved, process paralleling, in part, proximal portion of clasping apparatus. Clasping apparatus with horizontal and vertical rami merging gradually; horizontal ramus little longer than vertical, with entire external border divided into fairly straight proximal portion and slightly convex distal portion, the two portions meeting somewhat abruptly; regularly curved internal border bearing, in most cases, four well spaced teeth; clasping apparatus becoming narrowed distally and distal margin bearing two or three very small teeth more or less continuous with teeth of internal border.

FEMALE: Shell length 500-520 μ (average, 510 μ); height, 290-310 μ (average, 300 μ). Shell similar to that of male except that in triunguis female posterior-ventral region of shell slightly swollen.

Relationships. — Superficially, the copulatory complex of this ostracod resembles that of *Dacytlocythere banana* (q.v.), but can easily be distinguished from it by the absence of a finger guard.

Type-locality. — Burrows in swamp on farm of Roy Foster, 3 mi W of Alpha, Henry Co., Illinois, U.S.A. (types, USNM: holotype, allotype, and paratype, 81557). Host, *Cambarus d. diogenes*.

Distribution (*in addition to type-locality*). —

United States

ILLINOIS — KNOX CO. — Abingdon.

INDIANA — COUNTY UNKNOWN — Kankakee River in NW Indiana. MARSHALL CO. — Lake Maxinkuckee.

IOWA — SCOTT CO. — Davenport.

MINNESOTA — OLMSTED CO. — 5.9 mi W of Eyota on rt 14.

In all of the collections listed above the host crayfish was *Cambarus d. diogenes*.

SAGITTOCYTHERE HART, 1962

Sagittocythere Hart, 1962: 125 (K), 135 (M, S, C), type-species, *Entocythere barri* Hart and Hobbs, 1961b. — Hart, 1963: 51 (R, D). — Hobbs III, 1965: 160 (K), 163 (R). — Hart and Hart, 1969a: 185-187 (D, E, R). — Hart and Hart, 1969b: 158 (M, R). — Hobbs and Hobbs, 1970: 2 (K).

Diagnosis. — Male copulatory complex equipped with finger guard; clasping apparatus heavy, well developed, and prominently grooved. Female genitalia in triunguis specimens consisting of rod-like body superior to amiculum. Second antenna lacking accessory structure at distal end of ultimate podomere; two or three terminal claws present. Posteroventral portion of peniferum of male copulatory complex extended to form an acute projection barbed on its anterior surface.

Type-species. — *Sagittocythere barri* (Hart and Hobbs, 1961).

Distribution. — Caves in Alabama, Tennessee, Kentucky, and Indiana, U.S.A.

Hosts. — Freshwater crayfishes of the family Astacidae.

Figure 36. — Distribution of the genus *Sagittocythere*.

Sagittocythere barri (Hart and Hobbs, 1961)

Plate XXXIII, figures 8-9
Plate LIII

Entocythere barri Hart and Hobbs, 1961b: 174 (K), 175-177 (M, R, D, H), figs. 1-8.
Sagittocythere barri. — Hart, 1962: 135-136 (C, D, S). — Hobbs III, 1965: 163 (R). — Hart and Hart, 1966a: 7-9 (H, A, D). — Barr, 1967: 160 (C, D, H), 190-192 (Z). — Ferguson, 1968: 502 (C, D). — Hobbs and Barr, 1972: 81 (D, H).

Diagnosis. — MALE: Shell length 440-500 μ (average, 470 μ); height, 260-290 μ (average, 281 μ). Eyeless. Shell subelliptical with posterior portion only slightly higher than anterior; margins entire, lacking spines or distinct emarginations. Marginal setae absent dorsally, sparse ventrally; a few present on anterior margin, several evenly spaced on posterior margin.

Copulatory complex with peniferum not extending distally beyond border of clasping apparatus; posteroventral portion of peniferum extended to form acute projection, barbed on its anterior surface; anteroventral portion produced in spine-like process directed ventrally along base of aforementioned barb. Accessory groove present and extending proximally to level of loop of spermatic duct. Finger guard bifid distally. Dorsal finger terminating in one or two setae. Clasping apparatus with distinct horizontal and vertical rami of subequal lengths; external borders of both rami entire; internal border of horizontal ramus with three or four teeth variously spaced; distal extremity with two teeth.

TRIUNGUIS FEMALE: Shell length 360-470 μ (average, 441 μ); height, 230-320 μ (average, 283 μ). Triunguis females with J-shaped rod and amiculum, characteristic of *Sagittocythere* and *Dactylocythere* females.

Remarks. — This ostracod has been found associated with crayfishes in some 25 collections, representing 22 caves in Alabama, Tennessee, Kentucky, and Indiana.

Of the 11 supposedly troglobitic entocytherid species that have been reported so far from the United States (Klie, 1931; Walton and Hobbs, 1959; Hart and Hobbs, 1961; Hart and Hart, 1966a), *S. barri* is represented in more collections than any of the others, and without exception has been found only in subterranean waters. Furthermore, its range corresponds with that of the hypogean representatives of the crayfish genus *Orconectes* — see Hobbs and Barr (1960: 12) — and it appears likely that this ostracod is found exclusively on this group of crayfishes.

At the time of its description, it was assumed that variations in the penifera of this species, in which the peniferum is highly variable, formed a gradient — with the specimens from the northern localities exhibiting more truncate penifera than those from the south (Hart and Hobbs, 1961: 176; Hart, 1962: 147). Upon subsequent examination of additional specimens from all parts of the range, however, the apparent north-south gradient has been found to be non-existent; specimens from all parts of the range are highly variable and no distribution pattern related to the shape of the peniferum is evident.

Type-locality. — Cave Spring Cave, 6 mi N of New Hope, Madison Co., Alabama, U.S.A. (types, ANSP). Host, *Orconectes australis australis.* Hart and Hobbs (1961) confused the cave in Morgan Co. bearing the same name with that in Madison Co., and also erred in citing *Cambarus jonesi* as one of the hosts. These errors were pointed out to us by Martha R. Cooper.

Distribution (in addition to type-locality). —

United States

ALABAMA — JACKSON CO. — Blowing Cave, 7 mi N of Scotsboro (host, *Orconectes pellucidus australis*); Fern Cace, near Huntsville (host, *O. p. australis*). MADSON CO. — Shelta Cave, 1 mi N of Huntsville (host, *O. p. australis*).

INDIANA — CRAWFORD CO. — Seibert's Well Cave (host, *Orconectes pellucidus inermis*). HARRISON CO. — Bradford Cave (host, *O. p. inermis*). MONROE CO. — Eller's Cave, W of May's Cave (host, unknown); May's Cave, near Bloomington (host, *Orconectes pellucidus testii*).

KENTUCKY — BARREN CO. — Diamond Cave, N of Park City on rt 255 (host, *O. p. pellucidus*). BRECKINRIDGE CO. — Lockard Cave, 4 mi W of Big Spring (host, *Orconectes* sp.). EDMONSON CO. — Mammoth Cave, Styx and Echo Rivers and Lake Lethe (host, *O. p. pellucidus*). HART CO. — Buckner Hollow Cave (hosts, *O. p. pellucidus* and *Cambarus tenebrosus*); Rider's Mill Cave (host, *O. p. pellucidus*). LARUE CO. — Bland Cave, 7 mi W of Sonora (host, *O. p. pellucidus*). MEAD CO. — Rockhaven Cave, Otter Creek State Park (hosts, *C. tenebrosus* and *O. pellucidus inermis*). PULASKI CO. — Hyden's Cave, 1.3 mi NE of Blue John (host, *Orconectes pellucidus packardii*). ROCKCASTLE CO. — Teamer's Cave, 1.2 mi E of Mullins (hosts, *O. pellucidus packardii* and *O. p.* ssp.). TRIGG CO. — Cool Spring Cave, SE of Cadiz (host, *O. p. pellucidus*). WAYNE CO. — Horse Hollow Cave (hosts, *O. pellucidus packardii* and *Cambarus* nr. *distans*).

TENNESSEE — JACKSON CO. — Carter Cave (host, *O. p. pellucidus*). PUTNAM CO. — Blind Fish Cave, Long Hollow (host, *O. p. pellucidus*); Bridge Creek Cave, Calfkiller Valley (host, *O. p. pellucidus*).

Sagittocythere stygia Hart and Hart, 1966

Plate XXXIII, figure 10
Plate LIII

Sagittocythere stygia Hart and Hart, 1966a: 9 (M, R, H, D), figs. 12, 13. — Barr, 1967: 160 (C, D, H), 192 (Z). — Ferguson, 1968: 502 (C, D). — Hobbs and Barr, 1972: 81 (D, H).

Diagnosis. — MALE: Shell length 500 μ; height, 260 μ. Eyeless. Shell subelliptical in outline, with posterior portion only slightly higher than anterior; margins entire, lacking spines or distinct emarginations. Marginal setae absent dorsally and ventrally; few present on anterior and posterior margins.

Copulatory complex with peniferum not extending distally beyond external border of clasping apparatus; posteroventral portion of peniferum extended to form acute projection, barbed on its anterior surface; anteroventral portion cornified, rounded, and terminating dorsal to barbed projection. Accessory groove present and extending proximally to level of penis. Finger guard bifid distally. Dorsal and ventral fingers each terminating in one seta. Clasping apparatus with rami of subequal lengths; external borders of both rami convex, entire; internal border of horizontal ramus with two teeth proximal to midlength; distal extremity truncate; internal border of vertical ramus concave.

FEMALE: Unknown.

Relationships. — This ostracod bears a close resemblance to *S. barri,* but differs from it in the number of teeth on the internal border of the horizontal ramus (*S. barri* — 3 or 4; *S. stygia* — 2); the termination of the accessory groove (*S. barri* — at level of spermatic tube loop; *S. stygia* — at level of penis); and in the shape of the anteroventral portion of the peniferum (*S. barri* — a ventrally directed spine; *S. stygia* — a rounded and cornified projection).

Type-locality. — River Styx, Mammoth Cave, Edmonson Co., Kentucky, U.S.A. (types, USNM: holotype, 113338). Host, *Orconectes pellucidus* (Tellkampf).

Distribution. — Known only from type-locality.

SAUROCYTHERE HOBBS III, 1969

Saurocythere Hobbs III, 1969a: 167-168 (M), type-species, *Saurocythere rhipis* Hobbs III, 1969. — Hobbs and Hobbs, 1970: 2 (K).

Diagnosis. — Terminal tooth of mandible with cusps; no finger guard on copulatory complex of male; peniferum slender and elongate without accessory groove and extending ventrally beyond clasping apparatus; penis consisting of separate spermatic and prostatic elements, each one-fourth length of peniferum and reaching ventral extremity of peniferum; penis complex, three times longer than least diameter of peniferum; spermatic loop situated horizontally and supported by heavily sclerotized bar; clasping apparatus not clearly divisible into vertical and horizontal rami, external border sinuous, bearing small excrescence at base of distal fourth; internal border entire; apical portion of clasping apparatus slightly expanded and serrate (fan-like).

Type-species. — *Saurocythere rhipis* Hobbs III, 1969.

Distribution. — United States: west-central Alabama.

Hosts. — Freshwater crayfishes of the family Astacidae.

Figure 37. — Distribution of the genus *Saurocythere.*

Saurocythere rhipis Hobbs III, 1969

Plate XXXIV, figures 1-2
Plate LIV

Saurocythere rhipis Hobbs III, 1969a: 167-168 (M, H, R, A, D), figs. 1, 4, 5.

Diagnosis. — MALE: Shell length 442 μ; height, 294 μ. Eye pigmented, located approximately one-third shell length from anterior margin. Shell ovate with ventral margin entire, greatest height slightly posterior to midlength. Setae scarce, occurring irregularly but singly near anterior and posterior margins. Mandible with distal row of five multicuspid teeth, proximal tooth with two cusps, distal with five, others with three.

Copulatory complex: Described in generic diagnosis; in addition, dorsal finger long and slender, nearly one-half length of ventral finger and bearing two terminal setae, ventral finger slender and parallel to clasping apparatus ending in single terminal.

FEMALE: Unknown.

Relationships. — *Saurocythere rhipis* seems to be most closely allied to the members of the genus *Plectocythere,* Hobbs III, 1965. The principal resemblances are associated with the penis complex, which is composed of extremely long contiguous spermatic and prostatic elements in members of both genera. Although the peniferum of *Okriocythere cheia,* Hart, 1964, exhibits a strong superficial resemblance to that of *S. rhipis,* the penis is much shorter, the spermatic loop is vertical, and the form of the clasping apparatus in both genera is unique. The horizontally situated spermatic loop has not been observed to occur in any other entocytherid.

Type-locality. — Crayfish burrows in ditch along rt 14, 0.6 mi SE of Alabama-Mississippi state line, Pickins Co., Alabama, U.S.A. (types, USNM: holotype, 123880; paratype, 123881). Host, *Cambarus d. diogenes.*

Distribution. — Known only from type-locality.

THERMASTROCYTHERE HOBBS AND WALTON, 1966

Thermastrocythere Hobbs and Walton, 1966: 7-8 (M), type-species, *Entocythere harti* Hobbs and Walton, 1966 (= *Entocythere riojai* Hoff, 1943). — Hart and Hart, 1969a: 182-185 (D, E, R). — Hobbs and Hobbs, 1970: 3 (K), 16, 17 (M, S, C).

Diagnosis. — Terminal tooth of mandible pectinate. Male copulatory complex lacking finger guard and accessory groove; peniferum never extending ventrally beyond clasping apparatus, and, although slightly swollen ventrally, never with conspicuous terminal or subterminal process or barbed point; penis large, undulating, somewhat S-shaped, consisting of contiguous spermatic and prostatic elements, and directed anteroventrally; penis together with acute thickening of posteroventral margin of peniferum rendering forcipate appearance to ventral portion of peniferum. Clasping apparatus well developed with distinct vertical and horizontal rami; external border of both rami and internal border of vertical ramus entire, internal border of horizontal ramus with two widely separated teeth and apex of apparatus with four denticles. Triunguis female with second antenna lacking accessory pectinate process on distal podomere; genital complex consisting of slender ductule bearing hyaline cap suspended from sclerotized area with large, apparently delicate, tuberculiform lobe anteriorly.

Type-species. — *Thermastrocythere harti* Hobbs and Walton, 1966: 8 [= *Thermastrocythere riojai* (Hoff, 1943: 276)].

Remarks. — Recently, Dabney G. Hart and Horton H. Hobbs, Jr. examined some of the type-specimens of entocytherids deposited in the Smithsonian Institution by C. Clayton Hoff and discovered that among the paratypes of his *Entocythere riojai* are two species. Because the holotype was the male member of a copulating pair, certain features of the copulatory complex were somewhat obscured, and in the mount of Hoff's partially dissected male, from which his illustrations of the species were made, the penifera were not arranged so that a lateral view can be obtained, thus giving an erroneous impression of the details of the peniferum. Consequently, the previous concept of the species was based partially on the other two paratypic males in the series, both of which were members of an undescribed species of the genus *Uncinocythere.* The holotype and the partially dissected paratypic male were discovered to be almost identical to the holotype of *Thermastrocythere harti,* which, therefore, became a subjective junior synonym of Hoff's *Entocythere riojai.* Because the generic designation of Hobbs and Walton is appropriate, the synonymy for the single species of this genus (see below) is felt to be correct.

Distribution. — United States: New Mexico, Texas, Louisiana, and Alabama northward to Michigan, Wisconsin, and North Dakota. Canada: Manitoba.

Hosts. — Freshwater crayfishes of the family Astacidae.

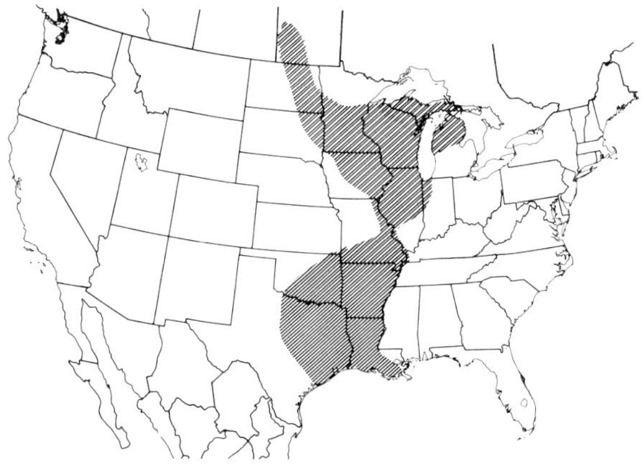

Figure 38. — Distribution of the genus *Thermastrocythere.*

Thermastrocythere riojai (Hoff, 1943)

Plate XXXIV, figures 3-5
Plate LIV

Entocythere (*Cytherites*) *riojai* Hoff, 1943: 276-282 (M, H, R, D, B), figs. 1a-1. — Rioja, 1943a: 565 (R). — Rioja, 1943b: 567, 582 (M, R), 583 (R).

Entocythere riojai. — Hoff, 1944b: 332 (K), 337 (R). — Rioja, 1945: 422 (R). — Tressler, 1947: 705 (C). — Rioja, 1949: 328 (R) (*sic riojae*). — Kozloff and Whitman, 1954: 162 (R). — Hobbs, 1955: 330 (R) (*sic roijai*). — Crawford, 1959: 167 (R). — Tressler, 1959: 730 (K), fig. 28.188. — Crawford, 1961: 238 (R). — Hart, 1962: 122 (C).

Uncinocythere riojai. — Hart, 1962: 138 (C, D, S).

Thermastrocythere harti. — Hobbs and Walton, 1966: 8-9 (M, H, R, A, D), figs. 2e-g. — Hart and Hart, 1969a: 185 (E, M, R). — Delorme, 1970: 1257, 1258 (C, H, D).

Thermastrocythere riojai. — Hobbs and Hobbs, 1970: 17 (R, S).

Diagnosis. — MALE: Shell length 338-390 μ (average, 362 μ); height, 188-218 μ (average, 202 μ). Eye present. Shell with highest portion slightly posterior to midlength and with dorsal margin more gently tapering posteriorly than anteriorly; ventral margin of shell entire; submarginal setae present anteriorly, posteriorly, and ventrally but absent dorsally, anterior and anteroventral setae closer together than posterior and posteroventral ones.

Copulatory complex as described in the diagnosis. In addition, dorsal finger slender with tip of apical seta reaching level of base of penis; ventral finger with moderate curve near base and a less prominent one short distance proximal to base of terminal seta.

FEMALE: Female unknown.

Relationships. — *Thermastrocythere riojai* has its closest affinities with members of the genus *Uncinocythere,* and is perhaps more closely allied to *U. simondsi* than to any other species; similarities exist in the armature of the clasping apparatus and in the distal portion of the peniferum; however, the penis in *riojai* reaches the anteroventral extremity of the peniferum — an arrangement that does not exist in any known species of the genus *Uncinocythere.*

Type-locality. — Stream in South Park, Quincy, Adams Co., Illinois, U.S.A. (types, USNM: holotype female, allotype male, and paratype, 81292). Host, *Orconectes virilis* (Hagen).

Distribution (in addition to type-locality). —

United States

ARKANSAS — BENTON CO. — Batey Ck, 0.9 mi NW of Maysville on rt 102 (host, *Orconectes neglectus chaenodactylus*); Little Sugar Ck N of Bentonville (hosts, *O. nais, O. macrus, O. n. neglectus*); Wildcat Ck, at Intersection of Ark 68 (hosts, *O. m. meeki, O. n. neglectus, O. nana*). BOONE CO. — Crooked River at Harrison (host, *O. hylas*). CLAY CO. — Jct of Hys 90 & 1, near Kennett (host, *O. p. palmeri*). LAWRENCE CO. — 5.5 mi NW of Powhatan Creek (host, *O. punctimanus*). LOGAN CO. — 7 mi W of Paris on rt 22 (host, *O. p. longimanus*). MARION CO. — Ash Creek in Flippin (host, *O. nais*). PERRY CO. — 2 mi W of Casa Mountain stream (host, *O. p. longimanus*); 5.5 mi S of Hallis on rt 7 (host, *O. p. longimanus*). POPE CO. — 6.4 mi N of Russellville on rt 7 (host, *O. p. palmeri*). PULASKI CO. — near crossroads on rt 10, 20 mi W of Little Rock (host, *O. meeki*). SCOTT CO. — 9 mi N of Waldron at jct of rt 23 & 71 (host, *O. p. longimanus*); 6.0 mi E of Cauthron on Ark 28 (host, *O. p. longimanus*). SHARP CO. — 0.7 mi NE of jct of 62 & 11 on US 62 (host, *O. punctimanus*). WASHINGTON CO. — Clear Ck about 4 mi NW of Fayetteville (hosts, *O. nana, O. n. neglectus*); Ozark Trout Farm near Fayetteville (host, *O. n. neglectus*); E fork of White R near rt 74 bridge, ¼ mi off rt 16 (host, *O. longidigitus*); S branch of Clear Creek, 3 mi N of Fayetteville, 1 mi E of rt 71 (hosts, *O. neglectus, O. nana*).

ILLINOIS — ADAMS CO. — Stream in South Park (host, *Orconectes virilis*). CHAMPAIGN CO. — Salt Fork at Homer Park, near Homer (host, *O. virilis*); Stream in Crystal Lake Park, Urbana (host, *O. propinquus*). MACOUPIN CO. — Staunton "Cahokiack" (host, *O. virilis*). OGLE CO. — White Pines State Forest (host, *O. virilis*). VERMILION CO. — Stony Creek, Oakwood (host, *O. propinquus*).

IOWA — DICKINSON CO. — Little Sioux R near Milford (host, *Orconectes virilis*).

KANSAS — LABETTE CO. — 1.5 mi N, 3.75 mi W of Oswego (host, *Orconectes nais*).

LOUISIANA — ALLEN PARISH — Castor Creek (host, *Orconectes hathawayi*); Bayou Blue (hosts, *O. lancifer, O. hathawayi, Procambarus clarkii, P. a. acutus*). OUACHITA PARISH — Bar pit on rt 165 4 mi N of rt 80 (hosts, *P. a. acutus, P. clarkii*). SABINE PARISH — Little Middle Ck (hosts, *P. natchitochae, O. difficilis*); Hapson Bayou (hosts, *P. dupratzi, O. p. longimanus, P. a. acutus*).

MICHIGAN — ARENAC CO. — Twinning, small str 2.6 mi N of jct of 65 & US 23 (host, *Orconectes virilis*); N Pine R, 1.7 mi NE of Standish (host, *O. propinquus*). IOSCO CO. — Trib of AuGres R, 3.1 mi W of jct of rts 55 & 65 (host, *O. propinquus*). OSCEOLA CO. — Middle branch of Muskegon R, 1.7 mi E of jct of rts 61 & 115 on 61 (host, *O. virilis*); Small branch, 0.6 mi E of Tustin on rt 63 (host, *O. virilis*).

MINNESOTA — BENTON CO. — St. Francis River at crossing of rt 23, 2.7 mi NE of Foley (host, *Orconectes virilis*). CARLTON CO. — S fork of Nemadji R at crossing of rt 23 (host, *O. virilis*); Nemadji River, S of Duluth at crossing of rt 23 (host, *O. virilis*). KANABEC CO. — Groundhouse River in Ogilvie at crossing of rt 23 (host, *O. virilis*). MORRISON CO. — Long Prairie River, at crossing of US 10, 7.6 mi N of Lincoln (host, *O. virilis*). OLMSTED CO. — 5.9 mi W of Eyota on rt 14 (hosts, *Cambarus diogenes, O. virilis*). OTTER TAIL CO. — Bluff Ck, at edge of Blufton on US 10 (host, *O. virilis*); Leaf River at crossing of US 10 between Blufton & Wadena (host, *O. virilis*). PINE CO. — Kettle River, at crossing of Minn 23, just NE of Sandstone, 1.7 mi W of jct of 23 & 123 (host, *O. virilis*). WASHINGTON CO. — Schettle, Stillwater.

MISSOURI — CARTER CO. — Van Buren, 8 mi NW of Butler-Carter Co. line on US 60 (host, *Procambarus a. acutus*). CRAWFORD CO. — Stream 1.5 mi N of Dillard, trib of Huzzah Ck & Meramec R (hosts, *Orconectes medius, O. punctimanus*); 2 mi E of Crawford-Phelps Co. line, 10 mi W of Steelville on MO. 8, Meramec River (host, *O. luteus*); 9 mi E of Steelville on MO 8, Huzzah Ck (host, *O. luteus*); 10 mi E of Steelville of MO 8 (host, *O. hylas*); 16 mi E of Steelville on MO 8, Curtois Ck (hosts, *O. luteus, O. hylas*). GREENE CO. — 6 mi E of Webster-Greene Co. line on US 66 (hosts, *O. nais, O. ozarkae.*) IRON CO. — Middlebrook Ck, 1 mi SW of Middlebrook (hosts, *O. hylas, O. quadruncus*); Trib of Clayton Ck, 2.5 mi N of Redmondville (host, *O. hylas*). KNOX CO. — N fk of South Fabius R, 1 mi N of Edina (host, *O. immunis*). LACLEDE CO. — Gasconade R, 7.2 mi W on back road (hosts, *O. luteus, O. punctimanus*);

15 mi NE of Lebanon on US 66, Gasconde R (hosts, *O. luteus, O. punctimanus*); 7 mi E of Lebanon on MO 32 (host, *O. punctimanus*); 11 mi E of Lebanon on MO 32 (host, *O. punctimanus*); 1 mi SE of Nebo on MO 32 (hosts, *O. luteus, O. punctimanus*). LAWRENCE CO. — Johnson Ck, near Halltown (host, *O. luteus*). MADISON CO. — Very rocky stream near Fredricktown (host, *O. punctimanus*); Very rocky stream near Fredricktown (hosts, *O. quadruncus, O. punctimanus*); Twelve Mile Ck, 7 mi S of Fredericktown (host, *O. luteus*); Twelve Mile Ck, 7 mi S of Fredricktown (host, *O. peruncus*). PHELPS CO. — 2 mi S of Rolla on US 63 (host, *O. punctimanus*); 8 mi N of Rolla on US 63 (host, *O. luteus*). REYNOLDS CO. — 2 mi SW of Reynolds-Iron Co. line, 2 mi NE of Lesterville (hosts, *O. hylas, O. ozarkae*); Ellington, 4 mi S of Reynolds-Iron Co. line on MO 21, East Fork Black River, 5 mi N of Lesterville (hosts, *O. hylas, O. ozarkae*); Ellington, 8 mi S of Centerville on MO 21, Sinking Ck (host, *O. hylas*). ST. FRANCOIS — St. Francois R at DeLassus (hosts, *O. quadruncus, O. punctimanus*); locality unknown (host, *O. medius*). ST. LOUIS — Small tributary of Moline Ck, 3 mi S of Blackjack (host, *O.* sp.). TEXAS CO. — 2 mi E of Texas-Laclede Co. line, 4 mi W of Plate on MO 32 (hosts, *O. luteus, O. n. neglectus*). WASHINGTON CO. — Stream at Irondale (host, *O. harrisoni*); Mill Ck at Mineral Point (hosts, *O. punctimanus, O. medius*); Clear Ck, 16 mi W of Potosi, 3 mi E of Washington-Crawford Co. line, on MO 8 (hosts, *O. hylas, O. luteus, O. quadruncus*); Shirley, 8 mi W of Potosi on MO 8 (hosts, *O. hylas, O. luteus, O. quadruncus*); Big River, 4 mi N of Caledonia (host, *O. hylas*). WAYNE CO. — McKenzie Ck at Piedmont (host, *O. hylas*); McKenzie Creek Bridge on MO 34 (hosts, *O. hylas, O. punctimanus*); 5 mi N of Piedmont on MO 49 (hosts, *O. hylas, O. punctimanus*). WEBSTER CO. — 16 mi NE of Webster-Greene Co. line on US 66 (host, *O. luteus*); 18 mi NE of Webster-Greene Co. line on US 66 (host, *O. nais*). WRIGHT CO. — Whetstone Ck, 5 mi W of Mountain Grove on US 60 (hosts, *O. luteus, O.* sp.).

OKLAHOMA — CHEROKEE CO. — McSpadin Falls, 10 mi NE of Tahlequah on rt 10 (hosts, *O. meeki brevis, O. nana, O. neglectus neglectus*); Ck about 10 mi E of Tahlequah (host, *O. n. neglectus*); about 10 mi N of Tahlequah, Spring fed ck, trib to Illinois R (host, *O. meeki brevis*). COMANCHE CO. — Cache Creek at White Wolf Crossing at Fort Sill (host, *O. nais*). LATIMER CO. — Turkey Ck, 8.7 mi E of Wilburton on US 270 (host, *O. p. longimanus*). SEQUOYAH CO. — Between Marble City & Standing Rock Elk Ck (host, *O. meeki*).

SOUTH DAKOTA — BEADLE CO. — 6.6 mi W of junction of US 14 & US 281 on 14 (host, *Orconectes immunis*).

TENNESSEE — OBION CO. — Turtle trap in Bayou du Chein (host, *Procambarus clarkii*).

TEXAS — MENARD CO. — Powell's Cave Downstream (200′ from well pipe) (host, *Procambarus clarkii*).

WISCONSIN — CHIPPEWA CO. — Duncan Ck, 4.1 mi N of Bloomer on US 53 (host, *Orconectes virilis*); Bob's Ck, 4.5 mi W of Cornell on rt 64 (host, *O. virilis*). COLUMBIA CO. — Crawfish Ck at intersection US 16 (host, *O. virilis*). DANE CO. — Six Mile Ck, near Waunakee (near Madison) (hosts, *O. propoinquus, O. virilis*); Yahara R near Sun Prairie where R crosses rt 19 near Madison (host, *O. propinquus*); small stream 13 mi W of Madison (hosts, *O. propinquus, O. virilis*); Yahara R at St 19 near Sun Prairie (host, *O. propinquus*). DOUGLAS CO. — Amnicon R 3.3 mi N of Middle R on US 53 (host, *O. virilis*). JEFFERSON CO. — (host, *O. propinquus*). LINCOLN CO. — Pine River, 8.4 mi W of Langlade Co. line (host, *O. virilis*). TAYLOR CO. — Trib to Yellow R 7 mi W of jct of rts 64 & 73 (host, *O. virilis*); Black River, 12.9 mi W of Medford on rt 64 (host, *O. virilis*). WASHBURN CO. — Spooner, Totagatic R, 2.8 mi N of Minong on US 53 (host, *O. virilis*).

Canada

MANITOBA — Town of Swan River (host, *Orconectes rusticus*).

UNCINOCYTHERE HART, 1962

Columbia Group. — Rioja, 1943a: 564, 565 (M, R). — Rioja, 1949: 328 (R). — Rioja, 1955: 196 (R). — Hart, 1959: 203 (M). — Crawford, 1959: 167 (M, R, D) (Rioja 1943, *not* Rioja 1949). — Walton and Hobbs, 1959: 118 (R). — Hobbs and Walton, 1960: 17 (R). — Crawford, 1961: 236, 238 (R) (*sic* Rioja 1949). — Hart and Hobbs, 1961a: 25 (D, H). — Hart and Hobbs, 1961b: 174 (K), 181 (C).

Uncinocythere Hart, 1962: 125 (K), 136 (M, S, C), type-species, *Entocythere columbia* Dobbin, 1941. — Hart, 1963: 51 (R, D). — Hart, 1965a: 190 (D). — Hobbs III, 1965: 159 (R), 161 (K). — Frantz and Cordone, 1966: 6 (D). — Hart and Hart, 1966a: 3 (A). — Hart and Hart, 1969a: 182-185 (D, E, R). — Hart and Hart, 1969b: 158 (M, R). — Hobbs and Hobbs, 1970: 2 (K).

Diagnosis. — Terminal tooth of mandible pectinate. Copulatory complex of male without finger guard or accessory groove; spermatic and prostatic elements of penis contiguous and extending anteroventrally from base. External border of clasping apparatus entire; internal border with one to four teeth; apex with two to five denticles. Females without amiculum or J-shaped rod, and antenna of triunguis female without comblike appendix at base of apical claws.

Type-species. — *Entocythere columbia* Dobbin, 1941; designated by Hart, 1962: 136.

Distribution. — According to the present state of knowledge, this genus occupies an apparently discontinuous range; Washington, Oregon, northern California, to Idaho, Utah, and western Wyoming; South Dakota, Minnesota, Wisconsin eastward to western Pennsylvania, West Virginia, western Virginia, and southward to northern Florida; Louisiana, and Texas; Hidalgo, Puebla, and Veracruz, Mexico.

With additional field data the western segment of the range may well be joined with the eastern one by way of the northern United States and Canada.

Hosts. — Freshwater crayfishes of the family Astacidae and (for *U. zaruri*) a freshwater crab of the family *Pseudothelphusidae.*

Figure 39. — Distribution of the genus *Uncinocythere.*

Key to the Species of the Genus Uncinocythere

(From Hobbs, Jr., 1970)

1	Internal border of horizontal ramus of clasping apparatus with four teeth	2
1′	Internal border of horizontal ramus of clasping apparatus with fewer than four teeth	3
2(1)	Apex of clasping apparatus with four denticles	*U. columbia*
2′	Apex of clasping apparatus with two or three denticles	*U. cuadricuspide*
3(1′)	Internal border of horizontal ramus of clasping apparatus with three teeth	4
3′	Internal border of horizontal ramus of clasping apparatus with one or two teeth	15
4(3)	Ventral margin of peniferum entire	5
4′	Ventral margin of peniferum cleft	7

5(4) Apex of clasping apparatus with two or three denticles .... *U. dobbinae*
5′ Apex of clasping apparatus with four denticles 6
6(5′) Vertical ramus of clasping apparatus longer than horizontal ramus .... *U. clemsonella*
6′ Vertical ramus of clasping apparatus shorter than horizontal ramus .... *U. columbia*
7(4′) Shell with posteroventral spine 8
7′ Shell without posteroventral spine 10
8(7) Posteroventral spine arising anterior to posteriormost margin of shell .... *U. thektura*
8′ Posteroventral spine arising at posteriormost margin of shell 9
9(8′) Peniferal cleft more than ½ width of peniferum; clasping apparatus with three apical denticles .... *U. caudata*
9′ Peniferal cleft less than ½ width of peniferum; clasping apparatus with five apical denticles .... *U. cassiensis*
10(7′) Apex of clasping apparatus with four or five denticles 11
10′ Apex of clasping apparatus with three denticles 14
11(10) Apex of clasping apparatus with four denticles .... *U. warreni*
11′ Apex of clasping apparatus with five denticles 12
12(11′) Apical portion of peniferum constricted ventrally .... *U. neglecta*
12′ Apical portion of peniferum not constricted ventrally 13
13(12′) Peniferal cleft almost closed ventrally by posterior curve of anteroventral spine .... *U. ericksoni*
13′ Peniferal cleft not almost closed ventrally .... *U. holti*
14(10′) Peniferum with simple rounded cleft 26
14′ Peniferum with cleft other than rounded 24
15(3′) Ventral margin of peniferum cleft with anteroventral spine 16
15′ Ventral margin of peniferum entire 20
16(15) Horizontal ramus of clasping apparatus with two teeth on internal border and two or four denticles on apex 17
16′ Horizontal ramus of clasping apparatus with one tooth on internal border and three apical denticles 19
17(16) Apex of clasping apparatus with two denticles .... *U. zaruri*
17′ Apex of clasping apparatus with four denticles 18
18(17′) Clasping apparatus C-shaped; peniferal cleft shallow .... *U. occidentalis*
18′ Clasping apparatus L-shaped; peniferal cleft deep .... *U. simondsi*
19(16′) Peniferal cleft very shallow .... *U. ambophora*
19′ Peniferal cleft deep 25
20(15′) Clasping apparatus with two teeth on internal border 21
20′ Clasping apparatus with one tooth on internal border 23
21(20) Teeth on internal border subequal in size .... *U. bicuspide*
21′ Teeth on internal border distinctly unequal in size 22
22(21′) Clasping apparatus L-shaped; eye absent .... *U. lucifuga*
22′ Clasping apparatus C-shaped; eye present .... *U. equicurva*
23(20′) Anteroventral angle of peniferum acute, its anterior side simple .... *U. lucifuga*
23′ Anteroventral angle of peniferum 90°, its anterior side rolled or thickened .... *U. xania*
24(14′) Anteroventral peniferal projection heavy and rounded distally .... *U. spathe*
24′ Anteroventral peniferal projection spiniform .... *U. stubbsi*
25(19′) Anteroventral peniferal projection extending well beyond posteroventral projection .... *U. allenae*
25′ Anteroventral peniferal projection subequal in length to posteroventral projection .... *U. zancla*
26(14) Clasping apparatus with vertical ramus longer than horizontal ramus .... *U. pholetera*
26′ Clasping apparatus with vertical ramus subequal in length to horizontal ramus .... *U. xena*

Uncinocythere allenae (Hart and Hart, 1971)

Plate XXXIV, figures 6-8
Plate LV

Uncinocythere allenae Hart and Hart, 1971: 116-117 (M, H, R, A, D), figs. 10a-c.

Diagnosis. — MALE: Shell length 320 μ; height, 190 μ. Shell subovoid, venter straight, dorsum broadly arched. Greatest height ⅓ shell length from posterior end. Eye pigmented, situated ⅕ shell length from anterior end.

Peniferum slender, of approximately equal width throughout, extending to level of internal border of horizontal ramus of clasping apparatus. Penis mostly horizontal, occupying ½ to ¾ peniferum width in holotype, situated ⅕ peniferum length from ventral extremity. Peniferum bifid ventrally with concavity in middle, tapering to sharp, acute point anteriorly.

Horizontal and vertical rami of clasping apparatus of subequal length, meeting at angle of approximately 80°. External border of vertical ramus almost straight; internal border also straight but diverging anteriorly at base. External border of horizontal ramus convex; internal border with one large tooth at midlength, and one small one near proximalmost of two or three terminal denticles.

FEMALE. — Unknown.

Relationships. — *Uncinocythere allenae* appears to be closely related to *U. zancla,* but differs from it in having a sharp, tapering anterior point on the base of the peniferum rather than the flat base of *U. zancla.* The clasping apparatus most resembles that of *Uncinocythere lucifuga.*

Type-locality. — Tributary to Horse Creek, 11.4 mi E of Savannah on Route 64, Hardin Co., Tennessee, U.S.A. (types, USNM: holotype, 138456). Host, *Orconectes compressus.*

Distribution. — Known only from the type-locality.

Uncinocythere ambophora (Walton and Hobbs, 1959)

Plate XXXIV, figures 9-11
Plate LV

Entocythere ambophora Walton and Hobbs, 1959: 115-118 (M, H, R, A, D), figs. 1-6. — Hart and Hobbs, 1961b: 173 (B, C), 174 (K), 181 (R). — Hart, 1962: 123 (C).

Uncinocythere ambophora. — Hart, 1962: 136 (C, D, S). — Hobbs and Walton, 1963a: 457 (R). — Ferguson, 1968: 502 (C, D). — Hart and Hart, 1969a: 157 (M).

Diagnosis. — MALE: Shell length 305-310 μ (average, 307 μ); height, 160-175 μ (average, 165 μ). Shell subovate with posterior portion slightly higher than anterior; margins entire, lacking emarginations and spines, with few widely dispersed submarginal setae on valves.

Copulatory complex (figs. 1, 2), except for lacking finger guard, consisting of usual elements. Distal portion of peniferum with distinct ridge along cephalodistal margin, terminating in acute spine. Distal and proximal portions of peniferum not clearly distinguishable. Horizontal and vertical rami of clasping apparatus merge imperceptibly; distal extremity of latter ramus with three teeth, external border entire, and internal border with single tooth.

TRIUNGUIS FEMALE: Shell length 340-360 μ (average, 347 μ); height, 180-210 μ (average, 189 μ). Shell similar to that of male but larger. Third claw on antenna apparently frequently vestigial or absent. Small papilla near posterior extremity of body, but no amiculum or J-shaped rod.

Type-locality. — Palm Springs, Seminole Co., Florida, U.S.A. (types, USNM: holotype, 105953; allotype, 105954; morphotype, 105955; paratypes, 108186, 108187, 108188, 108189, and 108190). Host, *Procambarus acherontis.*

Distribution. — Known only from the type-locality.

Uncinocythere bicuspide (Rioja, 1943)

Plate XXXIV, figures 12-15
Plate LVIII

Entocythere (*Cytherites*) *dobbini bicuspide* Rioja, 1943a: 565-566 (M, H, R, A, D), fig. 22. — Rioja, 1943b: 576 (M, R).

Entocythere dobbinae bicuspis (*sic*). — Hoff, 1944b: 332 (K) (Rioja *1943,* not Rioja 1944).

Entocythere dobbini bicuspide. — Rioja, 1945: 420 (M), 422 (R), fig. 1. — Kozloff and Whitman, 1954: 162 (R). — Hart, 1962: 122 (C).

Entocythere uncinata. — Rioja, 1949: 326-328 (M, H, R, D), figs. 27, 28. — Hart, 1962: 122 (C).

Entocythere bicuspide. — Rioja, 1949: 322, 323 (R, S, D, H), figs. 17-20. — Crawford, 1959: 167 (R). — Hart, 1962: 122 (C).

Uncinocythere bicuspide. — Hart, 1962: 136 (C, D, S). — Hobbs, Holt, and Walton, 1967: 6 (B, C). — Ferguson, 1968: 502 (C, D). — Hobbs, 1971: 21-22 (K, D, A, M), fig. 8.

Diagnosis. — MALE: Shell length 290-310 μ (average, 300 μ); height, 160 - 190 μ (average, 180 μ). Copulatory complex with clasping apparatus comparatively heavy and usually strongly curved with angle formed by extensions of horizontal and vertical rami distinctly less than 90 degrees. Internal border of horizontal ramus with two teeth and with two, rarely three, small denticles at apex of ramus. Penis situated proximal to ventral margin of peniferum by distance no greater than its length. Ventral margin of peniferum always entire but often with anteroventral extremity produced.

FEMALE: Shell length 340 - 350 μ (average, 340 μ); height 200-220 μ (average, 210 μ).

Type-locality. - Huauchinango, Estado de Puebla, Mexico (types no longer extant). Host, *Procambarus riojai.*

Distribution (in addition to type-locality). — Mexico

HIDALGO — Arroyo de Puendó, 3 km E of Tenango de Doria (Rioja, 1949: 323) [host, *Procambarus erichsoni* (listed as "P. weigmanni?" by Rioja)]; Arroyo de San Bartolo, San Bartolo, 23 km NNW of Huauchinango (host, *P. erichsoni*); Arroyo de Puendó, Mamay y Bojoy, Tenango de Doria (Rioja, 1949:323), 16 km NW of Huauchinango (host, *P. erichsoni*).

PUEBLA — Tetela de Ocampo, La Cañada (Rioja, 1945: 422) (host, *Procambarus paradoxus*); Villa Juárez (Rioja, 1949: 323) (host, *P. hoffmanni*); Cumbres de Cuanepixca, 6 km NE of Zihuateutla (Rioja, 1949: 323) (host, *P. hoffmanni*); Arroyo de San Diego, 36 km NE of Villa Juárez (Rioja, 1949: 323) (host, *P. contrerasi*); El Coyular, 4 km NE of Zihuateutla (Rioja, 1949: 323) (hosts, *P. contrerasi* and *P. zihuateutlensis*); Arroyo que va a El Coyular, distrito de Zihuateutla (Rioja, 1949: 327), 29 km NE of Huauchinango (host, *P. erichsoni*); *Zihuateutla* (host, *P. hortonhobbsi*).

?VERACRUZ — Two km N of Coyutla (host, *P. hoffmanni*); Tomatlán (host, *P. aztecus*); this locality cannot be found on a map, and should be confirmed.

Thus, the range of this species lies on the eastern slopes of the plateau in the states of Hidalgo, Puebla, and Veracruz, and, except for the record on *Procambarus aztecus* at Tomatlán, lies wholly north of the Cordillera Volcánica Transversal.

Entocythere uncinata Rioja, 1949, is unquestionably a morphological variant of *U. bicuspide.* The only distinction made between them was in the curvature of the clasping apparatus, and a complete series of intermediate forms have been observed among the specimens cited above.

Uncinocythere cassiensis Hart, 1965

Plate XXXIV, figures 16-19
Plate LV

Uncinocythere cassiensis Hart, 1965a: 191, 192 (M, H, D), 195 (R), figs.1-5. — Ferguson, 1968: 502 (C, D). — Hobbs and Walton, 1968: 251 (R). — Hart and Hart, 1969a: 183 (D). — Hart and Hart, 1969b: 154 (M).

Diagnosis. — MALE: Shell length 420-480μ (average, 450 μ); height, 200-250 μ (average, 230 μ). Shell subelliptical in outline; posteroventral margin of each valve with conspicuous hook-like prominence directed posteriorly; dorsal margin of valves distinctly heavier than ventral margin; anterodorsal margin slightly concave immediately anterior to eye spot; greatest shell height at approximately midlength. Marginal setae absent dorsally and ventrally; a few present on anterior and posterior margins. Eye spot approximately ¼ shell length from anterior margin.

Copulatory complex with peniferum extending only slightly beyond external border of clasping apparatus; distal extremity bifurcate with apices apposed at right angles to one another. Clasping apparatus with distinct vertical and horizontal rami; horizontal ramus slightly longer than vertical. Internal border of horizontal ramus provided with three evenly spaced teeth; distal extremity with four or five denticles; external border evenly rounded or slightly undulate.

TRIUNGUIS FEMALE: Shell length 500-560 μ (average, 530 μ); height, 260-300 μ (average, 280 μ). Shell similar to that of male, but larger. Posteroventral hook-like prominence more pronounced than that of males or biunguis females, and on some specimens cornified; angle prominence makes with ventral shell margin usually less in triunguis females than in males or biunguis females.

Variations. — Variations in the peniferum are noticeable, but not extreme. In some specimens the penis is directed almost straight downward rather than having a horizontal position.

Relationships. — *Uncinocythere cassiensis* appears to have its closest affinities with *U. caudata, U. thektura,* and *U. holti.*

It differs from *U. caudata, U. thektura,* and *U. holti* in that the distal extremity of the clasping apparatus is provided with four or five denticles (*U. caudata* has three, *U. thektura* and *U. holti* have five or six), and the posteroventral hook-like projection forms a distinct obtuse angle with the ventral margin of the shell (there is little or no angle in *U. caudata;* the projection is distinctly smaller in *U. thektura* than in *U. cassiensis* and the angle is slight; and there is no projection in *U. holti*).

The points of the distal portion of the peniferum are close to one another in *U. cassiensis*, while those of *U. caudata* are relatively widely separated.

Uncinocythere cassiensis, *U. thektura*, and *U. holti* are similar in three respects: (1) the clasping apparatus of each possess three evenly spaced teeth on the internal border of the horizontal ramus, (2) the distal portion of the peniferum is bifurcate with apices at right angles to one another, and (3) a depression, or concavity, occurs in the anterodorsal margin of the shell immediately anterior to the eye.

Type-locality. — Raft River between Bridge and Almo, Cassia Co., Idaho, U.S.A. (types, USNM: holotype and allotype, 110778; paratype, 110779). Host, *Pacifastacus gambelii.*

Distribution (in addition to type-locality). —

United States

IDAHO — FREMONT CO. — Teton Basin on W side of Tetons (host, *Pacifastacus gambelii*); E of Lorenzo (host, *P. gambelii*). JEFFERSON CO. — Bannock Jim's Slough, trib to Snake River near Roberts (host, *P. gambelii*). TWIN FALLS CO. — Shoshone Falls (host, *P. gambelii*).

OREGON — MALHEUR CO. — Jordan River (host, *P. gambelii*).

WYOMING — LINCOLN CO. — Crow Creek 2 mi inside Wyoming from Utah, trib of Salt River (host, *P. gambelii*). TETON CO. — Polecat Creek (host, *P. gambelii*). UINTA CO. — Bear River on highway just E of Evanston (host, *P. gambelii*); Bear River 10 mi N of Evanston on rt 89 (host, *P. gambelii*). YELLOWSTONE NATIONAL PARK CO. — Yellowstone National Park (host, *P. gambelii*).

Uncinocythere caudata (Kozloff, 1955)

Plate XXXV, figures 1-2
Plate LV

Entocythere caudata Kozloff, 1955: 156-158 (M, H, R, B, D), figs. 1-12. — Hobbs, 1955: 330 (R). — Ferguson, 1958: 198 (C). — Westervelt and Kozloff, 1959: 243 (R). — Hart and Hobbs, 1961b: 182 (R). — Hart, 1962: 122 (C).

Uncinocythere caudata. — Hart, 1962: 136 (C, S, D). — Hart, 1965a: 190, 191, 195 (R). — Hobbs and Walton, 1966: 10, 11 (R). — Hobbs, Holt, and Walton, 1967: 6 (B, C). — Hobbs and Walton, 1968: 251 (R). — Hart and Hart, 1969a: 183 (D) (*sic candata*). — Hart and Hart, 1969b: 154 (M).

Diagnosis. — MALE: Shell length 350-390 μ (average, 370 μ); height, 160-180 μ (average, 170 μ). Shell with valves evenly rounded along dorsal margin; anterior end acutely rounded. Posterior end with distinctive tail-like projection. Ventral margin nearly straight, but bearing slight concavity near midlength, and also immediately anterior to posterior projection. Height of shell greatest near posterior one-third of body. A number of fine setae scattered over outside surface of each valve, principally close to margins. Eye situated dorsally near end of first one-fifth of body.

Copulatory complex with distal portion of peniferum bifurcate; penis vertically oriented in distal one-third. Clasping apparatus with distinct horizontal and vertical rami, subequal in length to one another; internal border of horizontal ramus with three poorly developed teeth; three dorsally directed terminal denticles.

TRIUNGUIS FEMALE: Shell length 370-450 μ (average, 410 μ); height, 180-220 μ (average, 200 μ).

Type-locality. — Spring near OO Ranch corral, Malheur National Wildlife Refuge, Harney Co., Oregon, U.S.A. (types, USNM: holotype, 95682; allotype, 95683; paratypes, 95684 and 95685). Host, *Pacifastacus gambelii.*

Distribution (in addition to type-locality).—

United States

CALIFORNIA — SHASTA CO. — Fall River Mills, where Fall River joins Pit River (host, *Pacifastacus fortis*).

OREGON — HARNEY CO. — Malheur River near Drewsey (host, *P. gambelii*).

Uncinocythere clemsonella (Crawford, 1961)

Plate XXXV, figures 3-7
Plate LV

Entocythere clemsonella Crawford, 1961: 236-238 (M, H, R, A, D), figs. 1-8. — Hart, 1962: 123 (C).

Uncinocythere clemsonella. — Hart, 1962: 136 (C, D, S). — Ferguson, 1968: 502 (C, D).

Diagnosis. — MALE: Shell length 345-375 μ (average, 361 μ); height, 210-225 μ (average, 214 μ). Shell suboval in lateral aspect with anterior and ventral margins smoothly rounded. Dorsal margin slightly arched along posterodorsal edge. Setae sparsely scattered on lateral surface of valve and irregularly dispersed along free margin. Eye located about one-sixth of valve length from anterior end.

Copulatory complex with peniferum terminating in single small ventrally directed point or tubercle. Rami of clasping apparatus forming angle of about 70 degrees. Flexor surface of horizontal ramus bearing three teeth on distal half, decreasing in prominence distally. Entire extensor margin of horizontal ramus smooth. Four prominent denticles borne on distal end of horizontal ramus.

TRIUNGUIS FEMALE: Shell length 395-420 μ (average, 410 μ); height, 240-270 μ (average, 258 μ). Shell suboval in lateral aspect with ventral margin bearing slight concavity just anterior to midlength. Dorsal and anterior margins evenly rounded, with posterior one being slightly arched. Free margins of valve are setiferous, and lateral surface sparsely so. Eye located one-fifth of valve length from anterior end.

Type-locality. — Intermittent stream on South Carolina rt 20, about 1.4 m N of Abbeville, Abbeville Co., South Carolina, U.S.A. (types, USNM: holotype, 107454; allotype, 107456; morphotype, 107457; paratype, 107455). Host, *Cambarus latimanus.*

Distribution (in addition to type-locality). —

United States

ALABAMA — ELMORE CO. — 21.6 mi S of Rockford on rt 231 [hosts, *Procambarus lophotus, C. latimanus,* and *C. (Depressicambarus)* sp.]. TALLADAGA CO. — 5.3 mi S of Monford on rt 21 [host, *C. (D.)* sp.].

GEORGIA — PUTNAM CO. — Tributary of Little River in Rock Eagle Park (host, unknown) (Crawford, 1961: 238).

SOUTH CAROLINA — CHEROKEE CO. — Skeleton Creek, near Gaffney (host, unknown) (Crawford, 1961: 238).

VIRGINIA — BOTETOURT CO. — Stream at Blue Ridge (host, *C. bartonii*).

Uncinocythere columbia (Dobbin, 1941)

Plate XXXV, figures 8-11
Plate LV

Entocythere columbia Dobbin, 1941: 184, 185 (M, H, R, D), figs. 1-5 (pl. 1). — Tressler, 1947: 705 (C). — Rioja, 1949: 328 (R). — Kozloff and Whitman, 1954: 159 (A, B, D). — Hobbs, 1955: 330 (R). — Kozloff, 1955: 156 (C), 158 (R). — Westervelt and Kozloff, 1959: 239 (A, C, H). — Tressler, 1959: 728 (K), fig. 28.182. — Evenson (*nee* Dobbin), 1969: 491, 493 (C).

Entocythere (Cytherites) columbia. — Hoff, 1942a: 66 (K), 69, 71 (R). — Rioja, 1942b: 688 (K), 689, 696 (R), fig. 2. — Rioja, 1943a: 564, 565 (R). — Rioja, 1943b: 469 (R), 576 (R), fig. 21.

Uncinocythere columbia. — Hart, 1962: 137 (C, D, S). — Hart, 1965a: 190, 195 (D). — Evenson (*nee* Dobbin), 1969: 491, 493 (C).

Diagnosis. — MALE: Shell length 410-510 μ (average, 457 μ); height, 220-290 μ (average, 252 μ). Shell with rounded protuberance on posteroventral margin. Peniferum tapering distally, terminating in bifurcation; penis directed ventrally, situated in distal ⅓ of peniferum. Clasping apparatus with curved vertical ramus and more-or-less straight horizontal ramus; external borders entire; internal border of clasping apparatus with four teeth; terminal portion with four denticles.

TRIUNGUIS FEMALE: Shell length 620-650 μ (average, 635 μ); height, 360-380 μ (average, 370 μ). Shell similar to that of male, but larger.

Type-locality. — Columbia River, Washington (probably in SW Washington), U.S.A. (lectotypes in Thomas Burke Memorial Washington State Museum, Seattle, Washington). Host, "crayfish branchiae," probably *Pacifastacus leniusculus klamathensis.*

Distribution (in addition to type-locality). —

United States

CALIFORNIA — SANTA BARBARA CO. — Santa Barbara (host, *Pacifastacus gambelii*).

OREGON — BENTON CO. — St. Mary's River at W edge of Philomath on rt 20 (host, *Pacifastacus l. klamathensis*). CLACKAMAS CO. — "Several localities" (host, *P. l. leniusculus, P. l. trowbridgii,* and *P. l. klamathensis*) (Kozloff and Whitman, 1954: 159). CLATSOP CO. — Astoria (host, *P. l. klamathensis*). JEFFERSON CO. — Deschutes River near Warm Springs (host, *P. l. klamathensis*). LINCOLN CO. — Creek near Burnt Woods Post Office on rt 20 (host, *P. l. klamathensis*); trib to Yaquina River, 14.4 mi E of Toledo on rt 20 (host, *P. l. klamathensis*). MARION CO. — Little Pudding Creek, 7.9 mi SW of Silverton at crossing of rt 213 (host, *P. l. klamathensis*); Battle Creek, 6 mi NW of Silverton on rt 213 (host, *P. l. klamathensis*). MULTNOMAH CO. — Crystal Springs Creek, Portland (host, *P. l. trowbridgii*). POLK CO. — Willamina, S of Yamhill River, 1.6 mi W of Valley Junction (host, *P. l. klamathensis*); Salt Creek, 3.5 mi NW of Dallas on rt 22 (host, *P. l. klamathensis*).

WASHINGTON — PACIFIC CO. — Naselle River (no host record) (see Dobbin, 1941: 185). SNO-

HOMISH CO. — Silver Lake (no host record) (see Dobbin, 1941: 185).

Uncinocythere cuadricuspide (Rioja, 1945)

Plate XXXV, figures 12-16
Plate LVIII

Entocythere dobbini cuadricuspide Rioja, 1945: 422 (M, R), fig. 3. — Rioja, 1949: 323-325 (S, H, R), figs. 21, 23. — Hart, 1962: 122 (C).

Entocythere dobbini multicuspide. — Rioja, 1945: 422 (M, R), fig. 4. — Hart, 1962: 122 (C).

Entocythere dobbini multidentata. — Rioja, 1949: 325 (R).

Entocythere cuadricuspide. — Rioja, 1949: 323, 325, 328 (M, H, R), figs. 21-23. — Crawford, 1959: 167 (R). — Hart, 1962: 122 (C).

Uncinocythere cuadricuspide. — Hart, 1962: 137 (C, D, S). — Ferguson, 1968: 502 (C, D). — Hobbs, 1971: 22-24 (K, D, A, M), fig. 10.

Diagnosis. — MALE: Shell length 280-290 μ (average, 285 μ); height, 170-180 μ (average, 175 μ). Clasping apparatus strongly curved, with angle formed by extension of horizontal and vertical rami highly variable but always less than 90 degrees. Internal border of horizontal ramus with four or more teeth and with two or three denticles at its apex. Ventral extremity of peniferum entire.

FEMALE: Shell length 330-340 μ (average, 345 μ); height, 210 μ.

Type-locality. — Tetela de Ocampo y La Cañada, Puebla, México (types no longer extant). Host, *Procambarus paradoxus.*

Distribution (in addition to type-locality). —

Mexico

PUEBLA — La Magdalena, 3 km N of Zihuateutla (hosts, *P. ortmanni* and *P. erichsoni*) (Rioja, 1949: 325); Los Estajos, 9-10 km E of Villa Juárez, Arroyo de Tlatentiloyan (host, *P. hoffmani*); Arroyo del Coyular, 7 km NE of La Union (host, *P. hortonhobbsi*); Las Cumbres de Cuanepixca, Municipio de La Unión (host, *P. zihuateutlensis*); Zihuateutla (host, *P. hortonhobbsi*); Mazacutlán, 16 km NE of Huauchinango (host, *P. caballeroi*).

Uncinocythere cuadricuspide, like the sympatric *U. dobbinae,* seems to be restricted to the gulf drainage systems of México north of the Cordillera Volcánica Transversal, where it is thus far known to occur only in Puebla.

Uncinocythere dobbinae (Rioja, 1943)

Plate XXXV, figures 17-21
Plate LVIII

Entocythere (Cytherites) dobbini. — Rioja, 1943a: 561-565 (M, H, R, A, D), figs. 11-21, 24. — Rioja, 1943b: 570, 586 (R), figs. 19, 20.

Entocythere dobbini. — Rioja, 1955: 419-423 (M, R), fig. 2. — Rioja, 1949: 322 (R, H), figs. 15, 1. — Kozloff and Whitman, 1954: 162 (R). — Hobbs, 1955: 330 (R). — Kozloff, 1955: 161 (R).

Entocythere dobbinae. — Hoff, 1944b: 332 (K) (Rioja 1943, not Rioja 1944), 337 (R) (*sic* Rioja 1944), 341 (R). — Tressler, 1947: 705 (C). — Tressler, 1954: 138 (D). — Tressler, 1959: 729 (K) (*sic* Rioja *1944*), fig. 28.187. — Crawford, 1959: 167 (R).

Entocythere recurvata Rioja, 1949: 325, 236, 328 (M, R, D, H), figs. 24-26. (Type-locality, "Los Estajos, a 6 Km NE de Zinhuateutla, Pue." Types no longer extant). — Hart, 1962: 122 (C).

Uncinocythere dobbinae. — Hart, 1962: 137 (C, D, S). — Hobbs, 1971: 24-26 (M, D, A, R, K), figs. 12, 13.

Diagnosis. — MALE: Shell length 290-320 μ (average, 300 μ); height, 180-200 μ (average, 180 μ). Clasping apparatus of male curved with angle formed by extension of horizontal and vertical rami distinctly less than 90 degrees. Internal border of horizontal ramus with three denticles that may or may not be well-developed and with two to four small denticles at apex of horizontal ramus. Ventral portion of peniferum entire.

Type-locality. — Huachinango, Estado de Puebla, México (types no longer extant). Hosts, *Procambarus riojai* and *Procambarus hoffmanni.*

Distribution (in addition to type-locality). —

Mexico

HIDALGO — Arroyo Puendó, 3 km E of Tenango de Doria (Rioja, 1949: 322) (host, *P. erichsoni*); Tenango de Doria (host, *P. erichsoni*); Arroyo de San Bartolo, San Bartolo (host, *P. erichsoni*); Arroyos Puendó, Mamay y Bojoy, Tenango de Doria (host, *P. erichsoni*).

PUEBLA — Villa Juárez (host, *P. hoffmanni*) (Rioja, 1943a: 565); Los Estajos, 6 km NE of Zihuateutla (hosts, *P. hoffmanni, P. ortmanni,* and *P. erichsoni*) (Rioja, 1949: 322); La Magdalena, 3 km N of Zihuateutla (host, *P. erichsoni*) (Rioja, 1949: 322); Camino de Villa Juárez a Coyutla (host, *P. hoffmanni*) (Rioja, 1949: 322); Arroyo del Coyular, 7 km NE of Unión, La Unión (host, *P. contrerasi*); Las Cumbres de Cuanepixa, La Unión (host, *P. hoffmanni*); Mazacuatlán, 16 km NE of Huauchinango (host, *P. caballeroi*); El Agengibre (host, *P.

contrerasi); Arroyo de San Diego, km 260 carretera México a Tuxpán (host, *P. contrerasi*); Chignautla, Teziutlan (host, *P. teziutlanensis*); La Junta (host, *P. contrerasi*); Arroyo de Apatlaco, Villa Juárez (host, *P. hoffmanni*); Barrio San Isidro, Villa Juárez (host, *P. caballeroi*); Lagunillas, Villa Juárez (host, *P. caballeroi*); Curva del Perro, km 172 carretera México a Tuxpán (host, *P. hoffmanni*); Tetela de Ocampo, La Cañada (host, *P. paradoxus*) (Rioja, 1945: 419); Barranca de Zacatlan, Río Hueyalpan (host, *P. riojai*); Arroyo de San Diego, 36 km NE of Villa Juárez (host, *P. contrerasi*) (Rioja, 1949: 326).

VERACRUZ — Tlapacoyan (host, *P. tlapacoyanensis*) (Rioja, 1949: 326 = host and locality cited for *E. recurvata*); km 260 carretera México — Tuxpán (host, *P. hoffmanni*) (Rioja, 1949: 326 = host and locality cited for *E. recurvata*); Sabana de El Poste, Miahuapan, 8-10 km N of Coyutla (Rioja, 1949: 322) (host, *P. hoffmanni*); 18 km SSE of Martínez de la Torre (host, *P. hoffmanni*); 2 km N of Coyutla (host, *P. hoffmanni*); 5 km E of Coyutla, Río Necaxa (host, *P. hoffmanni*) (Rioja, 1949: 322; note that this locality is in Veracruz, not in Puebla as indicated by Rioja); Tuxpán (host, *P. contrerasi*).

Uncinocythere dobbinae appears to be restricted to the Gulf Slope in Mexico north of the Cordillera Volcanica Transversal, in the states of Hidalgo, Puebla, and Veracruz.

Remarks. — *Entocythere recurvata* is considered to be a morphological variant of *Uncinocythere dobbinae,* for among the available specimens there is a complete series of intermediate forms between the *dobbinae*-type clasping apparatus and the strongly recurved *recurvata*-type. Hobbs (1971) has observed that in collections from several localities, both extremes were encountered, and, in addition, there are a number of specimens in which the clasping apparatus is neither strongly recurved nor quite so gently curved as in typical *dobbinae.*

The alteration of the original spelling of the name of this ostracod by Hoff (1944b) is entirely justified by the dedication of the name, "Nos complacemos en dedicar esta especie a la Dra. Catherine N. Dobbin, del Departamento de Zoologia de Smith College, Northampton, Mass. . . ." (Rioja, 1943a: 565).

Uncinocythere equicurva (Hoff, 1944)

Plate XXXVI, figures 1-3
Plate LV

Entocythere equicurva Hoff, 1944b: 332 (K), 337-341 (M, H, R, D), figs. 6-10. — Rioja, 1945: 422 (R). — Tressler, 1947: 705 (C). — Rioja, 1949: 328 (R). — Kozloff, and Whitman, 1954: 162 (R). — Hobbs, 1955: 332 (R). — Tressler, 1959: 729 (K), fig. 28.186. — Hart, 1959: 193 (D), 194 (K), 203-204 (M, H, D), fig. 14. — Walton and Hobbs, 1959: 114 (R, K), 118, 120 (R), fig. 17. — Crawford, 1961: 238 (R). — Hart, 1962: 122 (C).
Uncinocythere equicurva. — Hart, 1962: 137 (C, D, S).

Diagnosis. — MALE: Shell length 300-340 μ (average, 320 μ); height, 170-190 μ (average, 180 μ). Shell oval in form with the dorsal, posterior, and anterior margins rounded. Each margin blends into adjoining one without demarcation except for a slight angle at juncture of anterior and dorsal margins. Anterior and posterior ends of shell little different in shape and ventral margin nearly straight. Shell well pigmented, especially in central portion of dorsal one-half of each valve. Eye oval, deeply pigmented, and located within anterior one-fourth of body. Copulatory complex with peniferum terminating with distal apex directed anteroventrally. Clasping apparatus with two teeth situated at mid-length of internal border of horizontal ramus; horizontal ramus terminating in two dorsally directed denticles.

FEMALE: Shell length 340-370 μ (average, 355 μ); height, 210-250 μ (average, 230 μ). Shell of gravid female differing little from shell of male, except that posterior portion often somewhat swollen, resulting in slightly concave ventral margin. Size of shell about ten per cent greater than that of male.

Type-locality. — Stream 2.3 mi S of Woodbine, Camden Co., Georgia, U.S.A. (types, USNM: holotype, allotype, and paratype, 81558). Host, *Procambarus paeninsulanus.*

Distribution (in addition to type-locality). —

United States

ALABAMA — ? CO. — "and one (collection) from Alabama" (host, *Orconectes spinosus*) (Hoff, 1944b: 340). COOSA CO. — 8.8 mi S of Sylacauga on rt 231 (host, *C. latimanus*). HOUSTON CO. —locality unknown (hosts, one or more from among *P. spiculifer, P. paeninsulanus, C. d. diogenes, F. fodiens,* and *C. schmitti*).

FLORIDA — ? CO. — ". . . ten collections from Florida . . . ," [hosts, *Procambarus leonensis, P. fallax, P. kilbyi, P. lucifugus alachua* (see Walton and Hobbs, 1959: 118), and *P. paeninsulanus* (Hoff,

1944b: 340)]. ALACHUA CO. — Creek 1.5 mi S of LaCrosse on Gainesville Rd (host, *P. paeninsulanus*). FLAGLER CO. — 10.8 mi E of San Mateo, 3.9 mi N of rt 28 (host, *P. paeninsulanus*). GADSDEN CO. — 1 mi E of Torreya Pk line, Flat Ck (host, *P. spiculifer*). JACKSON CO. — Dry Creek, 5.5 mi S of Marianna on rt 84 (hosts, *P. spiculifer, C. d. diogenes*); stream 5.6 mi N of Altha on rt 71 (hosts, *P. spiculifer, C. d. diogenes*); 2.4 mi N of Campbellton on rt 231 (hosts, *P. spiculifer, P. paeninsulanus, Cambarellus schmitti*); 7.2 mi S of Campbellton on rt 231 (host, *P. paeninsulanus*). LIBERTY CO. — 1.6 mi N of Hosford on rt 65 (host, *P. spiculifer*). NASSAU CO. — Little St. Marys River, ¼ mi NW of jct of Hilliard-Becker Rd with King's Ferry Rd (host, *P. paeninsulanus*). TAYLOR CO. — Scanlan (host, *P. paeninsulanus*).

Georgia — CAMDEN CO. — Stream 2.3 mi S of Woodbine (host, *Procambarus paeninsulanus*) (Hoff, 1944b: 337). DECATUR CO. — Butler Creek, 5 mi from home of C. G. Duke on IPC property (host, *P. spiculifer*); Spring Creek, 4.8 mi E of Iron City on rt 84 (hosts, *P. spiculifer, C. d. diogenes*); Mosquito Creek, 2 mi S of rt 97 on country rd (host, P. spiculifer). EARLY CO. — 4.4 mi E of Blakely on rt 62 (hosts, *P. paeninsulanus, C. d. diogenes*); 1.8 mi NW of Jakin on rt 84 (host, *P. spiculifer*). LAURENS CO. — 15.3 mi NE of Cochran on rt 26 (host, *P. pubescens, P. spiculifer*). SEMINOLE CO. — Dry Creek, 3.1 mi N of Iron City (host, *P. paeninsulanus, Fallicambarus uhleri*). TWIGGS CO. — 0.5 mi E of Twiggs Co. line on rt 96 (hosts, *F. clypeata, P. a. acutus*).

Tennessee — OBION CO. — Burrows near Biological Sta (host, *C. diogenes*).

Texas — JEFFERSON CO. — Roadside ditch about ½ mi W of Nome on US 90 at Liberty Co. line (hosts, *P.* sp., *P. hinei*).

Uncinocythere ericksoni (Kozloff, 1955)

Plate XXXVI, figures 4-5
Plate LVII

Entocythere ericksoni Kozloff, 1955: 159-161 (M, H, R, D), 156 (B), figs. 13-24. — Hobbs, 1955: 330 (R). — Ferguson, 1958: 198 (C). — Westervelt and Kozloff, 1959: 243, 244 (R). — Hart and Hobbs, 1961b: 182 (R). — Hart, 1962: 122 (C).

Entocythere erichsoni. — Hobbs and Walton, 1960: 18 (R) (erroneous spelling).

Uncinocythere ericksoni. — Hart, 1962: 137 (C, D, S). — Hobbs and Walton, 1963: 457 (R). — Hobbs and Walton, 1966: 10, 11 (R).

Uncinocythere erichsoni. — Hobbs, Holt, and Walton, 1967: 6, 7 (B, C) (erroneous spelling). — Hobbs and Walton, 1968: 251 (R) (erroneous spelling).

Diagnosis. — Male: Shell length 360-390 μ (average, 375 μ); height, 190-210 μ (average, 200 μ).

Copulatory complex with incision at distal end of peniferum, and edges on either side of incision drawn out into short spine-like extensions. Horizontal ramus of clasping apparatus longer than vertical ramus. About one-half of total length of clasping apparatus extending beyond end of peniferum; tip divided into five teeth. Three teeth of approximately equal size on inner border of distal half of clasping apparatus. Middle tooth slightly nearer to more proximal of three teeth than to more distal one. Dorsal finger of copulatory complex little more than twice as long as ventral finger, but spines at tips of two fingers are of approximately equal length.

Female: Shell length 430-470 μ (average, 450 μ); height, 230-280 μ (average, 255 μ).

In copulating females, the ventral margin of the valves of the shell is nearly straight, with only a suggestion of the concavity which is well-marked in gravid females.

Type-locality. — Spring near OO Ranch corral, Malheur National Wildlife Refuge, Harney Co., Oregon, U.S.A. (types, USNM: holotype, 95686; allotype, 95687; paratypes, 95688 and 95689). Host, *Pacifastacus gambelii*.

Distribution (in addition to type-locality). —

United States

California — SHASTA CO. — Fall River Mills, where Fall River joins Pit River (host, *P. fortis*).

Idaho — CASSIA CO. — Marsh Ck, 1 mi NE of Albion (host, *P. gambelii;* Marsh Ck below Chapburn's Ranch, Albion (host, *P. gambelii*).

Oregon — BENTON CO. — Oliver Ck on Fred Cleveland farm near Dawson (host, *P. l. trowbridgii*). HARNEY CO. — Spring near OO Ranch corral, Malheur National Wildlife Refuge (host, *Pacifastacus gambelii*); Malheur River near Drewsey (host, *P. gambelii*); Warm Springs (host, *P. l. klamathensis*); Silver Creek at Double O Ranch near Harney Lake (host, *P. gambelii*). POLK CO. — Willamina, S of Yamhill R, 1.6 mi W of valley jct (host, *P. l. klamathensis*).

WASHINGTON — GRAY'S HARBOR CO. — Hogiam River at jct of rt 215 & 101, 14.5 mi S of Humptulips (host, *P. l. klamathensis*); Boulder Creek at jct of US 101, 18.5 mi SE of Queets (host, *P. l. klamathensis*). JEFFERSON CO. — Mill Creek 10.9 mi S of forks on rt 101 (host, *P. l. klamathensis*). WAHKIAKUM CO. — Peter-McKinnon Log Dump, tidal area of Lower Columbia River (host, *P. l. leniusculus*).

Uncinocythere holti Hart, 1965

Plate XXXVI, figures 6-8
Plate LVII

Uncinocythere holti Hart, 1965a: 193-195 (M, D, R), figs. 10-13. — Ferguson, 1968: 502 (C, D). — Hobbs and Walton, 1968: 251 (R).

Diagnosis. — MALE: Shell length 400-430 μ (average, 410 μ); height, 190-230 μ (average, 200 μ). Shell subelliptical in outline, posteroventral margin evenly rounded; slight concavity apparent in anterodorsal margin anterior to eye spot; greatest shell height ⅓ shell length from posterior end. Few marginal setae present on anterior and posterior ends. Eye spot approximately ⅕ shell length from anterior margin.

Copulatory complex with peniferum extending only slightly beyond external border of clasping apparatus; distal extremity bifurcate with apices apposed at right angles to one another. Penis small, when compared with those of *U. cassiensis* and *U. thektura*. Clasping apparatus with distinct vertical and horizontal rami; horizontal ramus longer than vertical; internal border of horizontal ramus provided with three evenly spaced teeth; distal extremity with five or six denticles; external border evenly rounded or slightly undulate.

FEMALE: Shell length 470-490 μ (average, 480 μ); height 270-300 μ (average, 290 μ). Shell of biunguis female similar to that of male, and roughly same size.

Shell of triunguis female similar to that of male, but larger.

Type-locality. — Cold stream in Wasatch Mtns., 2 mi E of Brigham City, Box Elder Co., Utah, U.S.A. (types, USNM: holotype and allotype, 110781; paratype, 110782). Host, *Pacifastacus gambelii.*

Distribution (in addition to type-locality). —

United States

IDAHO — BEAR LAKE CO. — Mt spring at Paris (host, *P. gambelii*). KOOTENAI CO. — Mouth of St Joe River, Le Coeur d'Alene (host, *P. gambelii*). TWIN FALLS CO. — Shoshone Falls (host, *P. gambelii*).

ILLINOIS — IROQUOIS CO. — 6.5 mi S of jct of US 24 & I11 highway running N from Stockland (host, *O. virilis*).

LOUISIANA — BEAUREGARD CO. — Bearhead Creek (hosts, *O. lancifer, O. hathawayi, P. dupratzi*).

MINNESOTA — STEARNS CO. — (host, *O. virilis*).

SOUTH DAKOTA — BEADLE CO. — 6.6 mi W of jct of US 14 & 281 on 14 (host, *O. immunis*).

UTAH — BOX ELDER CO. — Two miles E of Brigham City on rts 89 & 91 stream in Wasatch Mtns (host, *P. gambelii*). SAN JUAN CO. — Monticello (host, *P. gambelii*).

Uncinocythere lucifuga (Walton and Hobbs, 1959)

Plate XXXVI, figures 9-12
Plate LVII

Entocythere lucifuga Walton and Hobbs, 1959: 118-120 (M, H, R, D), figs. 7-13. — Hart and Hobbs, 1961b: 173 (B, C), 174 (K), 181 (R). — Hart, 1962: 123 (C).
Uncinocythere lucifuga. — Hart, 1962: 138 (C, D, S). — Hobbs and Walton, 1963a: 457 (R). — Ferguson, 1968: 502 (C, D).

Diagnosis. — MALE: Shell length 350-390 μ (average, 367 μ); height, 195-210 μ (average, 198 μ). Shell subovate with slight emargination, sometimes absent, near posteroventral extremity.

Peniferum terminating in single point directed anteroventrally; penis situated in distal ¼ of peniferum. Clasping apparatus with two teeth on internal border of horizontal ramus; three terminal denticles.

FEMALE: Shell length 410-425 μ (average, 417 μ); height, 220-250 μ (average, 238 μ).

Shell of triunguis female conspicuously higher posteriorly than that of male, and ventral margin with distinct concavity slightly anterior to midlength.

Shell of biunguis female similar to that of triunguis form, but lacking concavity on ventral margin; anterior portion proportionately higher.

Type-locality. — Hog Sink, 13 mi W of Gainesville, Alachua Co., Florida, U. S. A. (types, USNM:

holotype, 105957; allotype, 105958; morphotype, 105959; paratypes, 108191, 108192, 108193, and 108194). Host, *Procambarus lucifugus alachua.*

Distribution (*in addition to type-locality*). —

United States

ALABAMA — HOUSTON CO. — 6 mi S of Ashford on Cowarts Creek, (host, *P. spiculifer*).

FLORIDA — ALACHUA CO. — Hog Sink, Gainesville (host, *P. lucifugus alachua*); Hog Sink, about 13 mi W of Gainesville (host, *Procambarus lucifugus alachua*); (Walton and Hobbs 1959: 119); Squirrel Chimney, 12.5 mi NW of Gainesville (host, *Procambarus pallidus*); (Walton and Hobbs 1959: 119); Well several mi W of Gainesville (host unknown). CALHOUN CO. — 1.7 mi S of Chipola on rt 6 (host, *P. spiculifer*). DIXIE CO. — US 19, 8 mi SE of Taylor Co. line 3 mi SE of rt 358 (hosts, *P. paeninsulanus,* shrimp). GADSDEN CO. — 9.9 mi NW of Greenesboro on rt 269 (host, *P. spiculifer*); Ck 15.1 mi E of Chattahoochee (hosts, *P. fallax, P. spiculifer*). HOLMES CO. — 13.5 mi N of Ponce de Leon on rt 81 (host, *P. versutus, C. d. diogenes*). JACKSON CO. — Blue Springs Run E of Marianna (host, *P. spiculifer*); Chipola River at Merritts Bridge S of Marianna (host, *P. spiculifer*). LEVY CO. — Ck under US 19 & 98, 6 mi SE of Gulf Hammock on rt 19 (host, *P. paeninsulanus*); 13 mi NW of Inglis on US 98 [host, *P. paeninsulanus* (Faxon)]. PUTNAM CO. — 8 mi E of Hawthorne (host, *P. fallax*).

GEORGIA — BIBB CO. — Wesleyan Campus (hosts, *P. spiculifer, P. howellae*). BLECKLEY CO. — 7.9 mi S of Twiggs-Bleckley Co. line on rt 129 (hosts, *P. howellae, Faxonella clypeata*). BURKE CO. — Briar Creek at Heath's Bridge between Sardis and Girard (host, *C. troglodytes*); 5.1 mi W of Waynesboro on rt 24 (host, *C. latimanus*). CHEROKEE CO. — small stream 3 mi N of Canton, pool below bridge (hosts, *C. latimanus, P. spiculifer*). COLQUITT CO. — Ochlockonee R near Riverside (hosts, *P. paeninsulanus, P. spiculifer*). DODGE CO. — 4.6 mi W of Telfair-Dodge Co. line & 3.6 mi N of US 23 on GA 165 (host, unknown). McINTOSH CO. — 5.4 mi S of Eulonia (host, *P. troglodytes*). PIKE CO. — N limit of Zebulon, 0.4 mi N of center of town on rt 19 (host, *P. spiculifer*). RANDOLPH CO. — 11.8 mi S of Cuthbert on US 27 (host, *P. spiculifer*). STEWART CO. — 8.8 mi N of Lumpkin on US 27 (host, *P. spiculifer*). TWIGGS CO. — 2.9 mi S of jct of rts 129 & 96 on 129 (hosts, *C.* sp., *P. spiculifer, P. howellae*).

Uncinocythere neglecta (Westervelt and Kozloff, 1959)

Plate XXXVI, figures 13-15
Plate LVII

Entocythere neglecta Westervelt and Kozloff, 1959: 240-244 (M, H, R, A, D), figs. 1-14. — Hobbs and Walton, 1960: 18 (R). — Hart and Hobbs, 1961b: 182 (R). — Levinson, 1962: 94 (C). — Hart, 1962: 123 (C).

Uncinocythere neglecta. — Hart, 1962: 138 (C, D, S). — Hobbs and Walton, 1963a: 457 (R). — Hart, 1965a: 190 (R). — Hobbs and Walton, 1966: 10, 11 (R). — Hobbs, Holt, and Walton, 1967: 7 (B, C). — Ferguson, 1968: 502 (C, D).

Diagnosis. — MALE: Shell length 390-420 μ (average, 405 μ); height, 220-250 μ (average, 235 μ). Shell smoothly rounded on dorsal and anterior margins. Posterior margin showing only slight curvature. Ventral margin nearly straight, exhibiting shallow concavity in anterior half. Eye heavily pigmented and situated about one-fifth distance from anterior margin.

Copulatory complex with distal portion of peniferum tapering and extending to level of clasping apparatus horizontal ramus; distal extremity bifurcate, with anterior process spiculiform, posterior process rounded. Clasping apparatus curved throughout; external borders entire; internal border of horizontal ramus with three teeth; terminating in six denticles.

TRIUNGUIS FEMALE: Shell length 360-490 μ (average, 425 μ); height, 200-310 μ (average, 255 μ). Posterior margin of shell of gravid female more conspicuously rounded than in male, and concavity in ventral margin more pronounced, but in other respects shells similar.

In the copulating female, posterior and ventral margins of shells come together to form a rather sharp point.

Type-locality. — San Francisco, San Francisco Co., California, U.S.A. (types, MCZ: holotype, allotype, paratypes, 126676. USNM: paratypes, 102106). Host, *Pacifastacus nigrescens* (Stimpson).

Distribution (*in addition to type-locality*). —

United States

CALIFORNIA — SHASTA CO. — Fall River, trib to Pit River (host, *Pacifastacus nigrescens*).

WASHINGTON — CLALLAM CO. — Indian Ck at jct of rt 101, SW of Port Angels (host, *Pacifastacus l. klamathensis*). WALLA WALLA CO. — Mill Creek at Walla Walla (host, *P. l. klamathensis*).

WYOMING — UINTA CO. — Bear River, 10 mi N of Evanston on rt 89 (host, *Pacifastacus gambelli*).

Uncinocythere occidentalis (Kozloff and Whitman, 1954)

Plate XXXVI, figures 16-19
Plate LV

Entocythere occidentalis Kozloff and Whitman, 1954: 160-162 (M, H, R, D), 159 (A, B), figs. 1-13. — Hobbs, 1955: 330 (R). — Kozloff, 1955: 161 (R). — Ferguson, 1958: 198 (C). — Westervelt and Kozloff, 1959: 239 (A, C, H). — Hobbs and Walton, 1960: 18 (R). — Hart and Hobbs, 1961b: 182 (R). — Hart, 1962: 122 (C).

Uncinocythere occidentalis. — Hart, 1962: 138 (C, D, S). — Hart, 1965a: 195, 196 (D).

Diagnosis. — MALE: Shell length 380-400 μ (average, 390 μ); height, 210-220 μ (average, 215 μ).

Copulatory complex with peniferum tapering distally and terminating in two closely apposed processes. Clasping apparatus evenly curved throughout, with single tooth at midlength of internal border of horizontal ramus; four terminal denticles.

TRIUNGUIS FEMALE: Shell length 410-470 μ (average, 440 μ); height, 250-280 μ (average, 265 μ).

Type-locality. — Crystal Springs Creek, Multonomah Co., Oregon, U.S.A. (types, USNM: holotype, 93503; allotype, 93504; paratypes, 93505, 93506, and 93507). Host, *Pacifastacus leniusculus trowbridgii.*

Distribution (in addition to type-locality). —

United States

CALIFORNIA — CALAVERAS CO. — Ck into Melones Res on rd to Ghighlieri Ranch, E of Melones (host, *Pacifastacus* sp.). CONTRA COSTA CO. — ½ mi downstream from Twitchel Island, San Joaquin R (host, *P. l. leniusculus*). HOMBOLDT CO. — Briceland (host, *P. l. klamathensis*); S fork of Erl R, Sarberville (host, *P. l. klamathensis*); Trinity R at Hoopa (host, *P. l. klamathensis*). LAKE CO. — Clear Lake near Pirate's Cove (host, *P. l. leniusculus*). MARIN CO. — Muir Woods Nat. Forest (host, *P. l. leniusculus*); Papermill Ck at Samuel B. Taylor State Park (host, *P. leniusculus*). MODOC CO. — Lost R (host, *P.* sp.). SACRAMENTO CO. — American R near Sacramento (host, *P.* sp.). SAN JOAQUIN CO. — North fork of Mokelumne R (host, *P. l. leniusculus*). SANTA CRUZ CO. — Waddell Ck (host, *P. l. klamathensis*). SISKIYOU CO. — Klamath R (host, *P. l. klamathensis*); Buttonwood Ck near Hornbrook (host, *P. l. klamathensis*); Wagon Ck, trib to Sacramento R, about 1 mi N of Mt. Shasta State Fish Hatchery (host, *P. l. leniusculus*). TRINITY CO. — Trinity R, Lewiston (host, unknown). TUOLUMNE CO. — Stanislaus R at jct of rt 49 (host, *P. leniusculus*).

IDAHO — KOOTENAI CO. — Coeur d'Alene Lake, Coeur d'Alene City (host, *Pacifastacus l. klamathensis*).

NEVADA — WASHOE CO. — Tucker R at Reno (host, *Pacifastacus l. trowbridgii*).

OREGON — (County unknown) — Sikan Ck, Henshaw (host, *Pacifastacus l. klamathensis*). BENTON CO. — St. Mary's R at W edge of Plilomath on US 20 (host, *P. l. klamathensis*). CLACKAMAS CO. — "... several localities ...," (hosts, *P. l. leniusculus, P. trowbridgii,* and *P. l. klamathensis*). CLATSOP CO. — Astoria (host, *P. l. klamathensis*). CROOK CO. — Des Schutes R (host, *P. klamathensis*). DOUGLAS CO. — S fork of Umpqua R, 1.2 mi W of Tilleron on rt 42; 33.5 mi SE of Myrlle Ck (host, *P. l. klamathensis*). JACKSON CO. — Bear R, Medford (host, *P. l. klamathensis*); trib about 5 mi N of Rogue Elk on OR 42, 10.8 mi S of Douglas Co. line (host, *P. l. klamathensis*). KLAMATH CO. — Sprague R & Williamson's R, near Fort Klamath (host, *P. l. klamathensis*). LANE CO. — Jasper, Hill Ck, 100 yds above middle fork of Williamette R (host, *P. l. klamathensis*); Coast Fork of Willamotte R at jct of US 99, 15.2 mi S of Eugene (host, *P. l. klamathensis*). LINCOLN CO. — Ck next to Burnt Woods PO on US 20 (host, *P. l. klamathensis*); Trib to Yaquina R 14.4 mi. E of Toledo on US 20 (host, *P. l. klamathensis*); Trib to Yaquina R 8.6 mi NE of Toledo on US 20 (host, *P. l. klamathensis*). MALHEUR CO. — Jordon Ck, trib of Owyhee R (host, *P. gambelli*); Owyhee R at bridge on rt 201, 2.5 mi S of Adrian (host, *P. l. lenusculus*). MARION CO. — Little Pudding Ck, 7.9 mi SW of Silverton at jct of rt 213 (host, *P. l. klamathensis*); Battle Ck at Marion-Clackamas Co. line about 6 mi NW of Silverton on rt 213 (host, *P. l. klamathensis*). MULTNOMAH CO. — Crystal Springs Creek (host, *P. l. trow-*

bridgii); Johnson Ck, Portland (host, *P. l. klamathensis*). POLK CO. — Willamina, S of Yamhill R, 1.6 mi W of Valley jct (host, *P. l. klamathensis*); Salt Ck, 3.5 mi NW of Dallas on rt 22 (host, *P. l. klamathensis*). SHERMAN CO. — Barnum Pond at Grass Valley (host, *P. l. klamathensis*). WASHINGTON CO. — Glenwood Springs (host, *P. l. klamathensis*).

WASHINGTON — CLALLAM CO. — Indian Ck at jct of US 101 SW of Port Angels (host *Pacifastacus l. klamathensis*). GRAY'S HARBOR — Hogiam R at jct of rt 215 & 101, 14.5 mi S of Humptulips (host, *P. l. klamathensis*); 5 mi S of Rocky Mt, 3 mi S of Amanda Park on US 101 (host, *P. l. klamathensis*); Boulder Ck at jct of US 101, 18.5 mi SE of Queets (host, *P. l. klamathensis*). JEFFERSON CO. — Nolon Ck, 21.5 mi S of forks on US 101 (host, *P. l. klamathensis*); Mill Ck, 10.9 mi S of forks on US 101 (host, *P. l. klamathensis*). KITTITAS CO. — 7.2 mi E of Ellensburg on US 101 (host, *P. l. klamathensis*). PACIFIC CO. — Nasel R, Nasel (host, *P. l. klamathensis*); Willapa R, Holcomb (host, *P. l. klamathensis*). PIERCE CO. — Purdy Creek at Homewood, Purdy, (host, *P. l. klamathensis*); Minter Ck, 3.8 mi W of Wauna at jct of 14A (host, *P. l. klamathensis*); Purdy Pond at Homewood (host, *P. l. klamathensis*). SPOKANE CO. — Granite Lake, 17 mi N of Spokane (host, *P. l. klamathensis*); W branch of Little Spokane R at jct of US 2, 3.5 mi S of Spokane (host, *P. l. klamathensis*). THURSTON CO. — Littlerock (host *P. l. klamathensis*). WAHKIAKUM CO. — Peter-McKinnon log dump, tidial area of lower Columbia R (host, *P. l. lenuisculus*). WALLA WALLA CO. — Mill Ck at Walla Walla (host, *P. l. klamathensis*). WHITMAN CO. — Union Flat Ck about 8 mi W of Pullman (host, *P. l. klamathensis*).

WYOMING — TETON CO. — Polecat Ck (host, *Pacifastacus gambelli*). (County unknown) — North River (host, *P. l. klamathensis*).

Uncinocythere pholetera (Hart and Hobbs, 1961)

Plate XXXVII, figures 1-3
Plate LV

Entocythere pholetera Hart and Hobbs, 1961b: 174 (K), 181, 182 (M, H, R, D), figs. 25-28.
Uncinocythere pholetera. — Hart, 1962: 138 (C, D, S). — Hobbs and Bedinger, 1964: 14 (D, H). — Bedinger and Hobbs, 1965: 94 (D, H) (by implication). — Hobbs and Walton, 1966: 11 (R). — Ferguson, 1968: 502 (C, D). — Hobbs and Walton, 1968: 251 (R).

Diagnosis. — MALE: Shell length 350-370 μ (average, 360 μ); height, 220-240 μ (average, 230 μ). Eyeless. Shell ovate with greatest height near mid-length. Setae evenly spaced along anterior, ventral, and posterior margins.

Copulatory complex with ventral extremity of peniferum lying at level slightly dorsal to external border of clasping apparatus. Peniferum long, slender, and bifid apically with two rami subequal in length; posterior ramus with acute apex directed toward anterior ramus. Clasping apparatus L-shaped with vertical ramus distinctly longer than horizontal. Internal border of horizontal ramus with three teeth and distal extremity with three somewhat more prominent ones. External border of clasping apparatus entire.

TRIUNGUIS FEMALE: Shell length 280-380 μ (average, 335 μ); height, 170-260 μ (average, 216 μ). Triunguis female shell similar to that of male, but more variable in size. Small anteroventrally projecting papilla present near posterodorsal extremity.

Type-locality. — Lewis Cave, 15 mi NW of Doniphan, Ripley Co., Missouri, U.S.A. (types, ANSP). Host, *Cambarus hubrichti.*

Distribution. — Known only from type-locality.

Uncinocythere simondsi (Hobbs and Walton, 1960)

Plate XXXVII, figures 4-7
Plate LV

Entocythere simondsi Hobbs and Walton, 1960: 17, 18 (M, H, R, D), 19 (A), figs. 1-10. — Hart, 1962: 123 (C).
Uncinocythere simondsi. — Hart, 1962: 138 (C, D, S). — Hart and Hart, 1966a: 9 (A). — Hobbs and Walton, 1966: 11 (R). — Ferguson, 1968: 502 (C, D). — Hobbs and Walton, 1968: 251 (R). — Hart and Hart, 1969a: 183 (D). — Hobbs III, 1970: 181 (A). — Hobbs and Barr, 1972: 81 (D, H).

Diagnosis. — MALE: Shell length 325-350 μ (average, 342 μ); height, 200-205 μ (average, 200 μ). Shell elliptical with posterior portion slightly higher than anterior; margins entire, lacking spines or distinct emarginations. Marginal setae regularly spaced on both valves anteriorly, ventrally, and posteriorly and lacking dorsally.

Copulatory complex with peniferum extending distally slightly beyond external border of clasping apparatus, with undulating anterior margin, and distally bifid, apparently with ridges on antero- and posterodistal margins. Dorsal finger of usual type and directed ventrally; ventral finger strongly curved near base (extension of axes of rami forming angle of about 100 degrees) and second curve (about 140 degrees) near distal end. Clasping apparatus although not angular, with distinct horizontal and vertical rami of subequal lengths; external border of horizontal ramus entire; internal border with two teeth on distal half; and distal extremity with four denticles.

Type-locality. — Dunn Creek, 1.9 mi W of Fighting Town Creek on Hell's Hollow Road, Fannin Co., Georgia, U.S.A. (types, USNM: holotype and allotype, 105956; paratype, 108182). Hosts, *Cambarus b. bartonii* and *Cambarus* sp.

Distribution (in addition to type-locality). —

United States

ALABAMA — BLOUNT CO. — 3.7 mi SW of Hayden on rt 7 [host, *Cambarus* (*Depressicambarus*) sp.]; 5.1 mi E of Hayden on rt 7 [host, *C.* (*D.*) sp.]. CALHOUN CO. — Ft. McClellun [host, *C.* (*Puncticambarus*) sp.]. DeKALB CO. — Stream in Collinsville on rt 11, trib of Wills Ck (hosts, *C. longirostris, C. bartonii*). ETOWAH CO. — 10.4 mi E of Etowah Co. line on rt 278 [Hosts, *C.* (*Puncticambarus*) sp., *C.* (*Depressicambarus*) sp.]. FAYETTE CO. — 5 mi W of Fayette on rt 96 [hosts, *C.* (*Depressicambarus*) sp., *O.* sp.]. LOWNDES CO. — 32.4 mi S of Montgomery of rt 31 (hosts, *P. acutissimus, P. lewisi, C. latimanus, C. diogenes* subsp.). ST. CLAIR CO. — 4.1 mi E of Co. line on rt 231 [hosts, *C. diogenes* subsp., *C.* (*Puncticambarus*) sp., *C. jordani*]. TALLADEGA CO. — 12.3 mi N of Sylacauga on rt 231 [hosts, *P. a. acutus, C.* (*D.*) sp.]. TUSCALOOSA CO. — Ck at E city limits of Cottondale on rt 11, trib of Hurricane Ck (hosts, *C. jordani, C. latimanus, P. versutus*). WALKER CO. — 15.7 mi S of Double Springs on rt 195 [hosts, *C.* (*D.*) sp., *O.* sp.].

FLORIDA — LIBERTY CO. — near Blue Springs, Torreya State Park (host, *C. d. diogenes*).

GEORGIA — BALDWIN CO. — 7 mi. N of Milledgeville on US 441 (host, *P. spiculifer*). CARROLL CO. — 2.6 mi N of Villa Rica RR on rt 101 [hosts, *C. latimanus, C.* (*D.*) sp.]; 1.8 mi E of rt 101 on unpaved rd (hosts, *P. spiculifer, C. latimanus*); 9.0 mi S of Bremen on rt 27A (hosts, *C. latimanus, P. spiculifer*). FANNIN CO. — Dunn Creek, 1.9 mi W of Fighting Town Creek, Hell's Hollow Road (hosts, *C. bartonii, C.* sp.). FLOYD CO. — 9.7 mi N of Rome [hosts, *C. latimanus, C.* (*D.*) sp., *C. diogenes* subsp.]; 10.7 mi N of Rome on rt 27 [host, *C.* (*D.*) sp.]; 7.5 mi E of Alabama line on rt 20 [*C.* (*Puncticambarus*) sp., *C. latimanus*]. HARALSON CO. — rt 27 at Tallapoosa R (hosts, *C. halli, P. spiculifer*); 1.3 mi S of Tallapoosa R, ck parallel to rt 27 (host, *C. halli*). HART CO. — 7.5 mi S of Hartwell on rt 172 [hosts, *C.* (*Depressicambarus*) sp., *C. bartonii*]. HEARD CO. — 5.3 mi W of Newman on rt 34 (hosts, *C. latimanus, P. spiculifer*). MADISON CO. — 5.4 mi SE of Jackson Co. line on GA 98 [hosts, *C.* (*Depressicambarus*) sp., *P. raneyi*]. MONROE CO. — 3.4 mi N of Forsythe on rt 42 (hosts, *P. spiculifer, C. latimanus*). MURRAY CO. — 1 mi E of Chatsworth on rt 76 [hosts, *C.* (*Depressicambarus*) sp., *C.* (*Puncticambarus*) sp., *O. spinosus*]. NEWTON CO. — E of Starrsville on GA 213 (hosts, *P. spiculifer, C. latimanus*). PAULDING CO. — 12.8 mi N of Villa Rica RR on rt 101 (host, *C. latimanus*); rt 101 at Tallapoosa R (hosts, *C. latimanus, C. halli*). POLK CO. — Euharlee Ck [hosts, *C.* (*Puncticambarus*) sp., *C. latimanus, O. spinosus*]. STEWART CO. — 8.8 mi N of Lumpkin on US 27 (host, *P. spiculifer*). TROUP CO. — 3.2 mi S of Chattahoochee R on rt 219 (host, *C. latimanus*). WALKER CO. — Chickamauga Natl. Military Pk near Rossville [hosts, *P.* (*Ortmannicus*) sp., *O. erichsonianus, C.* (*Depressicambarus*) sp.].

ILLINOIS — EDGAR CO. — Twin Lakes, 1 mi N of Paris (host, *O. virilis*); Paris, Twin Lakes (host, *O. virilis*). MACOUPIN CO. — Staunton "Cahokiack" (host, *O. virilis*).

KENTUCKY — JESSAMINE CO. — 8 mi SE of Versailles on KY 169 (host, *O. rusticus*). MORGAN CO. — 4 mi E of Grassy Ck on US 460 (hosts, *Cambarus* sp., *O. juvenilis*). PULASKI CO. — Hydens Cave, 1.3 mi NE of Blue John, about 6 mi from Sloans Valley (host, *O. australis packardi*).

MISSISSIPPI — CHOCTAW CO. — 6 mi W of Oktibbeha-Choctaw Co. line on rt 12 [hosts, *P. a. acutus, P. viaeviridis, H. cristatus, C.* (*D.*) sp.]. OKTIBBEHA CO. — 2.7 mi W of Sturgis on rt 12 [hosts, *P. acutissimus, O.* sp., *C.* (*D.*) sp.].

TENNESSEE — DAVIDSON CO. — Richland Ck at Bellemeade W of Nashville (hosts, *O. rusticus* subsp., *O. placidus*). FENTRESS CO. — Poplar Cove Ck [hosts, *C.* (*Puncticambarus*) sp., *C. striatus*]. MEIGS CO. — 1.9 mi N of Georgetown (hosts, *C. longirostris, Orconectes* sp.]; 8.4 mi N of Decatur on TN 58 (hosts, *O.* sp., *O. forceps*). POLK CO. — 11.4 mi S of Benton [hosts, *C.* (*D.*) sp., *C.* (*Puncticambarus*) sp.]; 1.8 mi S of Benton (host, *Cambarus longirostris*); 0.2 mi W of Ocoee, (hosts, *Cambarus longirostris, Cambarus* (*D.*) sp., *Orconectes erichsonianus*).

Uncinocythere spathe Hart and Hart, 1971

Plate XXXVII, figures 8-10
Plate LV

Uncinocythere spathe Hart and Hart, 1971: 117-118 (M, H, R, A, D), figs. 11a-c.

Diagnosis. — MALE: Shell length 400-410 μ (average, 415 μ); height, 210-220 μ (average, 215 μ). Shell subovoid, with greatest height slightly posterior to midlength. Eye pigmented, situated ⅕ shell length from anterior end.

Peniferum distinctly bipartite; posterodistal portion slender, curving to sharp, almost spiculiform, point; anterodistal portion broad and spatulate. Penis situated ⅓ peniferum length from distal end.

Horizontal ramus of clasping apparatus approximately ⅔ length of vertical ramus, meeting at angle of approximately 80°. External and internal borders of vertical rami nearly straight. External border of horizontal ramus slightly convex; internal border with three teeth evenly spaced from midlength of ramus to proximal member of three terminal denticles.

FEMALE. — Unknown.

Relationships. — This species is unique in the deep division between the proximal and distal portions of the peniferum and the spathe-form proximal portion. *Uncinocythere simondsi* Hobbs and Walton has a division, but is not so deep as that of *U. spathe.*

Type-locality. — 7.5 mi SW of Athens on rt 11, McMinn Co., Tennessee, U.S.A. (types, USNM: holotype, 138467). Host, *Orconectes erichsonianus.*

Distribution. — Known only from type-locality.

Uncinocythere stubbsi Hobbs and Walton, 1966

Plate XXXVII, figures 11-13
Plate LVI

Uncinocythere stubbsi Hobbs and Walton, 1966: 9-11 (M, H, R, A, D), figs. 2a-d. — Hobbs and Walton, 1968: 239, 242 (A). — Hart and Hart, 1969a: 183 (D).

Diagnosis. — MALE: Shell length 270-293 μ (average, 281 μ); height, 158-165 μ (average, 161 μ). Eye present. Shell highest near midlength and with dorsal margin evenly contoured anteriorly and posteriorly; ventral margin convex without emargination; submarginal setae evenly but widely spaced anteriorly, ventrally, and posteriorly, absent dorsally.

Copulatory complex with posterior margin of peniferum entire, its ventral margin excised with anteroventral extremity directed ventrally and posteroventral extremity directed anteroventrally; anteroventral portion with a heavy, acute, dorsally directed penis guide. Penis of moderate size, length approximately one-half anterior-posterior dimension of distal portion of peniferum and situated far distal to base of clasping apparatus and within ventral one-third of area between dorsal margin of spermatic loop and ventral extremity of peniferum. Dorsal and ventral fingers slender, latter with strong caudal bend at distal end of proximal third. Clasping apparatus with vertical and horizontal rami of subequal lengths, and extensions of their rounded junction forming angle of less than 90°. Vertical ramus entire but with distal one-third curved posteriorly. External border of horizontal ramus gently rounded and entire; internal border with three teeth — largest immediately proximal to midlength and two smaller ones near distal extremity; terminal extremity with three teeth.

TRIUNGUIS FEMALE: Shell length 315-345 μ (average, 324 μ); height, 180-195 μ (average, 189 μ). Shell distinctly larger than that of male with its greatest height slightly posterior to midlength and distinctly lower anteriorly than posteriorly; ventral margin with emargination near midlength; submarginal setae as in male.

Relationships. — *Uncinocythere stubbsi* has its closest affinities with those species of the genus having a distinct bifid ventral (or posteroventral) margin of the peniferum: *U. ericksoni, U. caudata, U. pholetera, U. neglecta,* and *U. simondsi.* It differs from *U. caudata* in lacking a posteroventral acute projection of the shell; from *U. ericksoni* and *U. neglecta* in possessing only three teeth at the extremity of the ventral ramus of the clasping apparatus; from *U. pholetera* in having the vertical and horizontal rami of the clasping apparatus subequal in length and with the distal two teeth of the horizontal ramus closer together, the distal one more distally situated than in *pholetera;* and from *U. simondsi* in having the penis situated in the distal third of the area between dorsal margin of the spermatic loop and the ventral margin of the peniferum. It differs from all of its relatives in the arrangement of the penis guide.

Type-locality. — Tributary of Big Turnbull Creek on property of Bethany Hill Church Camp in S portion of Cheatham Co., Tennessee, U.S.A. (types, USNM: holotype and allotype, 111260; paratype, 111261). Host, *Orconectes rusticus* subsp.

Distribution (in addition to type-locality). —

United States

ALABAMA — COLBERT CO. — 12.6 mi S of Tuscumbia on US 43, Bear Ck (host, *Cambarus girardianus*). LAUDERDALE CO. — Second Ck TVA #5106 (host, *Orconectes compressus*); 18.3 mi W of Athens on US 72 (host, *O. spinosus*); 10.7 mi E of Florence on US 72 (host, *O. spinosus*); Cypress Ck, 3 mi N of Florence (hosts, *O. compressus, O. spinosus*). LIMESTONE CO. — 3¼ mi W of Athens on US 72 (host, *O. spinosus*). MADISON CO. — 5.9 mi W of Huntsville on US 72 (hosts, *O. spinosus, O. forceps*).

GEORGIA — TROUP CO. — 6.5 mi S of Chattahoochee R on rt 219 (hosts, *Procambarus spiculifer, C. latimanus*).

ILLINOIS — CHAMPAIGN CO. — Rocky Ck near Muncie (host, *O. propinquus*). CLARK CO. — 1.5 mi NE of Moriah (host, *O. immunis*). HARDIN CO. — Peters Ck, 2¼ mi E of Elizabethtown (host, *O. kentuckiensis*); Branch to Big Ck 3.5 mi NE of Elizabethtown (host, *O. r. rusticus*); Sugar Ck, 2 mi NE of Creal Springs (host, *O. indianensis*). McLEAN CO. — 2 mi NW of Bellflower (host, *C. diogenes* subsp.).

INDIANA — CLAY CO. — Birch Ck at jct of rt 59, 0.4 mi N of Ashboro (host, *O. virilis*). DeKALB CO. — 6.6 mi W of Butler on US 6 (host, *O. propinquus*). DUBOIS CO. — Birdseye, 4.0 mi N of Perry Co. line on rt 145 (hosts, *O. indianensis, O. immunis*). HENDRICKS CO. — Big White Lick Ck, 2 mi N of Brownsburg at PRR bridge (host, *O. r. rusticus*). MONROE CO. — Richland Creek (host, *C. laevis*). PARKE CO. — 5.6 mi W of Bellmore (jct US 36 & 39) on 36 (host, *O. immunis*). SCOTT CO. — Small stream on US 1, 3.6 mi S of Scottsburg (host, *O. sloani*). WHITE CO. — 2.6 mi N of jct of US 421 & 24 at Reynolds (host, *O. virilis*).

KANSAS — LABETTE CO. — 10.4 mi E of Coffeeville on US 166 (host, *O. nais*).

KENTUCKY — ADAIR CO. — 9.9 mi W of Columbia on rt 80, Hot water at temp. of warm bath water (host, *O. putnami*). ALLEN CO. — Trammel Fork at rt 240, Allen Springs (host, *Barbicambarus cornutus*). BARREN CO. — Skeggs Ck, 8.6 mi SW of Glascow on US 31E (host, *B. cornutus*). BATH CO. — 4.2 mi E of Owingsville on US 60 [hosts, *O. rusticus* a, *O. juvenilis* b, *C. (Depressicambarus)* sp.]. BOYLE CO. — Near Danville, North Rolling Fork (host, *O. juvenilis*). BREATHITT CO. — Boone Fork at Vox Cleane, 12.5 mi W of Jackson on rt 15 (host, *O. juvenilis*); Buckhorn Ck, Camp Robertson near Noble 5.2 mi NE of rt 15 on dirt rd [hosts, *C. (Jugicambarus)* sp., *O. juvenilis*]. BULLITT CO. — Trib to Floyds Fork Ck at Waterford Rd near Jefferson Co. line (host, *O. rusticus*). CARROLL CO. — Ellis Ck, 4.4 mi SW of Ky R bridge at Carrollton on Rt 42 (hosts, *C. ornatus* a, *O. rusticus*). CASEY CO. — South Fork Ck, 2 mi ENE of Dunville on US 127 (host, *B. cornutus*). CLARK CO. — (host, *O. rusticus*); 4 mi S of Winchester on US 227 (host, *O. rusticus*). CUMBERLAND CO. — 5 mi W of Marrow on St 90 (host, *O. placidus*). EDMONSON CO. — Dog Ck, 0.3 mi SW of Hart Co. line (host, *O. putnami*). FAYETTE CO. — Lexington stream (host, *O. rusticus*). FRANKLIN CO. — Ck at jct of US 60 & Ky 151 (hosts, *O. rusticus, C. ornatus*). GRANT CO. — Rattlesnake Ck, 4.3 mi NE of Holbrook on rt 22 (host, *O. rusticus*). HART CO. — Stream in Hidden River Cave, Horse Cave (host, *O. pellucidus*). JEFFERSON CO. — McNeely Lake (host, *O. juvenilis*); Long Run, 1.9 mi N of rt 60, 5.6 mi E of Middletown (hosts, *O. rusticus, C. laevis*). JESSAMINE CO. — 8 mi SE Versailles on Ky 169 (host, *O. rusticus*). KNOTT CO. — 3.7 mi NW of Knott-Letcher Co. line on Ky 15 Breeding Creek (host, *O. juvenilis*). MADISON CO. — (host, *O. rusticus*); 3 mi S of Clark-Madison Co. line on US 227 (host, *O. juvenilis*); Trib of Ky River 8 mi N of Richmond on rt 223 (host, *O. juvenilis*). MEADE CO. — Doe Run (host, *O. rusticus*). MORGAN CO. — 4 mi E of Grassy Ck on US 460 (hosts, *C.* sp., *O. juvenilis*). NEL-

SON CO. — Bloomfield (host, *O. juvenilis*). OLDHAM CO. — Harrods Ck, 6.7 mi SW of Sligo, 0.9 mi E of US 42 (hosts, *O. rusticus, C. diogenes* subsp.). OWEN CO. — Trib of Eagle Ck 2.3 mi E of Worthville on US 227 (host, *O. rusticus*). POWELL CO. — Natural Bridge State Park (host, *O. juvenilis*). PULASKI CO. — S outskirts of Somerset on US 27 (host, *O.* sp.). SHELLY CO. — Guists Ck, 3.4 mi E of Shelbyville on rt 60 Salt R drainage (host, *O. rusticus*). WARREN CO. — 7 mi S of Bowling Green on rt 71, Middle Ck (hosts, *O. putnami, O. barrenensis*).

MICHIGAN — ALGER CO. — Small pool below dam on AuTrain R, 0.2 mi W of Forest Lake on St 94 (host, *O. propinquus*); Prairie Ck, 2.3 mi W of Schoalcraft Co. line of St 28 (hosts, *O. propinquus, O. virilis*). ARENAC CO. — N Pine R, 1.7 mi NE of Standish (host, *O. propinquus*). CHIPPEWA CO. — 0.5 mi E of Luce Co. line (host, *O. propinquus*). CLARE CO. — Muskegon R just NW of Temple on Mich 61, 619 mi SE of Osceola Co. line (host, *O. virilis*). DELTA CO. — Bank R, 2.3 mi N of Menominee Co. line, just above lamprey control station (host, *O. propinquus*); Days R, 1.5 mi S of Masonville of US 41 & 2 (host, *O. propinquus*). IOSCO CO. — Trib to AuGres R, 3.1 mi W of jct of Rts 55 & 65 (host, *O. propinquus*). LAKE CO. — Twin Ck, 3.3 mi E of jct of rts 63 & 37 on 63 (host, *O. propinquus*). LUCE CO. — W. Prairie Sage R, 7.5 mi W of Chippewa Co. line (host, *O. propinquus*). MONROE CO. — Raisin R, St. Mary's Academy (host, *C. diogenes* subsp.). MONTMORENCY CO. — 5.4 mi W of jct of rts 32 & 33 (host, *O. propinquus*). NEWAYGO CO. — White R at jct of rt 3 in White Cloud (host, *O. virilis*). OSCEOLA CO. — Middle Branch of Muskegon R, 1.7 mi E of jct of rts 61 & 115 on 61, (host, *O. virilis*); Ck 0.6 mi E of Tustin on rt 63 (host, *O. virilis*).

OHIO — ASHLAND CO. — Mohican State Park (hosts, *O. s. sanborni, C. robustus*). CRAWFORD CO. — (host, *O. rusticus*). MIAMI CO. — Sec 35, Union Township (host, *O. rusticus*). PIKE CO. — Swift Ck at E city limits of Beaver (host, *O. s. sanborni*). SCIOTO CO. — Scioto Brush Ck near Rushtown on rt 104 (host, *O. rusticus*). STARK CO. — Beech Ck, 3.8 mi W of Alliance on US 62 (host, *O. obscurus*). WARREN CO. — Spring on Hillside about 1 mi E of S. Lebanon (host, *C. bartonii*).

PENNSYLVANIA — CRAWFORD CO. — Conneautville (host, *O. obscurus*). WASHINGTON CO. — Ten mile Ck, 5.5 mi N of Rut Ck (jct US 19 & Pa 22) on 19 (host, *O. obscurus*).

SOUTH CAROLINA — HAMPTON CO. — 1.4 mi SE of Early Branch on rt 28; 13.8 mi SE of Vannville (hosts, *P. lunzi, P. barbatus*).

SOUTH DAKOTA — BEADLE CO. — 6.6 mi W of jct of US 14 & 281 on 14 (host, *O. immunis*).

TENNESSEE — ANDERSON CO. — (host, *O. forceps*). CANNON CO. — (host, *O.* sp.). CHEATHAM CO. — Trib of Big Turnbull Creek on the property of Bethany Hills Church Camp (host, *O.* sp.); 18 mi W of Nashville on US 70 (host, *O.* sp.); 1mi W of Davidson Co. line W of Belleview (host, *O.* sp.); Nashville, Ck at US 70 (host, *O. rusticus* subsp.). CLAY CO. — Big Tree Ck, Hermitage Springs (hosts, *O. compressus, O. putnami*); Oak Grove W of Celina (host, *O. putnami*); (host, *O. placidus*); Ohio drainage (hosts, *O. compressus, O. putnami*); 8.7 mi N of Jackson Co. line on rt 53 (host, *O. placidus*); (host, *O. rusticus* subsp.). DAVIDSON CO. — (host, *O.* sp.); Highland Ck at Nashville (host, *O. placidus*); (host, *O. placidus*); 3 mi S of Nashville (host, *O. shoupi*); Trib of Harpeth R on Old Hickory Blvd. near Hillsboro Pike (host, *O.* sp.); Seven Mile Ck N of Oglesby (host, *O.* sp.); Mill Ck near Antioch Pike to Cumberland, 10 mi S of Nashville (host, *O.* sp.); 14 mi W of Nashville at jct of US 70S & 70N, Ck along highway (host, *O.* sp.); Richland Ck. at Bellemeade W of Nashville (hosts, *O.* sp., *O. placidus*); Richland Ck at Bellemeade W of Nashville (host, *C. tenebrosus*); Buffalo Ck at jct of 70S & 70N S of Nashville (host, *O. r.* subsp.); Ck at Buffalo School (host, *O.* sp.); small trib to Overall Ck, Hy 70N (host, *O.* sp.); Overall Ck on old Hickory Blvd. (host, *O.* sp.); Wilkerson Branch Nashville (host, *O.* sp.); Otter Ck at Hillsboro Pike (hosts, *O.* sp., *O. rhoadesi*); Mill Ck at Antioch Pike (host, *O. rhoadesi, O. shoupi*); Seven Mile Ck at Antioch Pike (host, *O. shoupi*); Richland Ck, Nashville (hosts, *O.* sp., *O. placidus*). DeKALB CO. — Sink Ck, Blue Springs Comm. (host, *O.* sp.). DICKSON CO. — Montgomery Bell State Park (host, *O.* sp.). FENTRESS CO. — Poplar Cove Ck, [hosts, *C.* (*Puncticambarus*) sp., *C. striatus*]. FRANKLIN CO. — Trib to Elk R 11.4 mi W of Winchester on rt 64 (hosts, *O. spinosus, O. mirus*); Trib to Elk R 11.4 mi W of Winchester on rt 64 (hosts,

C. girardianus, C. striatus). GILES CO. — 10.6 mi N of Pulaski on US 31A (host, *O. spinosus*); Ck 4.5 mi E of Pulaski on rt 64 (hosts, *O. spinosus, O. mirus*); Ck 4.5 mi E of Pulaski on rt 64 (hosts, *C. striatus, C. tenebrosus*); 1.5 mi E of Lawrence Co. line on rt 64, Walker Spg (hosts, *O. spinosus, O. mirus*). GRAINGER CO. — Indian Ck, 11.4 mi SE of Tazewell on rt 25E (host, *C. bartonii*). HAMILTON CO. — Wolftever Ck, Collegedale (host, *O. erichsonianus*). HICKMAN CO. — (hosts, *O. compressus, O.* sp., *O. placidus*). HUMPHREYS CO. — Hurricane Ck, S of Waverly (hosts, *O. compressus, O.* sp., *O. placidus*); Hurricane Ck 10.2 mi E of Waverly city limits of US 70 (host, *O. compressus*). LAWRENCE CO. — 1.4 mi E of Lawrenceburg on rt 64 Shoal Ck (host, *O. spinosus*); 1.4 mi E of Lawrenceburg on rt 64 (host, *O. spinosus*); Peter Cave Ck (host, *O. compressus*). LINCOLN CO. — Swan Ck about 11 mi W of Fayetteville on rt 64 (hosts, *O. erichsonianus, O.* sp.); (host, *O.* sp.); (host, *O.* sp.); Branch of Swan Ck 6.7 mi E of McBurg on Rt 64 (hosts, *O. mirus, O. erichsonianus*); Branch of Swan Ck 6.7 mi E of McBurg on rt 64 (hosts, *O.* sp., *C. striatus, C. tenebrosus*). MACON CO. — Red Boiling Springs (hosts, *Cambarus striatus, O. compressus, O. putnami*). MARION CO. — Battle Ck, 9 mi SE of Monteagle on rt 41 (host, *O. forceps*); 1.2 mi S of US 72 on rt 134, near R (hosts, *O. forceps, O. erichsonianus, C.* sp.). MARSHALL CO. — East Rock Ck 6 mi NE of Lewisburg on rt 64 (host, *O. placidus*); (host, *O.* sp.); (hosts, *O. rhoadesi, O. mirus*). MEIGS CO. — Goodfield Ck 1 mi N of Decatur on rt 58 (host, *O. erichsonianus*). OBION CO. — Samburg, Indian Ck (host, *O. p. palmeri*); Indian Ck, near Old Samburg (host, *O. p. palmeri*). OVERTON CO. — Eagle Ck just above jct with Little Eagle Ck (host, *O. placidus*); Little Eagle Creek 0.5 mi above confluence with Eagle Creek (hosts, *O.* sp., *O. placidus*); 3.0 mi S of Livingston on St 41 (host, *O.* sp.); (host, *O.* sp). PUTNAM CO. — Falling Water R below Burgess Falls Lake Dam (host, *O. placidus*); South Baxter on St 56 (host, *O. placidus*). RUTHERFORD CO. — Stone R where it goes under rt 70S 15 mi E of Murfreesboro (host, *O.* sp.); Murfreesboro (host, *O. placidus*). SMITH CO. — (host, *O.* sp.). WILLIAMSON CO. — Franklin-Harpeth River (hosts, *O.* sp., *O. placidus*); (host, *O.* sp.); (host, *O.* sp.); Franklin (host, *O.* sp.); Franklin (host, *O.* sp.); Fairview (host, *O.* sp.); (host, *O.* sp.). WILSON CO. — Barton Ck at W city limits of Lebanon (hosts, *O. rhoadesi, O. placidus*); Lebanon (host, *O. rusticus* subsp.).

VIRGINIA — BUCHANON CO. — Garden Ck at Skegge (host, *O. juvenilis*).

WEST VIRGINIA — GRANT CO. — S. fork of Potomac R, 1.9 mi NE of Petersburg on rt 220 (host, *O. obscurus*). POCAHONTAS CO. — Durbin, Greenbrier River (hosts, *C. robustus, C. chasmodactylus*).

WISCONSIN — COLUMBIA CO. — Crawfish Ck at jct with US 16 (host, *O. virilis*). OCONTO CO. — Oconto R at jct of Waupee R, 0.4 mi W of jct of rts 64 & 32 on 64 (host, *O. propinquus*).

Canada

ONTARIO — HASTINGS CO. — Salmon R at Shannonville (host, *O. propinquus*).

Uncinocythere thektura Hart, 1965

Plate XXXVII, figures 16-18
Plate LVI

Uncinocythere thektura Hart, 1965a: 192-193 (M, H, D), 195 (R), figs. 6-9. — Ferguson, 1968: 502 (C, D). — Hobbs and Walton, 1968: 251 (R). — Hart and Hart, 1969a: 183 (D). — Hart and Hart, 1969b: 154 (M).

Uncinocythere thecktura, Hobbs, Jr., 1971: 21 (K) (erroneous spelling).

Diagnosis. — MALE: Shell length 380-410 μ (average, 390 μ); height, 200-220 μ (average, 200 μ). Shell subelliptical in outline; posteroventral margin of each valve with small posteriorly directed hook-like prominence situated slightly anterior of extreme posterior limit of shell; dorsal margin of valves heavier than ventral margin; anterodorsal margin slightly concave immediately anterior of eye spot; greatest shell height at approximately midlength. Marginal setae absent dorsally; few present on anterior and posterior margins. Eye spot approximately ¼ shell length from anterior margin.

Copulatory complex with peniferum not extending beyond external border of clasping apparatus; distal extremity bifurcate with points apposed at right angles to one another. Clasping apparatus with distinct vertical and horizontal rami; horizontal ramus longer than vertical. Internal border of horizontal ramus provided with three evenly spaced teeth; distal extremity with five or six denticles; external border evenly rounded or slightly undulate.

TRIUNGUIS FEMALE: Shell length 470-490 μ (average, 480 μ); height 230-270 μ (average, 260 μ). Shell similar to that of male, but larger. Ventral shell margin undulate instead of simply convex as in male.

Shell of biunguis female similar to that of male, but slightly larger.

Relationships — See discussion under *Uncinocythere holti* (above).

Type-locality. — Cold spring in Snake River Canyon near Hagerman, Gooding Co., Idaho, U.S.A. (types, USNM: holotype and allotype, 110780). Host, *Pacifastacus gambelii* × *P. connectens.*

Distribution (in addition to type-locality). —

United States

IDAHO — GOODING CO. — Spring in Snake River canyon near Hageman (host, *P. gambelii*). KOOTENAI CO. — Mouth of St. Joe River, Le Coeur d'Alene (host, *P. gambelii*).

OREGON — MALHEUR CO. — Jordan Creek, trib to Owyhee River (host, *P. gambelii*).

SOUTH DAKOTA — BEADLE CO. — 6.6 mi W of jct of rts 14 and 281 on rt 14 (host, *Orconectes immunis*).

WYOMING — TETON CO. — Presidents' Camp, Snake River, S of Yellowstone Park (host, *P. gambelii*). YELLOWSTONE NATIONAL PARK CO. — Crayfish Creek, below Moose Falls (host, unknown).

Uncinocythere warreni Hobbs and Walton, 1968

Plate XXXVIII, figures 1-3
Plate LVI

Uncinocythere warreni Hobbs and Walton, 1968: 250, 251, (M, H, R, A, D), figs. 3g-j.

Diagnosis. — MALE: Shell length 350 μ; height, 180 μ. Eye without pigment, perhaps absent. Shell oblong ovate, highest at midlength and without concavity on ventral margin. Submarginal setae present except dorsally, and more abundant anteriorly and posteriorly than ventrally.

Copulatory complex with peniferum deeply bifid ventrally with cephalic extension spike-like and posterior one curved anteriorly, ending in short, acute protuberance. Penis situated horizontally and more than half as long as anteroposterior plane of peniferum at level of penis. Dorsal finger comparatively heavy with large setiform, simple terminal seta; ventral finger bent strongly at distal end of basal fourth and gently curved distally. Clasping apparatus with horizontal and vertical rami situated at angle of 80 to 85 degrees; vertical ramus slightly longer than horizontal; entire external border and internal border of vertical ramus entire; internal border of horizontal ramus with three evenly spaced teeth, distal one smallest; apex of apparatus with four anterodorsally directed teeth.

TRIUNGUIS FEMALE. — Eye without pigment, perhaps lacking. Shell highly arched dorsally and with distinct concavity on ventral margin anterior to midlength; length of three available females 0.39, 0.38, and 0.38 mm; respective heights, 0.21, 0.22, and 0.20 mm. Submarginal setae absent dorsally, otherwise almost evenly distributed. Genital area with deep (dorsally) fissure situated between finger-like anterior lobe and posterior slighted sclerotized prominence.

Relationships. — *Uncinocythere warreni* has its closest affinities with *U. cassiensis, U. caudata, U. erichsoni, U. holti, U. pholetera, U. simondsi,* and *U. thektura.* All of them share a deeply bifid ventral extremity of the peniferum, usually three teeth on the internal border of the horizontal ramus of the clasping apparatus, and three to five apical teeth. It differs from *U. cassiensis, U. caudata,* and *U. thektura* in that the shell lacks a posteroventral spine or prominence, from *U. ericksoni* and *U. holti* (the latter is possibly a synonym of the former) in having a longer horizontal ramus of the clasping apparatus and a more deeply bifid ventral portion of the peniferum, and from *U. simondsi* in possessing three teeth on the internal border of the clasping apparatus instead of two. It shares with *U. pholetera* the absence of a pigmented eyespot.

Type-locality. — Climax Cave, 3 mi N of Climax, Decatur Co., Georgia, U. S. A. (types, USNM: holotype and allotype, 123324). Host, *Cambarus cryptodytes.*

Distribution. — Known only from type-locality.

Uncinocythere xania (Hart and Hobbs, 1961)

Plate XXXVIII, figures 4-6
Plate LVI

Entocythere xania Hart and Hobbs, 1961b: 174 (K), 181 (M, H, R, D), figs. 21-24. — Bedinger and Hobbs, 1965: 94 (D, H) (by implication).

Uncinocythere xania. — Hart, 1962: 138-139 (C, D, S). — Hobbs and Walton, 1963: 457 (R). — Hobbs and Bedinger, 1964: 14 (D, H). — Ferguson, 1968: 502 (C, D).

Diagnosis. — MALE: Shell length 420 μ; height, 250 μ. Eyeless. Shell ovate, with greatest height near midlength. Setae widely spaced along anterior, ventral, and posterior margins.

Copulatory complex with ventral extremity of peniferum reaching level of internal border of clasping apparatus. Cephaloventral portion of peniferum reflexed. Clasping apparatus with distinct vertical and horizontal rami, former slightly longer than latter. Internal border of horizontal ramus provided with submedian prominent tooth proximal to which is row of fine serrations and distally by small tooth; distal extremity provided with three denticles. External border evenly rounded or undulate but without distinct eminence. Distal tooth on internal border almost obsolete in some specimens.

FEMALE. — Shell of biunguis female similar to that of male, but smaller.

Type-locality. — Smallins Cave, 7.4 mi SE of Galloway, Christian Co., Missouri, U. S. A. (types, ANSP). Host, *Cambarus setosus.*

Distribution (in addition to type-locality). —

United States

MISSOURI — GREEN CO. — Moore's Cave, 3 mi SW of Springfield (host, *Cambarus setosus*).

Uncinocythere xena Hart and Hart, 1971

Plate XXXVIII, figures 7-9
Plate LV

Uncinocythere xena Hart and Hart, 1971: 118 (M, H, R, A, D), figs. 12a-c.

Diagnosis. — MALE: Shell length 360 μ; height, 210 μ. Shell subovoid, with greatest height ⅓ shell length from posterior end. Eye pigmented, situated ¼ shell length from anterior end.

Peniferum slender, strongly convex posteriorly, tapering distally, and extending ventrally to internal border of horizontal ramus of clasping apparatus. Anterodistal portion of peniferum terminating in sharp point, concavity almost closed by acute forward-curving, posterior projection. Penis almost vertical.

Horizontal and vertical rami of clasping apparatus subequal in length, meeting at angle of approximately 80°. External border of vertical ramus evenly convex; internal border evenly concave. External border of horizontal ramus almost straight with broad concavity slightly distal to midlength. Internal border with two teeth near base of distal half and one tooth near three terminal denticles.

FEMALE. — Unknown.

Relationships. — This species is most closely allied to *Uncinocythere simondsi* and *Uncinocythere stubbsi.* However, the placement of the teeth on the clasper is different, with *U. simondsi* and *U. stubbsi* having one tooth near the midlength of the horizontal ramus and *U. xena* having two teeth.

Type-locality. — Richland Creek at Bellemeade, Davidson Co., Tennessee, U. S. A. (types, USNM: holotype, 138466). Hosts, *Orconectes* sp. and *O. placidus.*

Distribution. — Known only from type-locality.

Uncinocythere zancla Hobbs and Walton, 1963

Plate XXXVIII, figures 10-12
Plate LVII

Uncinocythere zancla Hobbs and Walton, 1963a: 456-457 (M, H, R, A, D), figs. 1-3. — Hart and Hart, 1966a: 8 (A). — Ferguson, 1968: 502 (C, D). — Hobbs and Barr, 1972: 81 (O, H).

Diagnosis. — MALE: Shell length 326-348 μ (average, 336 μ); height, 178-192 μ (average, 186 μ). Eyes present. Shell with dorsal margin rising gently from anterior extremity to slightly posterior to midlength; ventral margin with shallow broad emargination immediately anterior to midlength; marginal setae evenly, but sparsely spaced, anteriorly, posteriorly and along anterior half of ventral margin.

Copulatory complex with caudal margin of peniferum undulating, its ventral margin truncate with distinct emargination, caudodistal margin of latter projecting anteriorly into emargination. Penis, of moderate size, situated far distal to base of clasping apparatus and entirely within penultimate fourth of area between dorsal margin of spermatic loop and ventral extremity of peniferum. Both dorsal and ventral fingers slender with their principal axes approximately at right angles; ventral finger with distinct angular bend short distance from base.

Clasping apparatus L-shaped and clearly divisible into vertical and horizontal rami of subequal lengths. Vertical ramus entire and joined to horizontal ramus in gentle arc, extensions of which forming angle of slightly less than 90 degrees. External border of horizontal ramus gently rounded and entire; internal border with single blunt tooth near midlength; terminal extremity with three teeth.

Female. — Not definitely correlated with male.

Relationships. — *Uncinocythere zancla* probably has its closest affinities with *U. ambophora, U. lucifuga,* and *U. xania.* In all of these the internal border of the clasping apparatus bears but a single tooth. The cleft peniferum of *U. zancla* distinguishes it from the latter three species in which the penifera are entire or bear a terminal projection. The cleft peniferum resembles those of *U. ericksoni* and *U. neglecta,* both of which have three teeth on the internal border of the clasping apparatus.

Type-locality. — Hurricane and Tumbling Creeks, SE of Waverly, Humphreys Co., Tennessee, U. S. A. (types, USNM: holotype, 108033; paratype, 108034). Hosts, *Orconectes compressus, O. placidus, O.* sp., and *Cambarus striatus.*

Distribution (in addition to type-locality). —

United States

GEORGIA — FAYETTE CO. — 11.2 mi NE of Fayette Co. line on rt 54 (host, *Procambarus spiculifer*).

KENTUCKY — GRANT CO. — Rattlesnake Ck, 4.3 mi NE of Holbrook on rt 22 (host, *O. rusticus*). HARDIN CO. — Creek from cave at Crystal Lake 2 mi N of Muldraugh off rt 31 (host, *Cambarus* sp.); Bland Cave, 7 mi W of Sonora (host, *O. i. inermis*). HART CO. — Riders Mill Cave (host, *O. i. inermis*).

NORTH CAROLINA — ALLEGHENY CO. — 16.9 mi E of jct of parkway & rt 18 (hosts, *C. robustus, C. bartonii*). DURHAM CO. — Plot 60, Durham Div. Duke Forest, "Crow's Nest Spring" (host, *C. bartonii*). ORANGE CO. — Trib of New Hope Ck, E of Piney Mt (host, *C. bartonii*).

TENNESSEE — HUMPHREYS CO. — Hurricane and Tumbling Creeks SE of Waverly (hosts, *O. compressus, O. placidus, O.* sp., *C. striatus*). CANNON CO. — (host, *O.* sp.). DAVIDSON CO. — Highland Ck at Nashville (host, *O. placidus*); Richland Ck at Bellemeade W of Nashville (host, *C. tenebrosus*); Buffalo Ck at jct of 70S & 70N S of Nashville (host, *O.* sp.). DeKALB CO. — Smithville (hosts, *O.* sp., *O. placidus*). DICKSON CO. — Montgomery Bell State Park camping area (host, *O.* sp.). FENTRESS CO. — Popular Cove Ck, W of Jamestown [host, *C.* (*Puncticambarus*) sp.]. FRANKLIN CO. — Trib to Elk R 11.4 mi W of Winchester on rt 64 (hosts, *O. spinosus, O. mirus*); Trib to Elk R 11.4 mi W of Winchester on rt 64 (hosts, *C. girardianus, C. striatus*). HICKMAN CO. — (hosts, *O. compressus, O.* sp., *O. placidus*). HUMPHREYS CO. — Hurrican & SE of Waverly (hosts, *C. striatus, O. compressus, O. placidus, O.* sp.); Tumbling Ck, E of Bold Spring Coll. (hosts, *O. compressus, O. placidus, O.* sp.). LAWRENCE CO. — 1.4 mi E of Lawrenceburg on rt 64 Shoal Ck. (host, *O. spinosus*). LINCOLN CO. — Swan Ck, about 11 mi W of Fayetteville on US 64 (hosts, *O. erichsonianus, O.* sp.). MARSHALL CO. — East Rock Ck. 6 mi NE of Lewisburg on St 64 (host, *O. placidus*). MAURY CO. — Mt. Pleasant (hosts, *O. compressus, O. placidus, O. rusticus* subsp); Little Bigby Ck at Mt. Pleasant (hosts, *O. placidus, O. r.* subsp.). PERRY CO. — (host, *O.* sp.). PUTNAM CO. — Falling Water R immediately below Burgess Falls Lake Dam. Coney Fork of Cumberland (host, *O. placidus*); South Baxter on rt 56 (host, *O. placidus*). RUTHERFORD CO. — Murfreesboro (host, *O. placidus*). WAYNE CO. — 7.0 mi E of Waynesboro on US 64 Forty-eight Ck, trib to Buffalo R (hosts, *O.* sp., *O. compressus, O. spinosus, O. r. forceps*). WILLIAMSON CO. — (host, *O. rusticus* subsp.). WILSON CO. — Barton Ck at W city limits of Lebanon (host, *O. rusticus* subsp.).

Uncinocythere zaruri Hobbs, 1971

Plate XXXVIII, figures 13-16
Plate LVIII

Uncinocythere zaruri Hobbs, 1971: 26-28 (M, H, R, A, D, K), figs. 9, 14.

Diagnosis. — MALE: Shell length 310-330 μ (average, 320 μ); height, 180-190 μ (average, 185 μ). Eye pigmented situated about one-fifth shell length from anterior extremity. Shell ovate with greatest height almost at midlength; ventral margin gently curved or almost straight; submarginal setae present anteriorly, posteriorly, and ventrally.

Copulatory complex with peniferum rounded posteriorly and distal extremity directed anteroventrally, bearing shallow notch just posterior to small acute anteroventral angle. Penis considerably longer than half anteroposterior diameter of peniferum at level of penis and curved with proximal and distal ends disposed at angle of approximately 120 degrees. Dorsal and ventral fingers comparatively stout and both terminating in simple seta; ventral finger somewhat sinuous proximally but strongly curved at end of proximal fourth and more gently in distal half. Clasping apparatus with vertical and horizontal rami disposed at angle of 60 to 80 degrees; vertical ramus about one-tenth longer and wider than tapering horizontal ramus; external border of both rami and internal border of vertical ramus entire; internal border of horizontal ramus with two teeth near midlength; apex with two denticles. (Hobbs, 1971).

TRIUNGUIS FEMALE: Shell length 340-360 μ (average, 350 μ); height 220-230 μ (average, 225 μ). Eye pigmented, situated as in male. Shell subreniform, more highly vaulted than in male, highest posterior to midlength, and with distinct concavity anterior to midlength; submarginal setae present except anterodorsally.

Genital complex consisting of prominent, slender, weakly sclerotized papilla projecting posteroventrally between anterior slender lappet and heavier posterior one.

Relationships. — *Uncinocythere zaruri* is most closely allied to *U. bicuspide* but may be readily distinguished from it by the slenderer clasping apparatus which bears but a single tooth on the internal border of the horizontal ramus, and by the shell which is usually slightly larger.

Remarks. — According to Hobbs (1971), except for the questionable isolated record cited above of *U. bicuspide* occurring on *Procambarus aztecus* at Tomatlán, Veracruz, *U. zaruri* is the only member of the genus which ranges south of the Cordillera Volcánica Transversal. Furthermore, he says it is the only member of the subfamily known to be associated with a non-crayfish host.

The only known host of this ostracod is the freshwater crab *Pseudothelphusa veracruzana.* Specimens were found in sediment in a jar containing specimens of *Atya Scabra* collected from Tapalapan, Santiago Tuxtla, but it is probable that *P. veracruzana* had been preserved in the same bottle. Examination of Atya collections in the Instituto de Biología at the University of México disclosed no entocytherids.

Type-locality. — Río Tapalapa, Santiago Tuxtla, Veracruz, México (types, USNM: holotype and allotype, 128827). Host, *Pseudothelphusa veracruzana.*

Distribution. — Known only from type-locality.

Incertae Sedis

Entocythere insignipes (Sars, 1926)

Cytherites insignipes Sars, 1926: 10, 11 (M, H, R, D), figs. 1-13. — Hoff, 1942a: 63, 64 (C). — Rioja, 1943b: 569, 584, 585 (C, R). — Hobbs, Holt, and Walton, 1967: 6 (C).

Entocythere (Cytherites) insignipes. — Hoff, 1942a: 66, 69, 71 (C). — Rioja, 1942b: 687 (K). — Rioja, 1943b: 569 (R), 584 (R). — Howe, *in* Moore and Pitrat, 1961: 300 (C).

Entocythere insignipes. — Hoff, 1944b: 328 (R), 331 (K, R, M). — Tressler, 1947: 705 (C). — Rioja, 1949: 328 (R). — Tressler, 1959: 731 (K), fig. 28.192. — Hart, 1962: 139 (C, S, D).

Type-locality. — This species was described on the basis of three female specimens sent to Sars from Canada. Locality, date, collector, and male specimens are unknown. Because of the lack of adequate specific characters in female entocytherid ostracods, this name should be considered *incertae sedis.*

Subfamily Microsyssitrinae Hart, Nair, and Hart, 1967

Microsyssitrinae Hart, Nair, and Hart, 1967: 2 (M, R, H), type genus *Microsyssitria* Hart, Nair, and Hart, 1967. — Hart and Hart, 1969a: 182, 183 (R), fig. 5. — Hart and Hart, 1970: 283 (R).

Tribe Microsyssitrini. — Danielpol, 1971a: 190 (D).

Type of the subfamily. — *Microsyssitria indica* Hart, Nair, and Hart, 1967.

Diagnosis. — Antennule with five podomeres. Dorsal antennal claw subspiculiform, bearing setae on flexor face. Mandibular protopodite with distal row of five multicuspid teeth. Peniferum terminating in two movable claws. Penis weakly developed and situated in distal portion of peniferum. Commensal on marine wood-boring isopods.

Distribution. — INDIA: coastal waters of Kerala state.

MICROSYSSITRIA HART, NAIR, AND HART, 1967

Microsyssitria Hart, Nair, and Hart, 1967: 2 (M), type-species, *Microsyssitria indica* Hart, Nair, and Hart, 1967. — McKenzie, 1972: 38 (C).

Diagnosis. — See diagnostic characters enumerated above for the subfamily.

Type-species. — *Microsyssitria indica* Hart, Nair, and Hart, 1967.

Distribution. — Ayiramthengu, Kayankulam backwaters and Cochin backwaters, Kerala State, India.

Host. — Wood-boring marine isopods.

Figure 40. — Distribution of the genus *Microsyssitria.*

Microsyssitria indica Hart, Nair, and Hart, 1967

Page 175
Plate XXXIX, figures 1-5
Plate LIX

Microsyssitria indica Hart, Nair, and Hart, 1967: 2-11 (M, D, H, R, B), figs. 1-17. — Hart and Hart, 1969a: 180 (E), fig. 2. — Hart and Hart, 1969b: 157 (M), fig. 2. — Hart and Hart, 1970: 282, 283 (R). — Hart and Hart, 1971: 582 (R).

Diagnosis. — MALE: Shell length 380-420 μ (average, 395 μ); height, 210-260 μ (average, 238 μ). Copulatory complex with ventral portion of peniferum terminating in two movable claws and single fixed anterior spinous process. Movable claws apposed and apparently controlled by single muscle originating dorsally in peniferum. Penis short, weakly developed, originating and terminating between or slightly proximal to bases of movable claws. Mid-portion of peniferum tapering distally. Clasping apparatus subequal in length to entire peniferum, curved at right-angle in proximal $\frac{1}{4}$, mid-portion slightly sinuous, and tapering slightly distally, distal $\frac{1}{4}$ forming obtuse angle with mid-portion and terminating in six denticles. Dorsal and ventral fingers present; ventral finger approximately $\frac{1}{6}$ length of clasping apparatus; dorsal finger bifurcate (or double?) and shorter than ventral finger.

TRIUNGUIS FEMALE: Shell length 360-420 μ (average, 391 μ); height 210-270 μ (average, 243 μ). Shell approximately same size as adult males, but lacking conspicuous concavity in posterodorsal margin. Posterodorsal portion of triunguis specimens containing irregular, somewhat amorphous chitinous structure reminiscent of amiculum found in some North American genera.

Relationships. — This ostracod appears to have affinities with members of the Australian genera *Chelocythere* and *Lichnocythere* and with the New Guinea genus *Elachistocythere*. It differs from both the Australian and New Guinea forms in that its peniferum terminates in two movable claws (several Australian species appear to have at least one movable claw, the New Guinea species appears to have none.). In addition, its antennule is composed of five podomeres and the mandibular palp bears five teeth — as opposed to six antennular podomeres and six mandibular palp teeth in the Australian and New Guinea species.

It resembles the Australian and New Guinea species in having a weakly developed vertical penis.

Type-locality. — Ayiramthengu, Kayankulam backwaters, Kerala State, India (types, ANSP). Host, marine woodboring isopod *Sphaeroma terebrans*.

Distribution. — In addition to the type-locality, this ostracod has been found in the Cochin Backwater in Kerala State, where the host was also *Sphaeroma terebrans*.

Subfamily Notocytherinae Hart and Hart, 1967

Notocytherinae Hart and Hart, 1967a: (M, H, D), type genus *Notocythere* Hart and Hart, 1967. — Hart and Hart, 1969a: 182, 183 (R), fig. 5. — Hart and Hart, 1969b: 157 (M, R). — Hart and Hart, 1970: 283 (R).

Tribe Notocytherini. — Danielpol, 1971a: 190 (D).

Type of the subfamily. — *Notocythere syssitos* Hart and Hart, 1967.

Diagnosis. — Distal spine of mandibular palp spatulate; terminal portion rounded or truncate. Respiratory plate of maxilla lacking; masticatory lobes of maxilla present or absent. Peniferum highly variable; penis straight or sinuous and vertically positioned in distal portion of peniferum. Commensal on freshwater isopods and crayfishes.

Distribution. — Fresh waters of the Australian mainland and Tasmania, New Zealand (South Island), and New Guinea.

Hosts. — Freshwater crayfishes of the family Parastacidae and phreatoicoid isopods.

Key to the Genera and Species of Australasian Entocytherid Ostracods (the Subfamily Notocytherinae)

1 Clasping apparatus curved 7
1′ Clasping apparatus straight 2
2(1′) Peniferum with dagger-shaped device, similar to finger guard of some North American species, situated dorsal of fingers .... *Riekocythere* 4
2′ Peniferum without dagger-shaped device 3
3(2′) Dorsal claw of antenna longer than mesial claw, equipped with subterminal tuft of setae, and not sharply pointed; clasping apparatus long, extending beyond terminal portion of peniferum .... *Hesperocythere* 5
3′ Dorsal claw of antenna shorter than mesial claw, spiniform, not possessing terminal tuft of setae; clasping apparatus small, short, not extending half way to terminal portion of peniferum .... *Elachistocythere merista*
4(2) Peniferum terminating just distal to end of penis .... *Riekocythere cherax*
4′ Peniferum extending beyond penis for distance subequal to length of clasping apparatus .... *Riekocythere xenika*
5(3) Peniferum terminating in two apposed processes enclosing bulbous, reniform member .... *Hesperocythere klasteroides*
5′ No bulbous member at end of peniferum 6
6(5′) Clasping apparatus tapering toward end, bearing six teeth on internal border .... *Hesperocythere tallanalla*
6′ Clasping apparatus not tapering, bearing four teeth on internal border .... *Hesperocythere xiphoides*
7(1) Distal portion of peniferum terminating in two apposed claw-like processes, posteroventral process heavier than anteroventral process; dorsal antennal claw curved, crenulate on internal border, and terminating in digitiform projection .... *Lichnocythere* 8
7′ Distal portion of peniferum not as above; dorsal antennal claw not crenulated on internal border 10
8(7) Clasping apparatus with six or seven teeth on internal border 9
8′ Clasping apparatus with four teeth on internal border, proximal tooth heavier than others .... *Lichnocythere synethes*
9(8) Posteroventral peniferal process hammer-shaped, with acute angle on internal margin level with end of penis .... *Lichnocythere tubrabucca*
9′ Posteroventral peniferal process with smoothly convex external margin .... *Lichnocythere victoria*
10(7′) Distal portion of peniferum terminating in heavily sclerotized claw-like apparatus, composed of anteroventral and posteroventral processes, with apparent muscle insertion on base of anteroventral process .... *Chelocythere kalganensis*
10′ Distal portion of peniferum not terminating in claw-like apparatus; no apparent muscle attachment in distal portion of peniferum 11
11(10′) Distal portion of peniferum terminating in two poorly sclerotized processes, not extending beyond angle of clasping apparatus 12
11′ Peniferum narrowing abruptly in distal portion, usually terminating in one to three processes; dorsal antennal claw curved, without setae, and terminating abruptly .... *Notocythere* 18
12(11) Clasping apparatus with vertical and horizontal rami forming distinct angle with one another .... *Herpetocythere* 13
12′ Clasping apparatus with vertical and horizontal rami merging imperceptibly .... *Laccocythere aotearoa*
13(12) External border of clasping apparatus smoothly curved; one tooth on internal border of horizontal ramus .... *Herpetocythere acanthoides*
13′ Clasping apparatus with distinct horizontal and vertical rami; one to five teeth on internal border of horizontal ramus 14
14(13′) Anteroventral peniferal process extending distally beyond posteroventral process 15
14′ Anteroventral peniferal process not extending distally beyond posteroventral process .... *Herpetocythere labidioides*
15(14) Posteroventral peniferal process blunt, with tubercle near base .... *Herpetocythere gnoma*
15′ Posteroventral peniferal process blunt or sharp, without tubercle near base 16

16(15′) Posteroventral peniferal process directed anteriorly .... *Herpetocythere mackenziei*
16′ Posteroventral peniferal process directed ventrally 17
17(16′) Posteroventral peniferal process not well developed; five distinct teeth on horizontal ramus of clasping apparatus .... *Herpetocythere australensis*
17′ Posteroventral peniferal process well developed; three indistinct teeth on horizontal ramus of clasping apparatus .... *Herpetocythere bendora*
18(11′) Posteroventral peniferal process developed into posteroventrally directed flange 19
18′ Posteroventral peniferal process not strongly developed, comprised of one or two protuberances variously directed 21
19(18) External border of clasping apparatus smoothly convex 20
19′ External border of clasping apparatus distinctly concave .... *Notocythere rieki*
20(19) Distal extremity of clasping apparatus tapering, terminating in three denticles .... *Notocythere synomodites*
20′ Distal extremity of clasping apparatus not tapering, terminating in five denticles .... *Notocythere tasmanica*
21(18′) Posteroventral peniferal process bifid 22
21′ Posteroventral peniferal process not bifid 24
22(21) Clasping apparatus without distinct juncture between vertical and horizontal rami .... *Notocythere syssitos*
22′ Clasping apparatus divided into distinct vertical and horizontal rami 23
23(22′) Horizontal ramus of clasping apparatus not tapering at distal end; tip of horizontal ramus twice diameter of mid-length of ramus .... *Notocythere grampians*
23′ Horizontal ramus of clasping apparatus tapering slightly toward distal end .... *Notocythere mirranatwa*
24(21′) Distal portion of peniferum with two apposed processes .... *Notocythere erica*
24′ Distal portion of peniferum with only one protuberance 25
25(24′) Vertical and horizontal rami of clasping apparatus forming distinct right angle with one another .... *Notocythere blundelli*
25′ Vertical and horizontal rami of clasping apparatus not forming distinct right angle with one another .... *Notocythere antichthon*

CHELOCYTHERE HART AND HART, 1967

Chelocythere Hart and Hart, 1967a: 7 (D, R), 8 (K), 27 (M), type-species *Chelocythere kalganensis* Hart and Hart, 1967. — Hart, Nair, and Hart, 1967: 4 (R). — Hart and Hart, 1969b: 158 (M, R).

Diagnosis. — Distal portion of peniferum terminating in two well-developed claw-like processes — anteroventral process heavier than posteroventral process, and apparently hinged and equipped with muscles so that two processes may be apposed. Dorsal antennal claw curved, with terminal brush of setae directed distally.

Type-species. — *Chelocythere kalganensis* Hart and Hart, 1967a.

Distribution. — West Australia.

Hosts. — Freshwater crayfishes of the genus *Cherax*.

Figure 41. — Distribution of the genus *Chelocythere.*

Chelocythere kalganensis Hart and Hart, 1967

Plate XXXIX, figures 10-12
Plate LXI

Chelocythere kalganensis Hart and Hart, 1967a: 27 (M, R, A, D), 29, 31 (A), 8 (K), figs. 75-79. — Hart and Hart, 1969a: 180, 181 (M), fig. 3. — Hart and Hart, 1969b: 157 (M), fig. 3. — Hart and Hart, 1970: 283 (R). — Hart and Hart, 1971: 582 (R).

Diagnosis. — MALE: Eye spot present, situated approximately ¼ shell length from anterior end. Shell subelliptical in outline, with greatest height in posterior third. Five male specimens ranged between 220 and 260 μ in length (average, 238μ); between 120 and 130 μ in height (average, 126 μ).

Copulatory complex with peniferum terminating in well-developed claw-like apparatus; anteroventral process heavily sclerotized, curved posteriorly, and apparently hinged proximally apparently apposing distal peniferal process may be apposed; conspicuous muscle with origin in proximal portion of peniferum and insertion on base of anteroventral peniferal process; posteroventral process partially contiguous with peniferum, distal portion directed ventrally. Penis originating proximal to level of origin of clasping apparatus and extending ventrally to position of hinge of anteroventral peniferal process. Clasping apparatus with horizontal and vertical rami merging imperceptibly; horizontal ramus ⅔ length of vertical ramus and bearing three teeth on internal margin and seven terminal denticles. Dorsal and ventral fingers present.

FEMALE. — Unknown.

Type-locality. — Kalgan River, 10 mi W of Albany, Western Australia, Australia (types, AM).

Host. — Crayfish *Cherax* sp.

Distribution (in addition to type-locality). —

Australia

WEST AUSTRALIA — 5 mi NW of Shannon River Mill. Host, *Cherax* sp.

ELACHISTOCYTHERE HART AND HART, 1970

Elachistocythere Hart and Hart, 1970: 279-283 (M), type-species, *Elachistocythere merista* Hart and Hart, 1970.

Diagnosis. — Antennule with six podomeres. Dorsal antennal claw subspiculiform and devoid of setae. Mandibular protopodite with distal row of five teeth. Peniferum bifld. Penis straight, ventrally directed, and situated slightly below mid-length of peniferum.

Type-species. — *Elachistocythere merista* Hart and Hart, 1970.

Distribution. — New Guinea.

Hosts. — Freshwater crayfishes of the genus Cherax.

Figure 42. — Distribution of the genus *Elachistocythere*.

Elachistocythere merista Hart and Hart, 1970

Page 176
Plate XXXIX, figures 6-9
Plate LIX

Elachistocythere merista Hart and Hart, 1970: 279-283 (M, H, R, D), figs. 1-9.

Diagnosis. — MALE: Shell length 270-300 μ (average, 282 μ); height 140-190 μ (average, 159 μ). Shell subelliptical in outline with eyespot situated approximately ¼ shell length from anterior end.

Copulatory complex with ventral portion of peniferum terminating in two strongly sclerotized tapering processes, which may or may not be apposable. Penis straight, ventrally directed, and situated slightly below the mid-length of the peniferum. Clasping apparatus short, club-shaped, with seven or eight terminal denticles.

TRIUNGUIS FEMALE: Shell length 300-320 μ (average, 307 μ); height 160-190 μ (average, 177 μ). Shell similar to that of male, but averaging 25 μ more in length. Conspicuous triangular apparatus present in posterodorsal area, probably homologous with amiculum of genus *Dactylocythere* in North America (Hart, 1962).

Relationships. — This ostracod appears to have close affinities with members of the Australasian entocytherid subfamily Notocytherinae, as well as with those of the Indian subfamily Microsyssitrinae. It is similar to the Australian species in having six antennular podomeres (as opposed to five in the Indian), but similar to the Indian forms in having five teeth on the mandibular palp, (as opposed to six in the Australian). We feel, however, that the preponderance of evidence indicates it should be placed in the subfamily Notocytherinae.

Although the distal portion of the peniferum is bifid, we doubt that the two projections are apposable as they are in *Microsyssitria indica* and as they may be in *Chelocythere kalganensis.*

Type-locality. — Small stream near native village of Gariau, on Jomoer Lake, New Guinea (types, RMNH).

Host. — Crayfish *Cherax,* sp.

Distribution. — Known only from type-locality.

HERPETOCYTHERE HART AND HART, 1967

Herpetocythere Hart and Hart, 1967a: 6 (D, R), 8 (K), 18 (M), type-species, *Herpetocythere australensis* Hart and Hart, 1967.

Diagnosis. — Terminal portion of mandibular palp truncate. Maxilla with masticatory process. Dorsal antennal claw curved, with or without terminal setae. Distal portion of peniferum relatively wide in relation to proximal portion, terminating in two or three apposed acute prominences, none of which appear movable.

Type-species. — *Herpetocythere australensis* Hart and Hart, 1967.

Distribution. — Southeastern Australian mainland.

Hosts. — Crayfishes.

Figure 43. — Distribution of the genus *Herpetocythere.*

Herpetocythere acanthoides Hart and Hart, 1967

Plate XXXIX, figures 13-14
Plate LX

Herpetocythere acanthoides Hart and Hart, 1967a: 23 (M, R, A, D), 8 (K), 24 (A), figs. 61, 62.

Diagnosis. — MALE: Eye spot present. Shell subelliptical in outline with greatest height at mid-length. Three male specimens averaging 203 μ in length, 117 μ in height.

Copulatory complex with ill-defined posteroventral and anteroventral peniferal processes, giving distal portion of peniferum appearance of being bluntly rounded. Penis with origin above level of origin of clasping apparatus, extending sinuously ventrally to near base of posteroventral process. Clasping apparatus with juncture of horizontal and vertical rami indistinct; internal margins of horizontal ramus with one tooth situated in proximal half; horizontal ramus terminating in point, without denticles.

Variations. — Between specimens the two rami of the clasping apparatus may be more or less distinct. From observations of preserved copulating specimens, it appears that the clasping apparatus is very flexible.

FEMALE. — Shells of biunguis females similar to those of males, but smaller. Two biunguis specimens averaged 155 μ in length, and 80 μ in height.

Relationships. — This ostracod appears to have its closest affinity with *Herpetocythere gnoma,* but differs from it in that the terminal peniferal processes are even less well developed in *H. acanthoides* than in *H. gnoma,* and the clasping apparatus of *H. acanthoides* is distinct in its almost crescent shape and lack of terminal denticles.

Type-locality. — Mt. Victoria, Blue Mountains, New South Wales, Australia (types, AM). Host, *Euastacus nobilis.*

Distribution (in addition to type-locality). —

Australia

NEW SOUTH WALES — Bell. Host, *Euastacus spinifer.*

Herpetocythere australensis Hart and Hart, 1967

Plate XXXIX, figures 15-17
Plate LX

Herpetocythere australensis Hart and Hart, 1967a: 18, 19 (M, R, A, D), 8 (K), 6 (D), 19, 20, 22 (R), 19, 21, 26 (A), figs. 42-47.

Diagnosis. — MALE: Shell length 270-290 μ (average, 279 μ); height 150-190 μ (average, 167 μ). Shell subelliptical in outline with eye spot situated approximately ⅓ shell length from anterior end. Greatest shell height in posterior ⅓.

Copulatory complex with peniferum terminating in large recurved anteroventral process and smaller rounded posteroventral process; thin velum extending from posteroventral to anteroventral process. Penis originating near level of origin of clasping apparatus and extending ventrally to line between terminal peniferal processes. Clasping apparatus with distinct vertical and horizontal rami; vertical ramus smoothly convex on external margin, concave on internal margin; horizontal ramus almost straight on external margin, its internal margin with five teeth; terminal portion turned slightly dorsally, with four denticles. Dorsal and ventral fingers present; dorsal finger with diameter at least twice that of ventral finger.

FEMALE. — Shell of biunguis female similar to that of male, but smaller. The single specimen measured 230 μ in length; 130 μ in height.

Relationships. — This ostracod apparently has its closest affinities with *H. bendora.* It differs from *H. bendora,* however, in having a less well-developed posteroventral peniferal process and in having five well-developed teeth on the horizontal ramus instead of only two indistinct ones.

Type-locality. — Elaman Creek, Maleny, Queensland, Australia (types, AM). Host, *Euastacus hystericosus.*

Distribution. — Known only from type-locality.

Herpetocythere bendora Hart and Hart, 1967

Plate XXXIX, figures 18-19
Plate LX

Herpetocythere bendora Hart and Hart, 1967a: 20, 21 (M, R, A, D), 15 (A), 8 (K), 19 (R), figs. 50-53.

Diagnosis. — MALE: Shell length 220-240 μ (average, 229 μ); height 120-140 μ (average, 133 μ). Eye spot present. Shell subelliptical in outline with greatest height near mid-length. Copulatory complex with peniferum equipped with two terminal processes, anteroventral one longer than posteroventral and recurved at tip, posteroventral one ⅔ length of anteroventral one and terminating bluntly. Penis originating proximal to origin of clasping apparatus and extending vertically between bases of terminal peniferal processes. Clasping apparatus with horizontal ramus approximately ⅔ length of vertical ramus; vertical ramus almost straight, external margin slightly convex, internal margin concave; horizontal ramus with external margin convex, internal margin concave and bearing three almost imperceptible teeth, terminal portion with five indistinct denticles. Dorsal and ventral fingers present.

Variations. — In some specimens the peniferal processes are more apposed, or are bent closer together, than in others. Also some specimens appear to have three instead of two indistinct teeth on the internal margin of the clasping apparatus.

FEMALE. — The single biunguis female associated with this species was similar to the male, but slightly larger, measuring 260 μ in length; 160 μ in height.

Relationships. — This ostracod appears to have its closest affinities with *Herpetocythere australensis,* but differs from it in that the posteroventral peniferal process is developed to a greater extent in *H. bendora;* and the teeth of the horizontal ramus of *H. bendora* are quite indistinct, whereas in *H. australensis* they are quite evident.

Type-locality. — Bendora, Australian Capital Territory, Australia (types, AM). Host, *Euastacus crassus.*

Distribution. — Known only from type-locality.

Herpetocythere gnoma Hart and Hart, 1967

Plate XL, figures 1-6
Plate LX

Herpetocythere gnoma Hart and Hart, 1967a: 21, 22 (M, R, A, D), 8 (K), 17 (A), figs. 55-60.

Diagnosis. — MALE: Shell length 220-240 μ (average, 230 μ); height 120-140 μ (average, 128 μ). Eye spot apparently absent in some specimens; present in others. Shell reniform to subelliptical in shape, with greatest height at mid-length. Copulatory complex with peniferum terminating in posteroventral and anteroventral processes, more or less apposed to one another; anteroventral process larger than posteroventral process, pointed, bifurcate, and directed posteroventrally. Penis originating above level of origin of clasping apparatus and extending ventrally between bases of peniferal processes. Clasping apparatus with horizontal and vertical rami subequal in length and forming right angle with one another; single tooth on internal margin of horizontal ramus near juncture with vertical ramus; horizontal ramus terminating in five denticles. Dorsal and ventral fingers present.

Variations. — This ostracod appears to be somewhat more variable than most Australian entocytherids. The anteroventral peniferal process may or may not be bifurcate; the tooth on the internal margin of the horizontal ramus may be situated at the juncture of the two rami, or slightly distal to it; and the horizontal ramus may terminate in four or five denticles.

FEMALE. — Shells of biunguis female similar to males but smaller shell length 170-200 μ (average, 185 μ); height 100-120 μ (average, 110 μ).

Relationships. — This ostracod appears to have its closest affinities with *Herpotocythere labidioides.* It differs from it, however, in that the posteroventral peniferal process of *H. labidioides* is not bifurcate as it is in *H. gnoma,* and the horizontal ramus terminates in four or five denticles, rather than six.

Type-locality. — Twenty mi W of Dorrigo, New South Wales, Australia (types, AM). Host, *Euastacus neohirsutus.*

Distribution. — Known only from type-locality.

Herpetocythere labidioides Hart and Hart, 1967

Plate XL, figure 7
Plate LX

Herpetocythere labidioides Hart and Hart, 1967a: 21 (M, R, A, D), 6 (D), 8 (K), 19, 26 (A), 23 (R), fig. 54.

Diagnosis. — MALE: Eye spot absent from the single known specimen. Shell subelliptical in outline with greatest height in posterior ⅓. The single specimen measured 260 μ in length, 120 μ in height. Copulatory complex with anteroventral and posteroventral peniferal processes apposed, almost touching; posteroventral process sharply recurved. Penis originating proximal to level or origin of clasping apparatus, extending distally to line proximal to bases of peniferal processes. Clasping apparatus with rami subequal in length; vertical ramus convex on external margin, concave on internal margin; horizontal ramus slightly convex on external margin, with a single tooth situated at mid-length; terminal portion with six denticles. Dorsal and ventral fingers present.

FEMALE. — The single female specimen (biunguis), similar in shape to the male, measured 180 μ in length and 90 μ in height.

Relationships. — This ostracod probably has its closest affinities with *Herpetocythere mackenziei,* but differs from it in the position of the anteroventral peniferal process (pointing ventrally in *H. mackenziei,* posteroventrally in *H. labidioides*), and in the number of teeth on the internal margin of the horizontal ramus of the clasping apparatus (one in *H. labidioides,* four in *H. mackenziei*).

Type-locality. — Elaman Creek, Maleny, Queensland, Australia (types, AM). Host, *Euastacus hystericosus.*

Distribution. — Known only from type-locality.

Herpetocythere mackenziei Hart and Hart, 1967

Plate XL, figures 8-9
Plate LX

Herpetocythere mackenziei Hart and Hart, 1967a: 19, 20 (M, R, A, D), 8 (K), 6 (D), 21 (R), 19, 21, 26 (R), figs. 48, 49.

Diagnosis. — MALE: Shell subelliptical in outline with eye spot situated approximately ¼ shell length from anterior end; greatest shell height at approximately mid-length. Single male specimen measured 280 μ in length, 160 μ in height; single biunguis female, 230 μ in length, 130 μ in height.

Copulatory complex with posteroventral peniferal process sharply pointed and directed anteriorly. Clasping apparatus with distinct horizontal and vertical rami, approximately equal in length; external margin of horizontal ramus entire and almost straight; external margin of vertical ramus entire and slightly convex; internal margin of vertical ramus concave; internal margin of horizontal ramus with four evenly spaced teeth; terminal portion with four denticles.

FEMALE. — The single biunguis female associated with male was similar in shape to latter, but somewhat smaller (see above).

Relationships. — This ostracod has its closest affinities with *H. australensis,* but differs from it in the relatively greater development of the posteroventral peniferal process and the lesser number of terminal denticles and teeth on the horizontal ramus.

Type-locality. — Elaman Creek, Maleny, Queensland, Australia (types, AM). Host, *Euastacus hystericosus.*

Distribution. — Known only from type-locality.

HESPEROCYTHERE HART AND HART, 1967

Hesperocythere Hart and Hart, 1967a: 7, 29 (M), 7 (K), type-species, *Hesperocythere klasteroides* Hart and Hart, 1967.

Diagnosis. — Peniferum elongate (over 100 μ), tapering distally, terminating in one to three processes. Clasping apparatus straight, originating proximal to origin of penis. Dorsal and ventral fingers present; ventral finger slender and nearly straight; dorsal finger short, bifurcate. Dorsal claw of antenna curved, subequal in length to ventral claw, and bearing subterminal brush of setae.

Type-species. — *Hesperocythere klasteroides* Hart and Hart, 1967.

Distribution. — Southwestern Australian mainland.

Hosts. — Crayfishes.

Figure 44. — Distribution of the genus *Hesperocythere.*

Hesperocythere klasteroides Hart and Hart, 1967

Plate XL, figures 10-11
Plate LXI

Hesperocythere klasteroides Hart and Hart, 1967a: 29, 30 (M, R, A, D), 8 (K), 30, 31 (M), 31 (R), figs. 83-85.

Diagnosis. — MALE: Shell length 310-320 μ (average, 314 μ); height 190-200 μ (average, 196 μ). Eye spot present, situated approximately 1/3 shell length from anterior end. Shell subelliptical in outline, with greatest height at mid-length.

Copulatory complex with peniferum greatly elongate (when compared with other Australian entocytherids); smoothly convex on posterior margin, and bearing two simple terminal processes between which lies a loosely attached, reniform, bulbous device. Penis arising in distal 1/4 of peniferum and extending ventrally between bases of terminal peniferal processes. Clasping apparatus straight; external margin sinuous; internal margin with five teeth (distal two indistinct); terminating in five denticles. Dorsal and ventral fingers present; ventral finger straight; dorsal finger short, bifurcate.

FEMALE. — Unknown.

Relationships. — This ostracod probably has its closest affinities with *Hesperocythere xiphoides,* which it resembles by having two terminal peniferal protuberances. Its clasping apparatus, however, is more like that of *H. tallanalla* in that both taper distally, rather than becoming thicker as in *H. xiphoides.*

Type-locality. — Beedelup Falls, Western Australia, Australia (types, AM). Host, *Cherax* sp.

Distribution. — Known only from type-locality.

Hesperocythere tallanalla Hart and Hart, 1967

Plate XL, figures 12-14
Plate LXI

Hesperocythere tallanalla Hart and Hart, 1967a: 30 (M, R, A (*lapsus calami* — line 41, associate, should read *Riekocythere cherax*), D), 8 (K), 28 (A), 30 (R), figs. 86-88.

Diagnosis. — MALE: Shell length 360-390 μ (average, 375 μ); height 210-250 μ (average, 230 μ). Eye spot present, situated 1/4 shell length from anterior end. Shell subelliptical in outline with greatest height at mid-length.

Copulatory complex with peniferum greatly elongate; posterior margin convex proximally, concave distally; distal extremity of peniferum terminating in one anteroventrally-directed posteroventral process, subequal in length to peniferal width at its base. Anteroventral lip may be interpreted as a much-reduced process. Origin of penis indistinct, but probable in distal 1/4 of peniferum; extending vertically to base of posteroventral peniferal process. Clasping apparatus tapering distally; possessing convex flange in proximal 2/3 of external margin; six teeth on internal margin, diminishing in size with most distal indistinct; four terminal denticles. Dorsal and ventral fingers present; ventral finger slender; dorsal finger short, bifurcate.

FEMALE. — Unknown.

Relationships. — See discussion under *Hesperocythere klasteroides.*

Type-locality. — Tallanalla, Western Australia, Australia (types, AM). Host, *Cherax* sp.

In addition to the type-locality, this ostracod is also known from 15 mi W of Pemberton, Western Australia, Australia, where the host was *Cherax* sp.

Hesperocythere xiphoides Hart and Hart, 1967

Plate XL, figures 15-17
Plate LXI

Hesperocythere xiphoides Hart and Hart, 1967a: 31 (M, R, A, D), 8 (K), 30 (R), 27, 29 (A), figs. 89-91.

Diagnosis. — MALE: Shell length 310-320 μ (average, 317 μ); height 150-160 μ (average, 153 μ). Eye spot present, situated 1/4 shell length from anterior end. Shell subelliptical in outline with greatest height near mid-length.

Copulatory complex with greatly elongate peniferum; posterior margin convex in proximal 3/4, straight in distal 1/4; distal extremity bearing three processes — anteroventral process acute and curved posteriorly; posteroventral process blunt and scoop-shaped, and directed ventrally; poorly sclerotized blunt process at base of posteroventral process directed posteroventrally. Penis originating in distal 1/4 of peniferum and extending distally to base of posteroventral process. Clasping apparatus straight, concave on external margin, concave in proximal half of internal margin, bearing five teeth on distal half of internal margin (the distal two are present only as very indistinct ridges), and bearing four terminal denticles. Ventral finger slender; dorsal finger short and bifurcate.

FEMALE. — Unknown.

Relationships. — See discussion under *Hesperocythere klasteroides.*

Type-locality. — Kalgan River, 10 mi W of Albany, Western Australia, Australia (types, AM). Host, *Cherax* sp.

This ostracod is also known from the following locality: West Australia — 5 mi NW of Shannon River Mill (host, *Cherax* sp.).

LACCOCYTHERE HART AND HART, 1971

Laccocythere Hart and Hart, 1971: 579 (M), type-species, *Laccocythere aotearoa* Hart and Hart, 1971.

Diagnosis. — Antennule with six podomeres. Dorsal antennal claw truncate and devoid of setae. Mandibular protopodite with distal row of five teeth. Peniferum terminating bluntly with two possibly opposable processes. Penis short, ventrally directed, and situated entirely above mid-length of peniferum.

Type-species. — *Laccocythere aotearoa* Hart and Hart, 1971.

Distribution. — New Zealand: South Island.

Host. — Crayfishes.

Figure 45. — Distribution of the genus *Laccocythere.*

Laccocythere aotearoa Hart and Hart, 1971

Plate XL, figures 18-21
Plate LIX

Laccocythere aotearoa Hart and Hart, 1971: 579-583 (M, H, R, A, D), figs. 1-8.

Diagnosis. — MALE: Shell length 280-310 μ (average, 298 μ); height 160-190 μ (average, 170 μ).

Copulatory complex with ventral portion of peniferum blunt and terminating in two sclerotized, tapering, processes, which may or may not be apposable. Penis short, curved, ventrally directed, and situated above midlength of peniferum. Penis apparently introduced into well-defined duct leading antero-ventrally to anterior face of peniferum. Clasping apparatus falcate, bearing no teeth on external surface, four terminal denticles, and four irregular teeth on internal surface.

Variations. — On some specimens terminal denticles may be more or less indistinguishable from teeth on the internal margin. A paratypic specimen has four teeth that might be considered terminal denticles and two or possibly three that might be considered teeth on the internal margin. The most proximal of these teeth might be described as a large truncated tooth or as two distinct teeth — depending upon the whim or point of view of the describer.

TRIUNGUIS FEMALE: Shell length 320 μ; height, 180 μ. Shell similar to that of adult male in shape, but averaging 22 μ more in length. Conspicuous slender conical apparatus similar to that described in *Elachistocythere merista* from New Guinea (see p. 148). This apparatus probably homologous with *amiculum* of the genus *Dactylocythere* in North America.

Relationships. — This ostracod appears to have close affinities with the entocytherid subfamily Notocytherinae of Australia and New Guinea, as well as with the Indian subfamily *Microsyssitrinae.* It is similar to the Australian and New Guinea species in having six antennular podomeres (as opposed to five in the Indian), but similar also to the Indian and New Guinea forms in having five mandibular palp teeth (as opposed to six in the Australian). We feel, however, that it should be placed in the subfamily Notocytherinae.

Although the distal portion of the peniferum is apparently bifid, we doubt that the two projections are apposable as they are in *Microsyssitria indica* and as they may be in *Chelocythere kalganensis.*

Type-locality. — Pelorus Valley, South Island, New Zealand (types, USNM). Host, *Paranephrops planifrons.*

Distribution. — Known only from type-locality.

LICHNOCYTHERE HART AND HART, 1967

Lichnocythere Hart and Hart, 1967a: 6, 23, 24 (M), 7 (K, R), type-species, *Lichnocythere victoria* Hart and Hart, 1967. — Hart, Nair, and Hart, 1967: 4 (R). — Hart and Hart, 1969a: 158 (M, R).

Diagnosis. — Distal portion of peniferum terminating in two well-developed claw-like processes — posteroventral process heavier than anteroventral process, and possibly hinged so that the two may be movably apposed. Dorsal antennal claw curved, crenulate on flexor face and terminating in finger-like projection.

Type-species. — *Lichnocythere victoria* Hart and Hart, 1967a.

Distribution. — Eastern Australian mainland.

Host. — Crayfishes.

Figure 46. — Distribution of the genus *Lichnocythere.*

Lichnocythere synethes Hart and Hart, 1967

Plate XLI, figures 1-3
Plate LXI

Lichnocythere synethes Hart and Hart, 1967a: 25, 26 (M, R, A, D), 7 (M), 8 (K), 26 (R), figs. 67-69.

Diagnosis. — MALE: Shell length 230-250 μ (average, 245 μ); height 130-150 μ (average, 145 μ). Eye spot present. Shell subelliptical in outline, with greatest height at mid-length.

Copulatory complex with peniferum terminating in two well-developed processes; posteroventral process with convex posterior margin, external margin of posteroventral process with dorsally directed heel-like protuberance; base of posteroventral process extending mesially, heavily sclerotized, and possibly serving as a pivot on which process may be moved. Anteroventral process tapering to point. Penis with origin near level of origin of clasping apparatus, extending ventrally between bases of distal peniferal processes. Clasping apparatus with horizontal ramus $\frac{2}{3}$ length of vertical ramus; external margins of both rami convex, internal margins concave; internal margin of horizontal ramus with one large proximal tooth and three smaller teeth situated distal to it, four terminal denticles. Dorsal and ventral fingers present.

Variations. — Among the specimens examined, the terminal denticles of the clasping apparatus horizontal ramus varied — numbering either four or five.

FEMALE. — Shell of the single biunguis female was similar in shape to that of the males, but smaller in size. Length 220 μ; height 140 μ.

Relationships. — This ostracod has its closest affinities with *Lichnocythere tubrabucca,* but differs from it in having fewer teeth on the clasping apparatus horizontal ramus, and also in having a larger posteroventral peniferal process.

Type-locality. — Mt. Glorious, Queensland, Australia (types, AM). Host, *Euastacoides setosus.*

Distribution. — Known only from type-locality.

Lichnocythere tubrabucca Hart and Hart, 1967

Plate XLI, figures 4-6
Plate LXI

Lichnocythere tubrabucca Hart and Hart, 1967a: 26 (M, H, R, A, D), 6 (D), 7, 25 (R), 8 (K), 9, 21 (A), figs. 70-74. — Hart and Hart, 1969a: 180, 181 (M), fig. 4. — Hart and Hart, 1969b: 157 (M), fig. 4.

Diagnosis. — MALE: Shell length 240-280 μ (average, 258 μ); height 140-170 μ (average, 156 μ). Eye spot present, situated approximately $\frac{1}{4}$ shell length from anterior end. Shell subelliptical in outline, with greatest height at mid-length.

Copulatory complex with posteroventral peniferal process similar to that of *Lichnocythere synethes,* but shorter; anteroventral process truncate. Penis with origin on level of clasping apparatus origin, descending vertically to level of bases of distal peniferal processes. Clasping apparatus with horizontal and vertical rami merging imperceptibly; external margins of both rami convex; internal margins of both rami concave; internal margin of horizontal ramus with seven teeth, becoming smaller distally; terminal portion of horizontal ramus with three denticles.

FEMALE. — Shell of the single biunguis female specimen similar to that of the males, but smaller. Length 230 μ; height 140 μ.

Relationships. — This ostracod appears to have its closest affinities with *Lichnocythere synethes,* but differs primarily in the number and size of the teeth on the internal margin of the clasping apparatus horizontal ramus.

Type-locality. — Barrington Tops, Tubrabucca Creek, New South Wales, Australia (types, AM). Host, *Euastacus* sp.

Distribution (in addition to type-locality). —

Australia

QUEENSLAND — Cunninghams Gap (host: *Euastacus cunninghami.* Elaman Creek, Maleny (host: *Euastacus hystricosus*). Taromeo (host: *Euastacus cunninghami*).

Lichnocythere victoria Hart and Hart, 1967

Plate XLI, figures 7-10
Plate LXI

Lichnocythere victoria Hart and Hart, 1967a: 24, 25 (M, H, R, A, D), 8 (K), 25, 26 (R), 23 (A), figs. 63-66.

Diagnosis. — MALE: Shell length 300-330 μ (average, 308 μ); height 170-200 μ (average, 177 μ). Eye spot present, situated approximately $\frac{1}{4}$ shell length from anterior end. Shell subelliptical in outline, with greatest height at mid-length.

Copulatory complex with peniferum terminating in two well-developed claw-like processes; posteroventral process convex on posterior margin, concave on anterior margin, and terminating in anteriorly-directed point; anteroventral process tapering distally and truncate. Penis originating near level of origin of clasping apparatus and extending ventrally to point between bases of terminal peniferal processes. Clasping apparatus with juncture of horizontal and vertical rami indistinct; external margins of both rami convex, internal margins of both rami concave; internal margin of horizontal ramus bearing six evenly-spaced teeth and terminating in four denticles.

Variations. — Teeth on internal margin of clasping apparatus vary in number from six to seven; terminal denticles vary from three to four. Because of their proximity and size, it is often a matter of personal judgement as to where the teeth end and the denticles begin.

FEMALE. — Shell of single known biunguis female similar to that of male, and only slightly smaller. Shell length 280 μ; height, 180 μ.

Relationships. — This ostracod probably has its closest affinities with *Lichnocythere tubrabucca.* It differs from it, however, in that the posteroventral peniferal process is more heavily sclerotized in *L. tubrabucca* than in *L. victoria.*

Type-locality. — Mt. Victoria, Blue Mountains, New South Wales, Australia (types, AM). Host, *Euastacus nobilis.*

Distribution. — Known only from type-locality.

NOTOCYTHERE HART AND HART, 1967

Notocythere Hart and Hart, 1967a: 6 (D, R), 8 (K), 9 (M), type-species, *Notocythere syssitos* Hart and Hart, 1967. — Sohn, 1968: 441 (C).

Diagnosis. — Terminal portion of mandibular palp rounded or truncate. Dorsal antennal claw curved, without setae, and terminating bluntly. Peniferum narrowing abruptly in distal portion, terminating in one to three variously directed protuberances, none of which appears to be movable.

Type-species. — *Notocythere syssitos* Hart and Hart, 1967.

Distribution. — Southeastern Australia, including Tasmania.

Hosts. — Crayfishes and isopods.

Figure 47. — Distribution of the genus *Notocythere.*

Notocythere antichthon Hart and Hart, 1967

Plate XLI, figures 11-12
Plate LX

Notocythere antichthon Hart and Hart, 1967a: 16, 17 (M, H, R, A, D), 9 (K), 22 (A), figs. 36, 37.

Diagnosis. — MALE: Shell length 300 μ; height 150 μ (single specimen).

Eye spot present, situated $\frac{1}{4}$ shell length from anterior end of shell. Shell subelliptical in outline, with greatest height at mid-length.

Copulatory complex with terminal portion of peniferum devoid of acute prominences as in other members of genus; posteroventral prominence blunt and not heavily developed. Penis originating at level of origin of clasping apparatus; extending sinuously in vertical direction and terminating short distance from distal extremity of peniferum. Clasping apparatus without distinct juncture between horizontal and vertical rami; horizontal ramus with five teeth on internal border, three terminal denticles, and two almost imperceptible protuberances in proximal half of external margin. Dorsal and ventral fingers of single male specimen apparently broken off.

FEMALE. — No female specimens have been assigned to this species.

Relationships. — This ostracod appears to have its closest affinity with *N. erica.* It differs from *N. erica* in that it lacks sharp terminal peniferal protuberances, but is similar to it in that the unusual prominences on the external margin of the horizontal ramus of the clasping apparatus are common to both species.

Type-locality. — Twenty mi W of Dorrigo, New South Wales, Australia (types, AM). Host, *Euastacus neohirsutus.*

Distribution. — Known only from type-locality.

Notocythere blundelli Hart and Hart, 1967

Plate XLI, figures 13-16
Plate LX

Notocythere blundelli Hart and Hart, 1967a: 17, 18 (M, H, R, A, D), 9 (K), 14 (A), figs. 38-41.

Diagnosis. — MALE: Shell length 260-290 μ (average, 279 μ); height 140-170 μ (average, 156 μ). Eye spot present, situated approximately $\frac{1}{4}$ shell length from anterior end. Shell subelliptical in outline, with greatest height in posterior third.

Copulatory complex with peniferum terminating bluntly; posteroventral projection rounded. Penis originating on level with origin of clasping apparatus, and extending ventrally to point near distal extremity of peniferum. Clasping apparatus divided into horizontal and vertical rami; horizontal ramus $\frac{2}{3}$ length of vertical ramus; external margin of vertical ramus nearly straight, internal margin concave; external margin of horizontal ramus nearly straight internal margin with five teeth, distal extremity with five denticles.

FEMALE. — Shell of single biunguis female similar to that of male. Shell length 260 μ; height 130 μ.

Relationships. — This ostracod probably has its closest affinities with *N. antichthon,* but differs from it primarily in the shape of the clasping apparatus (smoothly curved in *N. antichthon;* distinct rami in *N. blundelli*).

Type-locality. — Condor Creek, Blundell's Farm, Australian Capital Territory, Australia (types, AM). Host, *Engaeus parvulus.*

Distribution. — Known only from type-locality.

Notocythere erica Hart and Hart, 1967

Plate XLI, figures 17-18
Plate LX

Notocythere erica Hart and Hart, 1967a: 15, 16 (M, H, R, A, D), 9 (K), 17 (R), figs. 34, 35.

Diagnosis. — MALE: Shell length 290-210 μ (average 282 μ); height 130-170 μ (average, 152 μ).

MALE. — Eye spot present, situated $\frac{1}{3}$ shell length from anterior end of shell. Shell subelliptical in outline; greatest height at mid-length; ventral margin slightly concave.

Copulatory complex with two terminal processes of peniferum apposed to one another. Penis originating near level of origin of clasping apparatus and extending vertically to level of terminal peniferal processes; distal extremity of penis located near circular or spiral corneous device probably serving as a guide. Clasping apparatus with distinct horizontal and vertical rami; vertical ramus smoothly convex on external margin, concave on internal margin; horizontal ramus only slightly curved, bearing four teeth on internal margin, six terminal denticles, and two almost imperceptible protuberances on proximal half of external margin.

FEMALE. — Shell of biunguis females essentially similar to those of males, but averaging slightly smaller. Shell length 240-280 μ (average, 256 μ); height 140-160 μ (average, 146 μ).

Relationships. — This ostracod appears to have its closest relationships with *Notocythere grampians* and *N. antichthon.* It is similar to *N. grampians* in the peculiar spiraled guide at the distal end of the penis, and resembles *N. antichthon* in the general plan of the clasping apparatus.

Type-locality. — Mt. Erica, Victoria, Australia (types AM). Host, *Engaeus affinis.*

Distribution. — Known only from type-locality.

Notocythere grampians Hart and Hart, 1967

Plate XLII, figures 1-3
Plate LX

Notocythere grampians Hart and Hart, 1967a: 12, 13 (M, H, R, A, D), 9 (K), 16 (R), figs. 16-20.

Diagnosis. — MALE: Shell length 340-380 μ (average, 354 μ); height 160-180 μ (average, 177 μ). Eye spot apparently absent; pigment possibly bleached in preservative. Shell subelliptical in outline with greatest height at mid-length.

Copulatory complex with ventral portion of peniferum terminating in two posteroventrally directed processes and rounded, anteriorly situated, projection; two posteroventral processes appearing as posterior angles of two truncated projections. Penis originating in lower ⅓ of peniferum, at level of origin of clasping apparatus; extending straight ventrally and terminating adjacent to rounded projection in ventral extremity of peniferum. Clasping apparatus composed of distinct vertical and horizontal rami forming obtuse angle with one another; external margin of vertical ramus slightly convex; external margin of horizontal ramus slightly concave; five evenly spaced teeth on internal margin of horizontal ramus; terminal portion of horizontal ramus swollen and bearing six denticles. Dorsal and ventral fingers present, similar to those of *N. syssitos.*

FEMALE. — Biunguis female specimens similar in shell shape to males, but averaging slightly smaller. Shell length 240-310 μ (average, 275 μ); height 130-150 μ (average, 138 μ).

Relationships. — This ostracod appears to have its closest relationships with *Notocythere mirranatwa.* It differs from that ostracod in that the dorsal claw of the antenna is less acutely curved, the external margin of the horizontal ramus of the clasping apparatus is concave instead of convex, and the distal portion of the clasping apparatus is slightly swollen.

Type-locality. — Hall's Gap, Grampians area, Victoria, Australia (types, AM). Host, phreatoicoid isopods.

Distribution. — Known only from type-locality.

Notocythere mirranatwa Hart and Hart, 1967

Plate XLII, figures H-S
Plate LX

Notocythere mirranatwa Hart and Hart, 1967a: 11-12 (M, H, R, A, D), 9 (K), 11 (A), 13 (R), figs. 14, 15.

Diagnosis. — MALE: Shell length 250-270 μ (average, 260 μ); height 130-150 μ (average, 143 μ).

Eye spot present. Shell essentially identical to that of *Notocythere syssitos.*

Copulatory complex with ventral portion of peniferum terminating in three ventrally directed protuberances. Penis originating ventral to origin of clasping apparatus in lower ⅓ of peniferum, extending ventrally as straight, apparently tubular organ, and terminating among three terminal protuberances of peniferum. Clasping apparatus composed of distinct horizontal and vertical rami; external margins of both rami convex and without teeth; internal margin of horizontal ramus with five teeth; horizontal ramus terminating in five denticles. Dorsal and ventral fingers present, similar to those of *N. syssitos.*

FEMALE. — Shell of one biunguis female similar to that of male; length 250 μ, height 130 μ.

Relationships. — This ostracod bears its closest relationships with *Notocythere syssitos.* It differs from it in that the penis originates below rather than above the origin of the clasping apparatus, the horizontal and vertical rami of the clasping apparatus form a right angle with one another, and the overall shell size appears to be smaller.

Type-locality. — Mirranatwa, Grampians area, Victoria, Australia (types, AM). Host, *Geocherax* sp.

Distribution (in addition to type locality). —

Australia

VICTORIA — Lake Learmonth (hosts, *Cherax* sp. and *Engaeus* sp.).

TASMANIA — NE of Mt. Bowes (Host, *Parastacoides* sp.).

Notocythere rieki Hart and Hart, 1967

Plate XLII, figures 6-7
Plate LX

Notocythere rieki Hart and Hart, 1967a: 15 (M, H, R, A, D), 9 (K), 14 (R), 20 (A), figs. 30-33.

Diagnosis. — MALE: Shell length 260-290 μ (average, 278 μ); height 150-180 μ (average, 160 μ).

Copulatory complex with two well developed terminal processes on peniferum, most anterior of which is pointed and apparently receives and serves as guide for terminal portion of penis; posteroventral protuberance strongly developed into two points. Penis originating at approximately level of origin of clasping apparatus and extending almost straight vertically to level of anterior peniferal protuberance. Clasping apparatus with distinct horizontal and vertical rami; horizontal ramus sinuously concave on external margin, bearing six teeth on internal margin and four denticles distally; vertical ramus concave on internal margin, convex on external margin. Dorsal and ventral fingers apparently somewhat poorly developed.

FEMALE. — No female specimens have been assigned to this species.

Relationships. — This ostracod is most closely related to *Notocythere synomodites,* but differs from it in the development of the anterior peniferal protuberance (more strongly developed in *N. rieki* than in *N. synomodites*) and in the shape of the clasping apparatus (smoothly curved in *N. synomodites,* horizontal and vertical rami distinct in *N. rieki*).

Type-locality. — Bendora, Australian Capital Territory, Australia (types, AM). Host, *Euastacus crassus.*

Distribution. — Known only from type-locality.

Notocythere synomodites Hart and Hart, 1967

Plate XLII, figures 8-10
Plate LX

Notocythere synomodites Hart and Hart, 1967a: 14 (M, H, R, A, D), 9 (K), 14 (R), 18 (A), figs. 26-29.

Diagnosis. — MALE: Shell length 350 μ; height 200 μ (single specimen). Eye spot present, situated approximately ⅕ shell length from anterior end. Shell subelliptical in outline, similar to that of *N. tasmanica.*

Copulatory complex with peniferum terminating in well developed posteroventral protuberance and pointed anterior protuberance. Penis originating proximal to origin of clasping apparatus, extending ventrally with only slight curve to level of anterior peniferal protuberance. Clasping apparatus without distinct horizontal and vertical rami; external margin convex and entire; distal half of internal margin with three indistinct denticles. Dorsal and ventral fingers similar to those of other members of genus.

FEMALE. — No female has yet been assigned to this species.

Relationships. — This ostracod bears its closest relationships with *Notocythere rieki* (see above). It differs from it in that the peniferal protuberances are not as heavily developed as in *N. rieki,* and in that the external margin of the horizontal ramus of the clasping apparatus is convex in *N. synomodites* and concave in *N. rieki.*

Type-locality. — Condor Creek, Blundell's Farm, Australian Capital Territory, Australia (types, AM). Host, *Engaeus parvulus.*

Distribution. — Known only from type-locality.

Notocythere syssitos Hart and Hart, 1967

Plate XLII, figures 11-20
Plate LX

Notocythere syssitos Hart and Hart, 1967a: 9-11 (M, H, R, A, D), 9 (K), 6 (D), 12, 13, 15, 24 (R), figs. 3-13.

Diagnosis. — MALE: Shell length 280-360 μ (average, 327 μ); height 160-190 μ (average, 175 μ). Shell subelliptical in outline with eyespot situated approximately ¼ shell length from anterior end.

Copulatory complex with ventral portion of peniferum terminating in three points or protuberances; posterior two directed posteroventrally, more anterior one directed ventrally. Penis originating above origin of clasping apparatus at approximately midlength of peniferum; extending ventrally as straight, apparently tubular organ, and terminating adjacent to three terminal points of peniferum. Clasping apparatus composed of horizontal and vertical rami, merging with one another without sharp line of demarcation; external margins of both rami convex and without teeth; internal margin of vertical ramus concave and bearing three teeth, proximal one being larger; horizontal ramus terminating in seven denticles. Ventral finger subequal in length to clasping apparatus, dorsal finger extending only slightly beyond base of clasping apparatus.

Variations. — The internal margin of the horizontal ramus of the clasping apparatus may have three to five teeth of varying size; the terminal denticles of the horizontal ramus vary from four to seven, but the matter of where the internal margin teeth end and the denticles begin may be subject to varying interpretations. In some specimens the break between the horizontal and vertical rami is more distinct than in others. The terminal points or protuberances of the peniferum may vary in direction; in some specimens the two posteroventral points are directed ventrally; in others, more posteriorly. The penis appears to vary from straight to somewhat sinuous in different specimens.

TRIUNGUIS FEMALE: Shell length 330-340 μ (average, 335 μ); height 150-160 μ (average, 155 μ).

Relationships. — This ostracod appears to have its closest affinities with *Notocythere grampians.* It differs from it in having a greater number of teeth on the internal margin of the clasping apparatus horizontal ramus and in having three peniferal protuberances rather than two.

Type-locality. — Lake Lilla, Cradle Mountain, Tasmania, Australia (types, AM). Hosts, *Astacopsis* sp. and *Parastacoides leptomerus.*

Distribution (in addition to type-locality). —

Australia

VICTORIA — Lake Learmonth (hosts: *Cherax* sp. and *Engaeus* sp.).

QUEENSLAND — Elaman Creek, Maleny (host: *Cherax punctatus*).

Notocythere tasmanica Hart and Hart, 1967

Plate XLIII, figures 1-2
Plate LX

Notocythere tasmanica Hart and Hart, 1967a: 13, 14 (M, H, R, A, D), 9 (K), 14 (R), figs. 21-25.

Diagnosis. — MALE: Shell length 310-330 μ (average, 320 μ); height 180-200 μ (average, 190 μ). Eye spot present situated approximately $\frac{1}{5}$ shell length from anterior end of shell. Shell subelliptical in outline, with greatest height at approximately mid-length.

Copulatory complex with peniferum terminating in two poorly defined protuberances, one directed ventrally, other posteroventrally. Penis originating in ventral half of peniferum above origin of clasping apparatus, and extending sinuously to a point between two terminal protuberances. Clasping apparatus divided into horizontal and vertical rami, merging imperceptibly with one another; entire external margin convex; internal margin of horizontal ramus with four evenly spaced teeth; horizontal ramus terminating in five denticles.

FEMALE: No female specimens have been definitely assigned to this species.

Relationships. — This ostracod appears to have its closest relationships with *Notocythere synomodites* (see above). It differs from it in that the two ventral protuberances of the peniferum are not as well developed in *N. tasmanica* as they are in *N. synomodites,* and the rami of the clasping apparatus of *N. tasmanica* form a more distinct angle with one another than they do in *N. synomodites.*

Type-locality. — Cuvier River, Lake St. Clair, Tasmania, Australia (types, AM). Host, *Parastocoides* sp.

Distribution. — Known only from type-locality.

RIEKOCYTHERE HART AND HART, 1967

Riekocythere Hart and Hart, 1967a: 4, 27, 28 (M), 7 (R, K, D), type-species, *Riekocythere cherax* Hart and Hart, 1967. — Hart and Hart, 1969b: 158 (M, R).

Diagnosis. — Peniferum tapering distally, terminating in single process. Clasping apparatus straight, originating slightly distal to proximal portion of penis. Dorsal and ventral fingers delicate and without pronounced bend. Arrow- or dagger-shaped device situated anterodorsally of fingers. Dorsal claw of antenna shorter than either mesial or ventral claws, devoid of setae, and tapering to sharp apex.

Type-species. — *Riekocythere cherax* Hart and Hart, 1967a.

Distribution. — Southwestern Australia.

Host. — Crayfishes.

Figure 48. — Distribution of the genus *Riekocythere.*

Riekocythere cherax Hart and Hart, 1967

Plate XLIII, figure 3
Plate LXI

Riekocythere cherax Hart and Hart, 1967a: 28 (M, H, R, A, D), 8 (K), 28 (R), 30 (A) (*lapsus calami*), fig. 80.

Atopocythere cherax. — Hart and Hart, 1967a: 30 (*lapsus calami*).

Diagnosis. — MALE: Eye spot present, situated approximately ¼ shell length from anterior end. Shell subelliptical in outline with greatest height near midlength. The single specimen measured 250 μ in length, 140 μ in height.

Copulatory complex with peniferum tapering to form a single, blunt process, adorned with two anteriorly-directed tubercles. Penis originating proximal to apparent origin of clasping apparatus, extending ventrally to within short distance of terminal portion of peniferum. Clasping apparatus composed of single ramus, bearing two teeth on its internal margin and eight terminal denticles; external margin convex proximally, concave distally. Ventral finger delicate and almost straight, dorsal finger heavier and ¼ length of ventral finger. Arrow-shaped corneous device of indeterminate origin situated dorsal of fingers (this possibly homologous with finger guard found in some North American species).

FEMALE: Unknown.

Relationships. — This ostracod has its affinities with *Riekocythere xenika,* but differs from it in having a shorter and less ornate peniferum, and in having three instead of two teeth on the clasping apparatus.

Type-locality. — Fifteen mi W of Pemberton, Western Australia, Australia (types, AM). Host, *Cherax* sp. (probably *C. crassimanus*).

Distribution. — Known only from type-locality.

Riekocythere xenika Hart and Hart, 1967

Plate XLIII, figures 4-5
Plate LXI

Riekocythere xenika Hart and Hart, 1967a: 28, 29 (M, H, R, A, D), 8 (K), 27, 31 (A), figs. 81, 82.

Diagnosis. — MALE: Shell length 250 μ; height, 160 μ.

Copulatory complex with peniferum extending distal of end of penis for distance subequal to length of penis. Penis with origin proximal to apparent origin of clasping apparatus; extending vertically for approximately 13 μ. Clasping apparatus straight; external margin sinuous; internal margin with five teeth, terminating in six denticles. Ventral finger slender and almost straight, dorsal finger short and bifurcate. Dagger-shaped corneous device of indeterminate origin situated dorsal of the fingers.

FEMALE: Unknown.

Relationships. — See discussion under *Riekocythere cherax.*

Type-locality. — Five mi NW of Shannon River Mill, Western Australia, Australia (types, AM). Host, *Cherax* sp.

Distribution. — Known only from type-locality.

Subfamily Sphaeromicolinae Hart, 1962

Sphaeromicolinae Hart, 1962: 139 (M, H, D), type-genus *Sphaeromicola* Paris, 1916. — Hart and Hart, 1969a: 182, 183 (R), fig. 5. — Hobbs, 1971: (M).
Tribe Sphaeromicolini. — Danielopol, 190 (R).

Diagnosis. — Distal podomere of mandibular palp spatulate in shape; respiratory plate of maxilla lacking; masticatory lobes of maxilla vestigial. Peniferum shortened, usually having highly convex posterior margin.

Type of subfamily. — *Sphaeromicola topsenti* Paris 1916.

Distribution. — Freshwater habitats of France, Italy, Yugoslavia, and Mexico. Hosts, isopods of the family Sphaeromatidae in Europe, and isopods of the family Cirolanidae in Mexico.

SPHAEROMICOLA PARIS, 1916

Sphaeromicola Paris, 1916: 307-309 (M, H, D), type-species, *Sphaeromicola topsenti* Paris, 1916. — Klie, 1931: — Stammer, 1935: 97 (H, R). — Rioja, 1950: 170 (M, R). — Delamare Deboutteville, 1960: 640 (E), 660 (E, H). — Hart, 1962: 121-124 (C), 124 (K), 139 (C, D, S). — Hartmann, 1963: 145 (C). — Rudjakov, 1963: 36 (B, H). — Vandel, 1964: 303 (C). — Ferguson, 1965: 12 (B). — Roelofs, 1966: 147-152 (B). — Hart and Hart, 1967a: 1, 6 (R). — McKenzie, 1967: 230 (B, H). — Hart and Hart, 1969a: 182 (D, E). — Hobbs, 1971: 44-45 (M, D, H). — Danielopol, 1971a: 190 (Z). — Danielopol, 1971b: 189-207 (M, D, H).

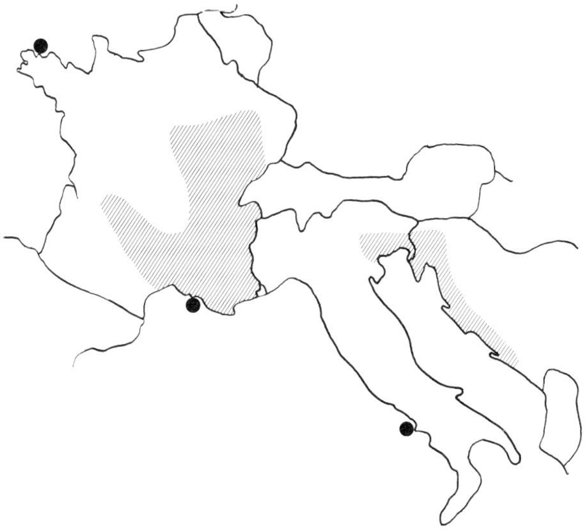

Figure 49. — Distribution of the genera *Sphaeromicola* and *Hartiella*. Diagonal hatching = *Sphaeromicola;* dots = *Hartiella.*

Diagnosis. — Same as that for subfamily (see above).

Type-Species. — *Sphaeromicola topsenti* Paris, 1916: 307.

Distribution. — Caves of France, Italy, Yugoslavia, and Mexico.

Hosts. — Fresh water isopods, marine amphipods.

Key to the Species of the Genus Sphaeromicola

(Modified from Hobbs, 1971)

1 Clasping apparatus with vertical and horizontal rami disposed at angle of 90° or less; its internal border with one or two teeth 2

1′ Clasping apparatus with rami (if at all distinct) disposed at angle much greater than 90° or curved throughout, its internal border lacking teeth 4

2(1′) Ventral finger of copulatory complex with single setiform tip 6

2′ Ventral finger of copulatory complex with two setiform tips 3

3(2′) Ventral surface of peniferum (masse pyriformis) with anteroventrally directed prominence .... *S. hamiger*

3′ Ventral surface of peniferum (masse pyriformis) with posteriorly directed prominence .... *S. cebennicus*

4(1′) Ventral surface of peniferum tapering, without acute or spiniform processes .... *S. stammeri*

4′ Ventral surface of peniferum with one or more spiniform or acute processes 5

5(4′) Clasping apparatus with six apical denticles; ventral surface of peniferum bearing long slender anteroventrally directed process and shorter acute projections .... *S. topsenti*

5′ Clasping apparatus with five apical denticles; ventral surface of peniferum lacking long slender anteroventrally directed process but with acute projection .... *S. sphaeromidicola*

6(2) Clasping apparatus with distinct tooth on internal border of horizontal ramus of clasping apparatus .... *S. cirolanae*

6′ Clasping apparatus with no tooth on internal border of horizontal ramus of clasping apparatus .... *S. coahuiltecae*

Sphaeromicola cebennicus Remy, 1948

Plate XLIII, figures 6-7
Plate LXII

Sphaeromicola cebennica Remy, 1948b: 129-132 (M, R, D, H), figs. 1-8. — Remy, 1951: 221 (D, H), 228 (C). — Hart, 1963: 139 (C, D, S). — Vandel, 1964: 303 (C). — Hobbs, 1971: 81, 45 (K).

Diagnosis. — MALE: Shell length 430-492 μ (average, 461 μ); height 246-261 μ (average, 253 μ). Copulatory complex with peniferum small (subequal) in size to clasping apparatus; "Piriforme mass" in distal portion of peniferum probably homologous with penis and associated terminal claw as in certain members of subfamily Notocytherinae. Clasping apparatus with horizontal and vertical rami forming acute angle with one another; horizontal ramus with single tooth at mid-length of internal border; six terminal denticles.

Type-locality. — Grotte de la Dragonière, commune de Labastide-de-Virac, canton de Vallon, Départment de l'Ardeche, France, about 60 km NW of Avignon (host, *Sphaeromides Raymondi*). Types not known.

Distribution (in addition to type-locality). —

France

DÉPARTMENT DE L'HERAULT, canton de Saint-Martin-de-Londres, résurgence des Cent-Fonts au Causse de la Salle, about 60 km W of Nimes (hosts, *Sphaeromides* sp. and possibly *Faucheria faucheri*) (Remy, 1951: 221).

Sphaeromicola cebennicus juberthiei Danielopol, 1971

Sphaeromicola cebennica juberthiei Danielpol, 1971b: 191, 201 (*nomen nudum*).

Sphaeromicola cirolanae Rioja, 1951

Plate XLIII, figures 8-14
Plate LXI

Sphaeromicola cirolanae Rioja, 1951: 170-179 (M, H, R, A, D), figs. 1-16. — Rioja, 1953: 290, 291 (C), 292 (R, E). — Levinson, 1959: 259 (C). — Hart, 1962: 122 (D), 139 (C, D, S). — Nicholas, 1962: 168 (B, C, D). — Vandel, 1964: 303 (C). — Hobbs, 1971: 45-47 (K, M, H, R, A, D), figs. 30-31. — Reddell and Mitchell, 1971: 142 (C).

Diagnosis. — MALE: Shell length 290-310 μ (average, 300 μ); height 160-180 μ (average, 170 μ). "Clasping apparatus (crochet) strongly bent with rami forming angle of less than 90 degrees with external border of bend rounded or produced in acute angle; internal border of horizontal ramus with one or two teeth and distal extremity with five or six denticles. Ventral surface of peniferum produced in retractable finger-like lobe bearing small, variable, subterminal projection. Ventral finger (flagellum) with two setiform processes at base, variously curved, and terminating in single setiform tip." (Hobbs, 1971).

Type-locality. — Cueva de los Sabinos, Valles, San Luis Potosí, Mexico (types, USNM 128952). Hosts, isopods *Speocirolana pelaezi* and *S. bolivari.*

Distribution (in addition to type-locality). —

Mexico

SAN LUIS POTOSÍ — La Cueva de Lagunita, 24 km NE of Ciudad de Maíz (host, *Speocirolana pelaezi*) (Hobbs, 1971: 46). Sótano de Pichijumo, about 9 km NE y Ciudad Valles (host, *S. pelaezi*) (Hobbs, 1971: 46). Sótano del Tigre, about 15 km NE of Ciudad Valles (host, *S. pelaezi*). Sótano del Arroyo, about 1.5 km SE of Los Sabinos (host, *S. pelaezi*).

TAMAULIPAS — Grutas de Quintero (= Cuevas de Quintero, Rioja, 1953: 178) about 11 km SW of Ciudad Mante (host, *S. pelaezi* and *S. bolivari*). Cueva de la Florida, about 15 km SW of Ciudad Mante (host, *S. pelaezi*).

Thus, as Hobbs (1971) points out, the known range of *Sphaeromicola cirolanae* is restricted to subterranean waters along the western flank of La Sierra de El Abra in the southeastern part of San Luis Potosí and southern Tamaulipas.

Remarks. — According to Hobbs (1971: 46-47), specimens of *Sphaeromicola* from Sótano del Arroyo possess a clasping apparatus identical to that of *S. cirolanae,* but because the ventral surface of the peniferum is so different from that of topoparatypes of the latter, they were originally believed to represent an undescribed species. When additional specimens from other localities became available, it was obvious that the differences observed were due to the retracted ventral fingerlike projection which in all of Rioja's specimens was presumably extended. The observed extreme state of retraction results in the exhibition of a flaplike flange which abuts the projecting penis.

Another rather surprising variation in the species, pointed out by Hobbs (1971), was in some specimens from Cueva de Lagunita, Sótano de Pichijumo, Sótano del Tigre, and Cueva de Florida. In this variant, the clasping apparatus is vastly different from that of the typical form of the species described and illustrated by Rioja. Instead of the clasping apparatus being bent at an angle, with the external border of the bend produced into an angular prominence, in this form, the bend is rounded. Furthermore, in none of the individuals possessing the rounded clasping apparatus is the ventral portion of the peniferum extruded. The combination of the rounded bend of the clasping apparatus and the retracted ventral extremity of the peniferum led Hobbs to conclude that two sympatric species, with apparently identical shell dimensions, were infesting the cirolanids of the region. The discovery of one specimen in which the dextral copulatory complex bore a clasping apparatus with a rounded bend and the sinistral one, an angular bend with a produced angular projection, indicated that *S. cirolanae* has a clasping apparatus which exists in two quite different forms, and that the variations noted in the ventral portion of the peniferum reflect nothing more than different degrees of its extension or retraction.

In only two localities, Sótano de Pichijumo and Sótano del Tigre, is the typical form not known to occur.

Sphaeromicola coahuiltecae Hobbs and Hobbs, 1973

Plate XLIII, figures 15-18
Plate LXI

Sphaeromicola coahuiltecae Hobbs and Hobbs, 1973: 39-41 (M, H, R, D), figs. 1c-g.

Diagnosis. — MALE: Shell length 300 μ; height 150 μ (2 specimens). Eye apparently absent. Shell elongate oval with ventral margin nearly straight.

"Copulatory complex with peniferum short, contracted posteriorly at base of ventral lobe; latter with anteroventral extremity forming angle of approximately 90° and prolonged posteriorly in conspicuous acute or subacute prominence; mesial surface of ventral lobe grooved vertically, receiving penis from main body of peniferum; anteroventral margin of main body of peniferum with prominent anteriorly directed spine situated immediately proximal to ventral lobe. Penis straight. Clasping apparatus L-shaped with vertical ramus distinctly shorter than horizontal ramus and extending far ventrally beyond peniferum; external border of vertical ramus with angular lobe near proximal end; internal border of vertical ramus and external border of horizontal ramus entire; internal border of horizontal ramus lacking teeth but undulating, high near midlength but comparatively low along distal third; apex of ramus with five denticles." (Hobbs, 1973).

FEMALE: Shell length 270-290 μ (average, 280 μ); height 140-150 μ (average, 145 μ). Eye, as in male, apparently absent. Shell similar to that of male.

Relationships. — According to Hobbs (1971), this ostracod has its closest affinities with the Mexican *S. cirolanae.* The clasping apparatus of *S. coahuiltecae* lacks a distinct tooth on the internal border of the horizontal ramus of the clasping apparatus, and it possesses a subangular lobe on the subproximal postaxial border of the vertical ramus. In *S. cirolanae* the horizontal ramus always bears a tooth on the internal border, and the vertical ramus lacks a lobe.

Type-locality. — Cueva del Huisache, 4 km NW of Micos, San Luis Potosí, Mexico (Host not known, but as all of the American members of this genus have been found only on cirolanids it is probable that the undescribed member of that family occurring in this cave series serves as the host.).

Distribution. — Known only from type-locality.

Sphaeromicola hamigera Remy, 1946

Plate XLIII, figure 15
Plate LXII

Sphaeromicola stammeri var. *hamiger* Remy, 1946: 7-9 (M, H), figs. 1-8. — Remy, 1948: 2-3 (B, D, H).
Sphaeromicola hamigera. — Remy, 1948b: 129 (M), 132 (R), fig. 9. — Remy, 1948c: 44, 45 (D, H), figs. 4, 1-4, 9. — Remy, 1951: 224, 225 (D, H), 228 (H). — Hart, 1962: 140 (C, D, S). — Vandel, 1964: 303 (C). — Illies, J., 1967: 171 (D). — Hobbs, 1971: 45 (K).

Diagnosis. — MALE: Shell length 400-480 μ (average, 440 μ); height 240-280 μ (average, 260 μ). Copulatory complex with peniferum reaching only to base of distal $\frac{1}{3}$ of vertical ramus of clasping apparatus. Peniferum terminating in anteriorly directed falcate process and posteroventrally directed rounded lobe. Clasping apparatus with distinct horizontal and vertical rami; external border of vertical ramus with two protuberances, one at mid-length, one in proximal quarter; horizontal ramus with single protuberance at mid-length of internal border; five terminal denticles.

Female: Shell length 390-490 μ (average, 420 μ); height 240-260 μ (average, 250 μ).

Type-locality. — Grotte de Sous-Roche, near Poncin (Department du Ain) at km 54,670 on route N 84 between Lyon and Geneve, France (types not known). Host, *isopod Caecosphaeroma (C.) virei.*

Distribution. — Known only from type-locality.

Sphaeromicola sphaeromidicola Hubault, 1938

Plate XLIV, figures 1-3
Plate LXII

Sphaeromicola sphaeromidicola Hubault, 1938: 11-24 (M, D, R), figs. 1-17. — Remy, 1948a: 1-4 (B, D, H). — Remy, 1948b: — Remy, 1951: 220, 221 (D, F, M, H), 228 (B, H). — Deboutteville, 1960: 658 (C, D, H). — Hart, 1962: 140 (C, D, S). — Vandel, 1964: 303 (C). — Vandel, 1964: 303 (C). — Illies, 1967: 171 (D). — Hobbs, 1971: 45 (K).

Diagnosis. — Male: Shell length 360 μ, height 214 μ. Copulatory complex with peniferum terminating in one, two, or three acute processes directed ventrally or slightly anteroventrally; penis straight, and directed anteroventrally. Clasping apparatus with indistinct juncture between horizontal and vertical rami; rami straight or slightly sinuous and lacking teeth on both margins; five or six terminal denticles.

Remarks. — It is entirely possible that the specimens described and figured by Hubault (1938) and Remy (1951) should be assigned to two or more species. However, because none of these specimens were available for study, we feel that decisions of this nature should await further study.

Type-locality. — Subterranean stream near Rovigno d'Istria, Croatia, Yugoslavia, about 70 km S of Trieste (types not known). Host, isopod *Sphaeromides verei.*

Distribution. — known only from type-locality.

Sphaeromicola stammeri Klie, 1930

Plate XLIV, figures 4-5
Plate LXII

Sphaeromicola stammeri Klie, 1930: 272-276 (M, R, D), figs. 1-8. — Stammer, 1932: 589, 590 (D, H). — Wolf, 1934-1938: (C). — Stammer, 1935: 94 (D, H).— Hubault, 1937: 366, 377 (H, M). — Klie, 1938: 317, 322 (R). — Hubault, 1938: 11, 12 (H), 18, 19 (R). — Remy, 1948a: 2, 3 (B, D, H). — Remy, 1948b: 132 (R). — Remy, 1948c: 44-45 (D, H). — Remy, 1951: 222-224 (D, H), 227, 228 (B, H). — Rioja, 1951: 178, 179 (R). — Deboutteville, 1960: 658 (C, D, S). — Vandel, 1964: 303 (C). — Illies, 1967: 171 (D). — Hobbs, 1971: 45 (K). — Danielopol, 1971b: 191 (R), fig. 5F.

Diagnosis. — Male: Shell length 342 μ; height 171 μ. Copulatory complex with peniferum terminating in two closely opposed acute projections. Clasping apparatus tripartite, with each of three parts subequal in length (Plate XLIV, fig. 5); or bipartite with short horizontal ramus and elongate, gently curving, vertical ramus (Plate XLIV, fig. 4); four or five small terminal denticles.

Female: Shell length 324 μ; height 162 μ.

Type-locality. — Monjalconebach (Gradiska), about 10 km N of Monjalcone and 25 km NW of Trieste, Venezia Guilia, Italy. Host, isopods *Monolistra (Typhlosphaeroma) recovitzai* and *Microlistria schollaender.*

Distribution (in addition to type-locality). —

Italy

Venezia Euganea — Caves of Monti Berici, about 10 km S of Vicenza (host, *Monolistra (T.) berica*) (Hubault, 1937: 367).

Yugoslavia

Herzegovine — The cave of "Vjetrenica pecina, a Zavala, Popovopolje," about 25 km NW of Dubrovnik (host, *Monolistra (P.) hercegoviniensis*) (Hubault, 1937: 367).

Slovenia — Podpeska jama (=Podpec-Höhle) at Podpec, 2 km de la gare de Dobre-Polje, between Ljubljana and Kocevje (host, *Monolistra (M.) caeca caeca*) (Remy, 1951: 222). Logarcek (=Graf Falkenhayn-Huhle) near Planina, N of Postojna (host, *M. (T.) racovitzai*) (Remy, 1951: 223). Krizna jama (=kreuzberghohle) near Loz (host, *M. (T.) racovitzai*) (Remy, 1951: 223).

Sphaeromicola topsenti Paris, 1916

Plate XLIV, figure 6
Plate LXII

"J'ai trouvé dans les trois tubes contenant les *Caecosphaeroma* que j'ai capturées moi-même, des Ostracodes de très petite taille, tout à fait transparents et aveugles, qui seront décrits ultérieurement. Il est probable que ces animaux vivent sur les gros Spheromiens et en sont les commensaux," Racovitza, 1910: 703 (H).

Sphaeromicola topsenti Paris, 1916a: 307-309 (M, H, B). — Paris, 1961b: 116 (D, H). — Paris, 1920: 475-487 (D, M, B, H), figs. 1-50. — Paris, 1925a: 449 (C). — Paris, 1925b: 454 (B). — Jeannel, 1926: 112 (D, H). — Spandl, 1926: 45-46 (B, D, H). — Chappuis, 1927: 35 (D), 38-40 (B, D, M). — Klie, 1930: 272 (D, H), 276 (R). — Stammer, 1932: 589 (R). — Wolf, 1934-1938: 43 (C). — Stammer, 1935: 94 (H, D). — Hubault, 1937: 366, 369, 371 (R). — Hubault, 1938: 11, 16, 18, 19 (R). — Klie, 1938: 317, 321 (R). — Wesenberg-Lund, 1939: 485 (H, C). — Remy, 1943: Rioja, 1943b: 567 (R). — Remy, 1946: 9 (R). — Remy, 1948a: 2-3 (B, D, H). — Remy, 1948b: 129 (C), 130 (D), 132 (R). — Remy, 1948c: 44-45 (H, D). — Kesling, 1951: 224-228 (H, D). — Rioja, 1951: 179 (R). — Deboutteville, 1960: 658 (H, C, D). — Hart, 1962: 140 (C, S, D). — Vandel, 1964: 303 (C). — Hart, Nair, and Hart, 1967: 5 (B, M). — Illies, 1967: 171 (D). — Hobbs, 1971: 45 (K). — Danielopol, 1971b: 191 (R).

Diagnosis. — MALE: Shell length 440 μ; height, 125 μ. Copulatory complex with trifid terminal portion of peniferum; narrow anterior terminal process longer than heavier middle process; small posterior acute process directed posteroventrally; penis straight and terminating between bases of first two processes. Clasping apparatus tripartite with proximal two parts subequal in length; terminal part short, with five terminal denticles. Dorsal and ventral fingers apparently fused in proximal ⅓, terminating in three tapering spines.

Remarks. — Remy (1938: 38) points out that slight differences exist between the ostracods assigned to this species, and that those from farther NE differ from those of the type-locality. He ascribes these differences to geographic isolation of the two groups.

Type-locality. — Grotte de la Douix à Darcey, Darcey, canton de Flavigny-sur-Ozerain, about 40 km NW of Dijon, départment de la Côte-d'Or, France (types not known). Hosts, *Caecosphaerum burgundum* and *C. virei.* (Paris, 1916a and 1920: 479.)

Distribution (in addition to type-locality). —

France

DÉPARTMENT DE LA CÔTE-D'OR — Carrière de Vry, Magny-les-Villers, Canton de Nuits-Saint-Georges, about 25 km SSW of Dijon (host, *C. burgundum*) (Paris, 1920: 479).

DÉPARTMENT DE LA DORDOGNE — La grotte de Beaussac, Canton de Mareuile-sur-Belle, à une trentaine de km au sud d'Yvrac-Malleyraut, about 35 km NW of Périgeux (host, *C. burgundum*) (Remy, 1943: 37-38 and Remy 1948b: 130).

DÉPARTMENT DE LA HAUTE-MEME — Trou-du-Bas (=Souffre-du-Bas) cummune de Mandres-la-Côte, canton de Nogent-en-Bassigny, about 90 km NE of Dijon (host, *C. burgundum*) (Remy, 1951: 225).

DÉPARTMENT DE LA HAUTE SAONE — . . . la résurgence de Veuvey à Calmoutier, canton de Noroy-le-Bourg, en aval de l'orifice, dans des trous d'eau du canon (host, *C. burgundum*) (Remy 1951: 226) [also Remy, 1948b: 130 (Footnote 3)]. Rivière soutérraine d'Oyrières, canton d'Autrey-les-Gray (host, *C. burgundum*) (Remy, 1951: 226). Grotte de Captiot (ou Capsio) près de Bucey-les-Gy, canton de Gy (host, *C. burgundum*) (Remy, 1951: 226).

DÉPARTMENT DU JURA — Grotte de Baume-les-Méssieurs, Baume-les-Méssieurs, Canton de Voiteur, about 80 km SW of Dijon (host, *C. burgundum*) (Paris, 1920: 479).

DÉPARTMENT DE MEURTHE-ET-MOSELLE — Grotte de Sainte-Reine near Pièrre-la-Treiche, canton de Toul, about 20 km W of Nancy (host *C. burgundum*) (Remy, 1951: 225).

DÉPARTMENT DE L'YONNE — Grotte d'Arcy-sur-Cure, Arcy-sur-Cure, Canton de Vermenton, about 110 km WNW of Dijon (host, *C. burgundum*) (Paris, 1920: 479). Puits à Ravières, Ravières, Canton d'Ancy-le-France, about 75 km NW of Dijon (host, *C. burgundum*) (Paris, 1920: 479).

Subfamily Hartiellinae Danielopol, 1971

Tribe Hartiellini Danielopol, 1971b: 189-207 (M, R, D). Subfamily Hartiellinae, Hart and Hart, 1974: 000 (S, D).

Type of the subfamily. — *Sphaeromicola dudichi* Klie, 1938.

Diagnosis. — Antennule with seven podomeres. Antennal claws spiculiform, bearing setae on flexor face. Mandibular protopodite with distal row of six multicuspid teeth. Penis weakly developed and situated in distal portion of peniferum.

Distribution. — On marine amphipods in the coastal waters of Italy and France.

HARTIELLA DANIELOPOL, 1971

Hartiella Danielopol, 1970: 14 (*nomen nudum*) (R). — Danielopol, 1971b: 190, 191, type-species *Sphaeromicola dudichi* Klie, 1938.

Diagnosis. — See diagnostic characters enumerated above for subfamily.

Type-species. — *Sphaeromicola dudichi* Klie, 1938.

Distribution. — See subfamily distribution above and Fig. 49.

Host. — Marine amphipods.

Hartiella dudichi (Klie, 1938)

Plate XLIV, figures 7-10
Plate LXII

Sphaeromicola dudichi Klie, 1938: 317-322 (M, D, R), figs. 1-11. — Hubault, 1938: 18, 19 (R). — Remy, 1946: 8 (R). — Remy, 1948a: 1-4 (B, D, H). — Remy, 1948b: 129 (B, C), 132 (R). — Kesling, 1951: 92 (B, D, H). — Rioja, 1951: 179 (R). — Deboutteville, 1960: 660 (C, D, H). — Hart, 1962: 140 (C, D, S). — Vandel, 1964: 303 (C, E). — Roelofs, J., 1966: 207-212 (M, D, H), figs. 1a-e. — Roelofs, H. M. A., 1966: 151 (B). — Hart, Nair, and Hart, 1967: 1 (R). — Roelofs, H. M. A., 1968: 39-51 (B, H, D), 16 figs. — Hobbs, 1971: 45 (K). — Danielopol, 1970: 14 (R). — Danielopol, 1971a: 181 (E).
Hartiella dudichi. — Danielopol, 1971b: 189-207 (M, H, R, A, D), figs. 1-8.

Diagnosis. — MALE: Shell length 320-350 μ (average, 335 μ); height, 250 μ. Copulatory complex with peniferum similar in terminal detail to that of *Notocythere erika* (see figure 18, Plate XLI) except that the vertically oriented straight penis is shorter in *S. topsenti*. Clasping apparatus gently curved throughout entire length and terminating in rounded lobe. Homolog of finger (probably dorsal finger) similar to clasping apparatus in shape and bulk, terminating in two ventrally directed setae. "Furcal lobe" of Danielopol (1971b: 197) possibly homologous with finger guard.

Type-locality. — Gulf of Naples, Italy. Host, *Chelura terebrans* (marine amphipod from tubes of *Limnoria lignorum* and *Teredo navalis*).

Distribution (in addition to type-locality). —

France

DÉPARTMENT DE N.-FINISTERE — Penpoull, East of St. Pol-de-Léon, 30 km NE of Brest (host, *C. terebrans*) (Roelofs, 1966a: 207).

DÉPARTMENT DE BOUCHES DU RHONE — Port Miou, Bay of Marseille (host, *C. terebrans*) (Roelofs, 1968: 40).

REFERENCES

ALLEN, S. A. 1933. Parasites and commensals of North Carolina crayfishes. — *Journal of the Elisha Mitchell Scientific Society,* 49: 119-121, pl. 3, figs. 1-3.

ANONYMOUS. 1964. Progress report, committee on recent Ostracods. — *Pubblicazioni della Stazione Zoologica di Napoli,* 33 (supplement): 588-590.

BAKER, JAMES H. 1969. On the relationship of *Ankylocythere sinuosa* (Rioja, 1942) (Ostracoda, Entocytheridae) to the crayfish *Procambarus simulans* (Faxon, 1884). — *Transactions of the American Microscopical Society,* 88(2): 293-294.

BARR, THOMAS C., JR. 1967. Ecological studies in the Mammoth Cave System of Kentucky, I: The Biota. — *International Journal of Speleology,* 3(1-2): 147-204, 28 pls.

BEDINGER, M. S. AND HORTON H. HOBBS, JR. 1965. Observations on a new troglobitic crayfish (with notes on the distribution of troglobitic crayfishes in the Ozark region). — *Bulletin of the National Speleological Society,* 27(3): 93-96.

BRONSTEIN, Z. S. 1947. Rakoobraznye (Crustacea), t. 2, pt. 1, Ostracoda presnykh vod. (fresh water). — *Fauna SSSR,* n.s., No. 31, 339 pp., 206 figs., 14 pls.

CHAPPUIS, P. A. 1927. *In* Thienemann, A., Die Tierwelt der Unterirdischen Gewässer. — *Die Binnengewasser, Stuttgart,* 3: 1-175.

CONDE, B. 1963. Notice sur Paul A. Remy et son oeuvre biospeologique. — *Annales de Speleologie,* 18(4): 463-471.

CRAWFORD, E. A. 1959. Five new ostracods of the genus *Entocythere* (Ostracoda, Cytheridae) from South Carolina. — *University of South Carolina Publications, Biology,* Ser. III, 2(4): 149-189, 5 pls.

————. 1960. The genus *Entocythere,* an ostracod epizooic on crayfishes (Abstract). — *The ASB Bulletin (Association of Southeastern Biologists),* 7(2): 26.

————. 1961. Three new species of the genus *Entocythere* (Ostracoda, Cytheridae) from North and South Carolina. — *American Midland Naturalist,* 65(1): 236-245, 21 figs.

————. 1965. Three new species of epizoic ostracods (Ostracoda, Entocytheridae) from North and South Carolina. — *American Midland Naturalist,* 74(1): 148-154.

CROCKER, D. W. AND D. W. BARR. 1968. Handbook of the crayfishes of Ontario. — *University of Toronto Press (for the Royal Ontario Museum),* 158 pp.

DANIELOPOL, DAN L. 1970. Une nouvelle espèce du genre *Darwinula* des eaux souterraines de Roumania et quelques remarques sur la morphologie des Darwinulidae (Ostracoda — Podocopida). — *Travaux de l'Institut de Spéologie "Émile Racovitza,"* 9: 135-149.

————. 1971a. Quelques remarques sur le peuplement ostracodologique des eaux douces souterraines d'Europe. — *Colloque sur la Paléoécologie des Ostracodes, Pau (France).* H. J. Oertli, ed., pp. 179-190.

————. 1971b. Définition de la tribu des Hartiellini nov. et remarques sur le poecilogynie de *Hartiella dudichi* (Ostracoda, Entocytheridae, Sphaeromicolinae). — *Travaux de l'Institut de Spéologie "Émile Racovitza,"* 10: 189-207.

DEBOUTTEVILLE, C. D. 1960. Biologie des eaux souterraines littorales et continentales. — *Hermann, Paris,* 740 pp.

DELORME, L. D. 1970. Freshwater ostracods of Canada. Part IV. Families Ilyocyprididae, Notodromadidae, Darwinulidae, Cytherideidae, and Entocytheridae. — *Canadian Journal of Zoology,* 48(6): 1251-1259, 6 pls.

DOBBIN, C. N. 1941. Fresh-water ostracoda from Washington and other western localities. — *University of Washington Publications, Biology,* 4: 174-246, 14 pls.

————. See also EVENSON, C. D.

EVENSON, C. D. (NEE C. N. DOBBIN). 1969. Designation of lectotypes of fresh-water species described by Dobbin, 1941 (Ostracoda, Crustacea). — pp. 491-494 IN *The Taxonomy, Morphology and Ecology of Recent Ostracoda (John W. Neale, editor). Oliver and Boyd, Ltd., Edinburgh.* 553 pp.

FAGE, L. 1931. Campagne speologique de C. Bolivar et R. Jeannel dans l'Amerique du Nord (1928). 5.

Crustaces Amphipodes et Decapodes. — *Biospeologica: Archives de Zoologie Experimentale et Generale,* 71(3): 361-374.

FERGUSON, E., JR. 1944. Studies of seasonal life history of three species of freshwater Ostracoda. — *American Midland Naturalist,* 32(3): 713-727.

————. 1958. A supplementary list of species and records of distribution for North American Ostracoda. — *Proceedings of the Biological Society of Washington,* 71: 197-202.

————. 1965. Morphology and taxonomy of freshwater Ostracoda (Abstract). — *Symposium on Crustacea, Marine Biological Association of India, Mandapam Camp, South India,* p. 12.

————. 1968. Recently described species and distributional records for North American freshwater Ostracoda. — *American Midland Naturalist,* 79(2): 499-506.

FRANTZ, T. C. AND A. J. CORDONE. 1966. A preliminary checklist of invertebrates collected from Lake Tahoe, 1961-1964. — *Occasional Papers of the Biological Society of Nevada,* 8: 1-12.

HARDING, J. P. 1964. Crustacean cuticle with reference to ostracod carapace. — *Pubblicazioni della Stazione Zoologica di Napoli,* 33 (supplement): 1-31.

HARLAND, W. B. ET AL. 1967. The fossil record: A symposium with documentation. — *Geological Society of London,* 827 pp.

HART, C. W., JR. 1959. The ostracods of the genus *Entocythere* from the lower Chattahoochee-Flint basin: with a review of the occurrence of the genus in Florida, and descriptions of two new species. — *Bulletin of the Florida State Museum, Biological Sciences,* 4(6): 193-204, figs. 1-19.

————. 1960. A new ostracod (Crustacea, Ostracoda) of the genus *Entocythere* from Pennsylvania. — *Notulae Naturae of the Academy of Natural Sciences of Philadelphia,* No. 335, 4 pp., 9 figs.

————. 1962. A revision of the ostracods of the family Entocytheridae. — *Proceedings of the Academy of Natural Sciences of Philadelphia,* 114(3): 121-147.

————. 1963. The status of entocytherid ostracod research in the United States and Mexico (Abstract). — *Proceedings of the XVI International Congress of Zoology, Washington, D.C.,* 1: 51.

————. 1964. Two new entocytherid ostracods from the vicinity of Washington, D.C. — *Proceedings of the Biological Society of Washington,* 77(28): 243-146.

————. 1965a. Three new entocytherid ostracods from the western United States with new locality data for two previously described western entocytherids. — *Crustaceana,* 8(2): 190-196.

————. 1965b. New entocytherid ostracods and distribution records for five midwestern states. — *Transactions of the American Microscopical Society,* 84(2): 255-259.

————. 1971. A new species of parasitic ostracod of the genus *Acetabulastoma* (Paradoxostomatidae, Paradoxostominae) with a discussion of the copulatory appendage homologies. — *Notulae Naturae, Academy of Natural Sciences of Philadelphia,* 442: 1-11.

———— AND DABNEY G. HART. 1966a. Four new entocytherid ostracods from Kentucky, with notes on the troglobitic *Sagittocythere barri.* — *Notulae Naturae of the Academy of Natural Sciences of Philadelphia,* No. 388, 10 pp.

———— AND ————. 1966b. New Australian entocytherid ostracods and their relation to North American species (Abstract). — *American Zoologist,* 6(4): 564.

———— AND ————. 1967a. The entocytherid ostracods of Australia. — *Proceedings of the Academy of Natural Sciences of Philadelphia,* 119(1): 1-51.

———— AND ————. 1967b. The copulatory apparatus of male entocytherid ostracods and its homologous structures in certain free-living marine ostracods (Abstract). — *Program of the Symposium on the Taxonomy, Morphology and Ecology of Recent Ostracoda, held at the University of Hull, England,* 10-14 July 1967.

———— AND ————. 1968. The sorting, storage, and preparation of microscopic crustaceans, with special reference to entocytherid ostracods. — *Crustaceana,* 14(3): 315-318.

———— AND ————. 1969a. Evolutionary trends in the ostracod family Entocytheridae, with notes on the distributional patterns in the Southern Appalachians. — pp. 179-190 IN *The Distributional History of the Biota of the Southern Appalachians. Part I. Invertebrates. (Research Division Monograph 1, Virginia Polytechnic Institute, Blacksburg.)* 295 pp.

———— AND ————. 1969b. The functional morphology of entocytherid ostracod copulatory appendages, with a discussion of possible homologues in other ostracods. — pp. 154-167 IN *The Taxonomy, Morphology and Ecology of Recent Ostracods (John W. Neale, editor). Oliver and Boyd, Ltd., Edinburgh.* 553 pp.

———— AND ————. 1971. A new ostracod (Entocytheridae, Notocytherinae) commensal on New Zealand crayfishes. — *Proceedings of the Biological Society of Washington,* 83(50): 579-584.

———— AND ————. 1974. Ostracoda: Podocopa: Family Entocytheridae. — *Catalogus Crustaceorum* (Amsterdam).

See also HART, DABNEY G. AND C. W. HART, JR.

HART, C. W., JR. AND HORTON H. HOBBS, JR. 1961a. Observations on the distribution and relationships of the troglobitic ostracods of North America (Abstract). — *The ASB Bulletin (Association of Southeastern Biologists),* 8(2): 25.

———— AND ————. 1961b. Eight new troglobitic ostracods of the genus *Entocythere* (Crustacea, Ostracoda) from the eastern United States. — *Proceedings of the Academy of Natural Sciences of Philadelphia,* 113(8): 173-185, 32 figs.

————, N. BALAKRISHNAN NAIR, AND DABNEY G. HART. 1967. A new ostracod (Ostracoda: Entocytheridae) commensal on a wood-boring marine ostracod from India. — *Notulae Naturae of the Academy of Natural Sciences of Philadelphia,* No. 409, 11 pp.

Hart, Dabney G. and C. W. Hart, Jr. 1970. A new ostracod (Entocytheridae, Notocytherinae) commensal on New Guinea crayfishes. — *Zoologische Mededelingen,* 44(19): 279-283.

———— and ————. 1971. New entocytherid ostracods of the genera *Ankylocythere, Dactylocythere, Entocythere, Geocythere,* and *Uncinocythere* — with a new diagnosis of the genus *Entocythere.* — *Proceedings, Academy of Natural Sciences of Philadelphia,* 123(5): 105-125.

See also Hart, C. W., Jr. and Dabney G. Hart.

Hartmann, Gerd. 1963. Zur phylogenie und systematik der ostracoden. — *Zeitschrift für Zoologische Systematik und Evolutionsforschung,* 1(1-2): 1-204.

————. 1964. Neontological and paleontological classification of Ostracoda. — *Pubblicazioni della Stazione Zoologica di Napoli,* 33 (supplement): pp. 550-587.

Hobbs, Horton H., Jr. 1941. A new crayfish from San Luis Potosi, Mexico (Decapoda, Astacidae). —*Zoologica, Scientific Contributions of the New York Zoological Society,* 26(1): 1-4.

————. 1953. The epizootic associates of the crayfishes of the New River system with particular reference to the ostracods (Abstract). — *Journal of the Tennessee Academy of Sciences,* 28(3): 180-181.

————. 1955. Ostracods of the genus *Entocythere* from the New River system in North Carolina, Virginia, and West Virginia. — *Transactions of the American Microscopical Society,* 74(4): 325-333.

————. 1956 (1957). Observaciones acerca de las especies del genero *Entocythere* (Crustaceos, Ostracodos) de Cuba. — *Anales del Instituto de Biologia, Mexico,* 27(2): 131-136, 8 figs.

————. 1966. An illustrated key to the species of the genus *Ankylocythere* with a description of a new species from Louisiana (Ostracoda, Entocytheridae). — *Proceedings of the Louisiana Academy of Sciences,* 29: 67-75.

————. 1967. A new genus and three new species of ostracods with a key to the genus *Dactylocythere* (Ostracoda: Entocytheridae). — *Proceedings of the United States National Museum,* 122(3587): 1-10.

————. 1971. The entocytherid ostracods of Mexico and Cuba. — *Smithsonian Contributions to Zoology,* 81: 1-55.

————. 1974. A checklist of the North and Middle American Crayfishes (Decapoda: Astacidae and Cambaridae). — *Smithsonian Contributions to Zoology,* No. 166.

Hobbs, Horton H., Jr. and Thomas C. Barr, Jr. 1972. Origins and affinities of the troglobitic crayfishes of North America (Decapoda: Astacidae) II. Genus *Orconectes.* — *Smithsonian Contributions to Zoology,* 105: 1-84.

Hobbs, Horton H., Jr. and M. S. Bedinger. 1964. A new troglobitic crayfish of the genus *Cambarus* (Decapoda, Astacidae) from Arkansas with a note on the range of *Cambarus cryptodytes* Hobbs. — *Proceedings of the Biological Society of Washington,* 77: 9-16.

Hobbs, Horton H., Jr. and C. W. Hart, Jr. 1966. On the entocytherid ostracod genera *Ascetocythere, Plectocythere, Phmocythere* (gen. nov.), and *Cymocythere,* with descriptions of new species. — *Proceedings of the Academy of Natural Sciences of Philadelphia,* 118(2): 35-61.

Hobbs, Horton H., Jr. and H. H. Hobbs III. 1970. New entocytherid ostracods with a key to the genera of the subfamily Entocytherinae. — *Smithsonian Contributions to Zoology,* 47: 1-19.

———— and ————. 1973. The genus *Sphaeromicola* (Ostracoda, Entocytherifae) in Mexico. — Studies on the Cavernicole Fauna of Mexico and Adjacent Regions. — *Association for Mexican Cave Studies, Bulletin 5, Speleo Press, Austin, Texas.*

Hobbs, Horton H., Jr., Perry C. Holt, and Margaret Walton. 1967. The crayfishes and their epizootic ostracod and branchiobdellid associates of the Mountain Lake, Virginia, region. — *Proceedings of the United States National Museum,* 123(3602): 1-84.

Hobbs, Horton H., Jr. and Frank O. Perkins. 1967. A new burrowing crayfish from North Carolina (Decapoda, Astacidae). — *Proceedings of the Biological Society of Washington,* 80(21): 141-146.

Hobbs, Horton H., Jr. and Alejandro Villalobos. 1958. The exoskeleton of a freshwater crab as a microhabitat for several invertebrates (Abstract). — *Virginia Journal of Science,* 9(4): 395-396.

Hobbs, Horton H., Jr. and Margaret Walton. 1960. Three new ostracods of the genus *Entocythere* from the Hiwassee drainage system in Georgia and Tennessee. — *Journal of the Tennessee Academy of Sciences,* 35(1): 17-23, 20 figs.

———— and ————. 1961. Additional new ostracods from the Hiwassee drainage system in Georgia, North Carolina, and Tennessee. — *Transactions of the American Microscopical Society,* 80(4): 379-384, 8 figs.

———— and ————. 1962. New ostracods of the genus *Entocythere* from the Mountain Lake region, Virginia (Ostracoda, Entocytheridae). — *Virginia Journal of Science,* 13(2): 42-48, 13 figs.

———— and ————. 1963a. Three new ostracods (Ostracoda, Entocytheridae) from the Duck River drainage in Tennessee. — *American Midland Naturalist,* 69(2): 456-461, 10 figs.

———— and ————. 1963b. Four new species of the genus *Donnaldsoncythere* (Ostracoda, Entocytheridae) from Virginia with a key to the species of the genus. — *Transactions of the American Microscopical Society,* 82(4): 363-370, 26 figs.

———— and ————. 1966. A new genus and six new species of entocytherid ostracods (ostracoda, Entocytheridae). — *Proceedings of the United States National Museum,* 119(3542): 1-12.

———— and ————. 1968. New entocytherid ostracods from the southern United States. — *Proceedings of the Academy of Natural Sciences of Philadelphia,* 120(6): 237-252.

———— and ————. 1970. New entocytherid ostracods from Tennessee and Virginia. — *Proceedings* of the Biological Society of Washington, 82(68): 851-864.

HOBBS, H. H. III. 1965. Two new genera and species of the ostracod family Entocytheridae, with a key to the genera. — *Proceedings of the Biological Society of Washington,* 78(19): 159-164.

————. 1968. The possibility of host specificity in the genus *Ornithocythere* (Ostracoda, Entocytheridae) (Abstract). — *The ASB Bulletin (Association of Southeastern Biologists),* 15(2): 40-41.

————. 1969a. A new genus and two new species of entocytherid ostracods from Alabama and Mississippi. — *Proceedings of the Biological Society of Washington,* 82(10): 167-170.

————. 1969b. Studies in ecological and host specificity in entocytherid ostracods (Ostracoda, Entocytheridae). — *Unpublished Master's Theses, Mississippi State University,* 93 pp.

————. 1970. New entocytherid ostracods of the genus *Ornithocythere* and the description of a new genus. — *Proceedings of the Biological Society of Washington,* 83(15): 171-182.

————. 1971. New entocytherid ostracods of the genera *Ankylocythere* and *Dactylocythere.* — *Proceedings, Biological Society of Washington,* 84(17): 137-146.

HOFF, C. CLAYTON. 1942a. The subfamily Entocytherinae, a new subfamily of freshwater cytherid ostracods, with descriptions of two new species of the genus *Entocythere.* — *American Midland Naturalist,* 27(1): 63-73, 13 figs.

————. 1942b. The ostracods of Illinois: their biology and taxonomy. — *Illinois Biological Monographs,* 19(1-2): 1-196, 9 pls., 135 figs.

————. 1943. Two new ostracods of the genus *Entocythere* and records of previously described species. — *Journal of the Washington Academy of Sciences,* 33(9): 275-286, 2 figs.

————. 1944a. The origin of nearctic fresh-water ostracods. — *Ecology,* 25(3): 369-372.

————. 1944b. New American species of the ostracod genus *Entocythere.* — *American Midland Naturalist,* 32(2): 327-357, 33 figs.

HOWE, HENRY V. 1955. Handbook of ostracod taxonomy. — *Louisiana State University Studies, Physical Sciences,* Series No. 1, 386 pp.

————. 1962. Ostracod taxonomy. — *Louisiana State University Press, Baton Rouge,* 366 pp.

HUBAULT, E. 1937. *Monolistra hercegoviniensis* Absolon; spheromien cavernicole d'Herzegovine et *Sphaeromicola stammeri* Klie, son commensal. — *Archives de Zoologie Experimentale et Generale,* 78(9): 357-373, pls. 5-8. (Biospeologica LXVI).

————. 1938. *Sphaeromicola sphaeromidicola* nov. sp., commensal de *Sphaeromides virei* Valle, en Istrie et considerations sur l'origine de diverses especes cavernicoles perimediterraneennes. — *Archives de Zoologie Experimentale et Generale,* 80 (Notes et Revue 1): 11-24, 17 figs.

HUTCHINSON, G. EVELYN. 1967. *A treatise on limnology. Vol. II. Introduction to lake biology and limnoplankton.* — *John Wiley & Sons, New York.*

ILLIES, JOACHIM (editor). 1967. *Limnofauna Europaea.* — *Gustav Fischer Verlag, Stuttgart,* 474 pp.

JEANNEL, R. 1926. Faune cavernicole de la France. — *Encyclopedie Entomologique,* Paris, 7: 1-334.

KAESTNER, A. 1970. *Invertebrate zoology, Vol. III (translated and adapted by Levi and Levi).* — *Interscience Publishers, J. Wiley & Sons, New York.*

KESLING, R. V. 1951. The morphology of ostracod molt stages. — *Illinois Biological Monographs,* 21(1-3): 324 pp., 96 pls., 36 figs., 5 charts.

KLIE, W. 1926. IN *Biologie der Tiere Deutschlands by Paul Schultz, Berlin,* 22(16): 1-56.

————. 1930. Uber eine neue Art der Ostracodengattung *Sphaeromicola.* — *Zoologischer Anzeiger,* 88 (9-10): 272-276, 8 figs.

————. 1931. Campagne speologique de C. Bolivar et R. Jeannel dans l'Amerique du Nord (1928). 3. Crustaces Ostracodes. — *Biospeologica: Archives de Zoologie Experimentale et Generale,* 71(3): 333-344, 20 figs.

————. 1938. *Sphaeromicola dudichi* n. sp. (Ostr.) ein kommensale des Bohramphipoden *Chelura terebrans.* — *Zoologischer Anzeiger,* 121(11-12): 317-322, 11 figs.

KOZLOFF, EUGENE N. 1955. Two new species of *Entocythere* (Ostracoda, Cytheridae) commensal on *Pacifastacus gambelii* (Girard). — *American Midland Naturalist,* 53(1): 156-161, 24 figs.

KOZLOFF, EUGENE N. AND DONALD C. WHITMAN. 1954. *Entocythere occidentalis* sp. nov., a cytherid ostracod commensal on western species of *Pacifastacus.* — *American Midland Naturalist,* 52(1): 159-163, 13 figs.

KÜKENTHAL, W. 1926, 1927. *Handbuch der Zoologie III,* 1. — *Berlin und Leipzig.* 1158 pp.

LEVINSON, STUART. 1959. Bibliography and index to new genera and species of ostracods for 1957. — *Micropaleontology,* 5(2): 241-260, 3 figs.

————. 1962. Bibliography and index to new genera and species of ostracods for 1958-1959. — *Micropaleontology,* 8(1): 77-105.

MARSHALL, W. S. 1903. *Entocythere cambaria* n. g. n. sp., a parasitic ostracod. — *Transactions of the Wisconsin Academy of Sciences, Arts, and Letters,* 14(1): 117-144, 4 pls., 30 figs.

MCGREGOR, DON L. AND ROBERT V. KESLING. 1969. Copulatory adaptations in ostracods, Part II. Adaptations in living ostracods. — *Contributions from the Museum of Paleontology, The University of Michigan,* 22(17): 221-239.

MCKENZIE, K. G. 1967. The distribution of Caenozoic marine Ostracoda from the Gulf of Mexico to Australia. — pp. 219-238 IN *Aspects of Tethyan Biogeography (C. G. Adams and D. V. Ager, editors). Systematics Association Publication* No. 7 (London).

MCKENZIE, K. G. (ed.). 1972. *Check list of ostracoda recorded from the Indian subcontinent and Ceylon (1840-1971).* — *Privately printed,* 59 pp.

MOORE, RAYMOND C. AND C. W. PITRAT. 1961. *Treatise on invertebrate paleontology, Part Q, Arthropoda 3: Crustacea, Ostracoda.* — *Geological Society of America, New York, and the University of Kansas Press.*

MORKHOVEN, F. P. C. M. VAN. 1962. *Post-Palaeozoic Ostracoda, their morphology, taxonomy, and economic use. Volume I, General.* — *Elsevier Publishing Co., Inc., New York.* 204 pp.

MULLER, G. W. 1912. *Ostracoda,* IN *Das Tierreich 31, Leiferung.* — *Berlin.* 434 pp.

NEALE, JOHN W. 1964. Some factors influencing the distribution of recent British Ostracoda. — *Pubblicazioni della Stazione Zoologica di Napoli,* 33 (supplement): 247-307.

NICHOLAS, BROTHER G. 1962. Checklist of troglobitic organisms of middle America. — *American Midland Naturalist,* 68(1): 165-188.

PARIS, P. 1916a. *Sphaeromicola topsenti* n. g. n. sp., Ostracode commensal d'isopodes troglobies du genre *Caecosphaeroma.* — *Comptes Rendus Hebdomadaires des Seances de l'Academie des Sciences, Paris,* 163: 307-309.

————. 1916b. Note sur quelques Entomostraces recoltes dans le Department de la Cote-d'Or. — *Bulletin de la Societe Zoologique de France,* 41: 112-116.

————. 1920. Ostracods (premiere serie). — *Biospeologica: Archives de Zoologica Experimentale et Generale,* 58: 475-487, 50 figs.

————. 1925a. Al faune cavernicola de la Cote-d'Or. — *Association Francaise pour l'Avancement des Sciences,* 49th session, pp. 449-450.

————. 1925b. Sur la bionomie de quelques crustaces troglobies de la Cote-d'Or. — *Association Francaise pour l'Avancement des Sciences,* 49th session, pp. 451-454.

PENNAK, T. W. 1953. *Fresh-water invertebrates of the United States.* — *Ronald Press Co., New York.* 769 pp.

PETERS, DANIEL J. 1971. Entocytherid ostracods of the Lower James-York peninsula, Virginia. — *Virginia Journal of Science,* 22(3): 100.

PRATT, H. S. 1948. *A manual of the common invertebrate animals, exclusive of insects.* — *P. Blakiston's Sons and Co., Philadelphia.* 845 pp.

PRINS, RUDOLPH. 1964. *Attheyella carolinensis* Chappuis (Copepoda, Harpacticoida) on freshwater crayfishes from Kentucky. — *Transactions of the American Microscopical Society,* 83(3): 370-371.

RACOVITZA, EMILE-G. 1910. Spheromiens (premiere serie) et revision des Monolistrini (Isopoda spheromiens). — *Biospeologica, No. XIII: Archives de Zoologie Experimentale et Generale,* 5th series, tome IV, pp. 625-758, 31 pls.

REDDELL, JAMES R. 1965. A checklist of the cave fauna of Texas. I. The invertebrates (exclusive of Insecta). — *Texas Journal of Science,* 17(2): 143-187.

REDDELL, JAMES R. AND ROBERT W. MITCHELL, (Eds). 1971. Studies on the cavernicole fauna of Mexico. — *Association for Mexican Cave Studies,* Bulletin 4, 239 pp. (The Speleo Press, Austin, Texas.)

REMY, P. A. 1943. Nouvelle station du sphaeromien troglobie *Caecosphaeroma* (V.) *burgundum* Dollfus var. *rupis fucaldi* Hubault et de l'ostracode commensal *Sphaeromicola topsenti* Paris. — *Bulletin de la Society Zoologique de France,* 68: 35-38.

————. 1946. *Sphaeromicola stammeri* Klie var. hamiger (sic) n. var. ostracode commensal du Sphaeromide obscuriole *Caecosphaeroma* (*C.*) *Virei* Dollfus. — *Collections du Musee de Zoologie, Nancy,* 1: 7-9, 8 figs.

————. 1948a. Sur la distribution geographique des Sphaeromidae, Monolistrini, et des *Sphaeromicola,* leurs ostracodes commensaux. — *Bulletin du Museum d'Histoire Naturelle de Marseille,* 8(1): 1-4.

————. 1948b. Description de *Sphaeromicola cebennica,* n. sp., ostracode cytheride commensal de l'isopode cirolanide cavernicola *Sphaeromides raymondi* Dollfus. — *Bulletin Mensuel de la Societe Linneenne de Lyon,* 17(7): 129-132, 9 figs.

————. 1948c. Sur quelques crustaces cavernicoles d'Europe. — *Notes Biospeologiques,* 3: 35-47, 4 figs. (Also as, *Publications du Museum d'Histoire Naturelle, Paris,* 11.)

————. 1951. Stations de crustaces obscurioles. — Biospeologica: *Archives de Zoologie Experimentale et Generale,* 88: 217-230, 2 figs.

————. 1959. *Notice sur les titres travaux scientifique de M. Paul A. Remy, professeur a la faculte de sciences de Nancy.* — *Mimeographed,* 59 pp.

RIEK, E. F. 1959. The Australian freshwater crustacea. — pp. 246-258 IN *Biogeography and Ecology in Australia, edited by A. Keast, R. L. Crocker and C. S. Christian (Uitgeverij Dr. W. Junk, Den Haag)* 640 pp.

RIOJA, ENRIQUE. 1940a. Notas acerca de los crustaceos del Lago Patzcuaro. — *Anales del Instituto de Biologia, Mexico,* 11(2): 469-475.

————. 1940b. Estudios carcinológicos. V. Morfología de un ostrácodo epizoario observado sobre *Cambarus* (*Cambarellus*) *montezumae* Sauss. de México, Entocythere heterodonta n. sp. y descripcion de algunos de sus estados lavarios. — *Anales del Instituto de Biologia, Mexico,* 11(2): 593-609, 6 figs.

————. 1940c. Un caso de biocenosis observado sobre *Cambarus* (*Cambarellus*) *montezumae* Saussure, de México. — *Revista Chilena de Historia Natural pura y Applicado,* 44: 57-59.

————. 1941a. Estudios carcinológicos. VI. Estudio morfológico del esquelto interno de apodemas quitinoso de *Entocythere heterodonta* Rioja (Crust. Ostracodos). — *Anales del Instituto de Biologia, México,* 12(1): 177-191, 9 figs.

————. 1941b. Estudios carcinológicos. VIII. Las especies del genero *Entocythere* (Crust. Ostracodos). — *Anales del Instituto de Biologia, México,* 12(1): 193-198, 1 fig.

————. 1942a. Descripción de una especie y una subspecie nuevas del genero *Entocythere* Marshall, precedentes de la Cueva Chica (San Luis Potosi, México). — *Ciencia* (*México*), 3(7): 201-204, 6 figs.

————. 1942b. Estudios carcinológicos. XIII. Consideraciones y datos acerca del género *Entocythere* (Crust. Ostracodos) y algunas de sus especies, con descripcion de una nueva. — *Anales del Instituto de Biologia, México,* 13(2): 685-697, 21 figs.

————. 1943a. Estudios carcinológicos. XIV. Nuevos datos acerca de los *Entocythere* (Crust. Ostracodos) de México. — *Anales del Instituto de Biologia, México,* 14(2): 553-566, 24 figs.

————. 1943b. Estudios carcinológicos. XV. Polimorfismo femenino en los ostracodos del genero *Entocythere.* — *Anales del Instituto de Biologia, México,* 14(2): 567-585, 39 figs.

————. 1945. Estudios carcinológicos. XVIII. Observaciones acerca de las variaciones de la pieza copuladora en *Entocythere dobbini,* Rioja (Crust. Ostracodos). — *Anales del Instituto de Biologia, México,* 16(2): 419-423, 4 figs.

————. 1949. Estudios carcinológicos. XXI. Contribución al conocomiento de las especies del genero *Entocythere* de México. — *Anales del Instituto de Biologia, México,* 20(1-2): 315-329, 28 figs.

————. 1951. Estudios carcinológicos. XXV. El hallazgo del genero *Sphaeromicola* en America (Ostracodos, Citeridos) y descripcion de una nueva especie. — *Anales del Instituto de Biologia, México,* 22(1): 169-179, 16 figs.

————. 1953. Los crustaceos cavernicolas de Mexico. IV. Centenario de la Universidad de Mexico (1151-1951). *Memoria del Congreso Cientifico Mexicano.* — *Ciencias Biologicas,* 7: 285-298.

————. 1955. Estudios carcinológicos. XXXII. Primeros datos acerca de las especies del genero *Entocythere* (Crustaceos, Ostracodos) de la Isla de Cuba. — *Anales del Instituto de Biologia, Mexico,* 26(1): 193-197, 9 figs.

ROELOFS, H. M. A. 1966. Verslag over het onderzoek naar biologie en voorkomen van met de mariene, houtborende amphipode *Chelura terebrans* Philippi geassocieerde ostracoden. — *Verslagen van de gewone vergadering der Afdeeling natuurkunde Kon. Nederlandse Akademie van wetenschappen, Amsterdam,* 75(9): 147-152.

————. 1968. Etude du developpement de l'ostracode marin *Sphaeromicola dudichi Klie,* 1938. — *Bulletin Zoologisch Museum Universiteit van Amsterdam,* 1(5): 39-51.

ROELOFS, H. M. A. — See also ROELOFS, JIET.

ROELOFS, JIET. 1966. Redescription de l'ostracode marin *Sphaeromicola dudichi* Klie, 1938, et sa presence dans l'Atlantique. — *Beaufortia* (*Amsterdam*), 13(161): 207-212.

ROELOFS, JIET — See also ROELOFS, H. M. A.

RUDJAKOV, I. A. 1963. On the fauna of cave ostracods in western Transcaucasia (Russian with English summary). — *Zoologicheskii Zhurnal,* 42(1): 32-40, 4 figs.

SARS, G. O. 1926. Fresh-water Ostracoda from Canada and Alaska. — *Report, Canadian Arctic Expedition,* 1913-1918, 7(I): 1-32, 5 pls.

SKOGSBERG, T. 1928. Studies on marine ostracods. Part II. External morphology of the genus *Cythereis* with descriptions of twenty-one new species. — *Occasional Papers of the California Academy of Sciences,* XV, 155 pp.

SOHN, I. G. 1968. Premature citations and zoological nomenclature. — *Science,* 159(3813): 441-442.

SPANDL, HERMANN. 1926. Die Tierwelt der Unterirdischen Gewässer. — IN *Kyrle, G., Spelaologische Monographien, Wein,* 11: 1-235.

STAMMER, HANS-JURGEN. 1930. Eine neue Höhlensphäromide aus dem Karst, *Monolistra* (*Typhlosphaeroma*) *schottlaenderi,* und die Verbreitung des Genus *Monolistra.* — *Zoologischer Anzeiger,* 88(910): 291-304.

————. 1932. Die Fauna des Timavo. — *Zoologische Jahrbucher, Jena,* 63(5-6): 521-656.

————. 1935. Untersuchungen über die Tierwelt der Karsthöhlengewasser. — *Verhandlungen der Internationalen Vereinigung für theoretische und angewandte Limnologie* (*Beograd*), 7(1): 92-99.

STAMPER, M. N. 1957. Incubation and hatching of the egg of a commensal ostracod, *Entocythere illinoisensis* Hoff, 1942. — *Journal of the Colorado-Wyoming Academy of Science,* 4(9): 50.

STRASKRABA, M. 1969. Lista de los crustaceos dulceacuicolas de Cuba y sus relaciones zoogeograficas. — *Academia de Ciencias de Cuba, Instituto de Biologia, Serie Biologica,* 8: 1-37.

TRESSLER, WILLIS L. 1947. A check list of the known species of North American freshwater ostracods. — *American Midland Naturalist,* 38: 698-707.

————. 1954. Fresh-water Ostracoda from Texas and Mexico. — *Journal of the Washington Academy of Sciences,* 44(5): 138-149, 48 figs.

————. 1959. IN *Fresh-water biology, 2nd edition, Ward, H. B. and Whipple, G. C. (edited by W. T. Edmondson), John Wiley and Sons, New York.* 1248 pp.

TSCHERNYSHEVA, E. N. (editor). 1961. Basic paleontology, Arthropoda volume. — *State Scientific-Technological Publishing House, Moscow. Yu. A. Orlov, director.* 515 pp.

VANDEL, A. 1964. *Biospeologie, la biologie des animaux cavernicoles.* — *Gauthier-Villars Editeur, Paris.* 619 pp.

WALTON, MARGARET AND HORTON H. HOBBS, JR. 1959. Two new eyeless ostracods of the genus *Entocythere* from Florida. — *Quarterly Journal of the Florida Academy of Sciences,* 22(2): 114-120, 20 figs.

———— AND ————. 1971. The distribution of certain entocytherid ostracods on their crayfish hosts. — *Proceedings of the Academy of Natural Sciences of Philadelphia,* 123(4): 87-103.

WESENBERG-LUND, C. 1939. *Biologie der susswassertiere; Wirbellose tiere.* — *Julius Springer, Wein.* 817 pp.

WESTERVELT, CLINTON A., JR. AND EUGENE N. KOZLOFF. 1959. *Entocythere neglecta* sp. nov., a cytherid ostracod commensal on *Pacifastacus nigrescens* (Stimpson). — *American Midland Naturalist,* 6(1): 239-244, 14 figs.

Wolf, B. (editor). 1934-1938. *Animalium cavernarum catalogus*. — *Junk, s'Gravenhage*, 3 volumes.
Young, Willard. 1971. Ecological studies of the Entocytheridae (Ostracoda). — *American Midland Naturalist*, 85(2): 399-409.
Zenker, W. 1854. Monographie der ostracoden. — *Archiv für Naturgeschichte, Band*, 1: 1-87.

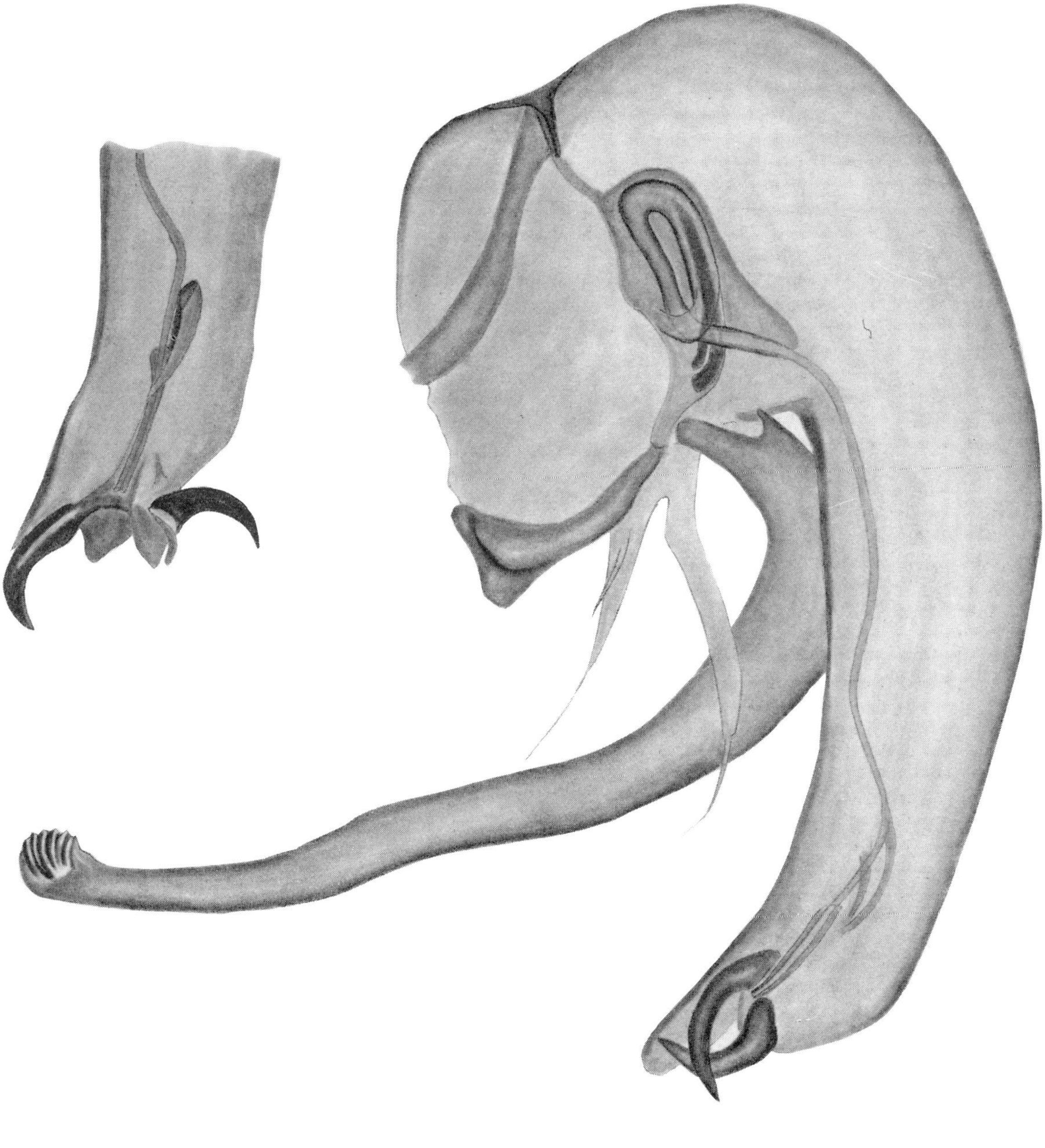

Copulatory apparatus of *Microsyssitria indica*

PLATES

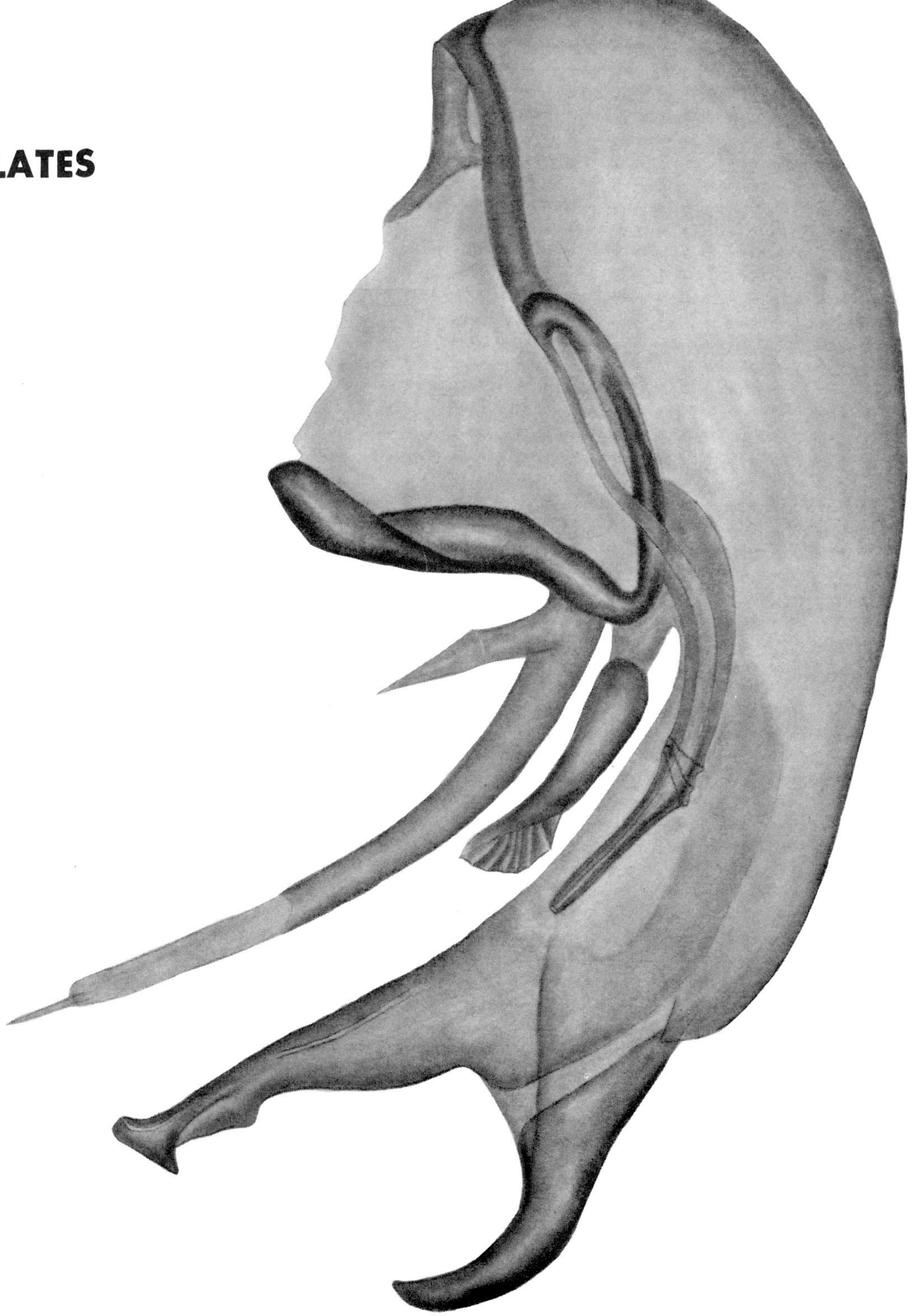

Copulatory apparatus of *Elachistocythere merista*

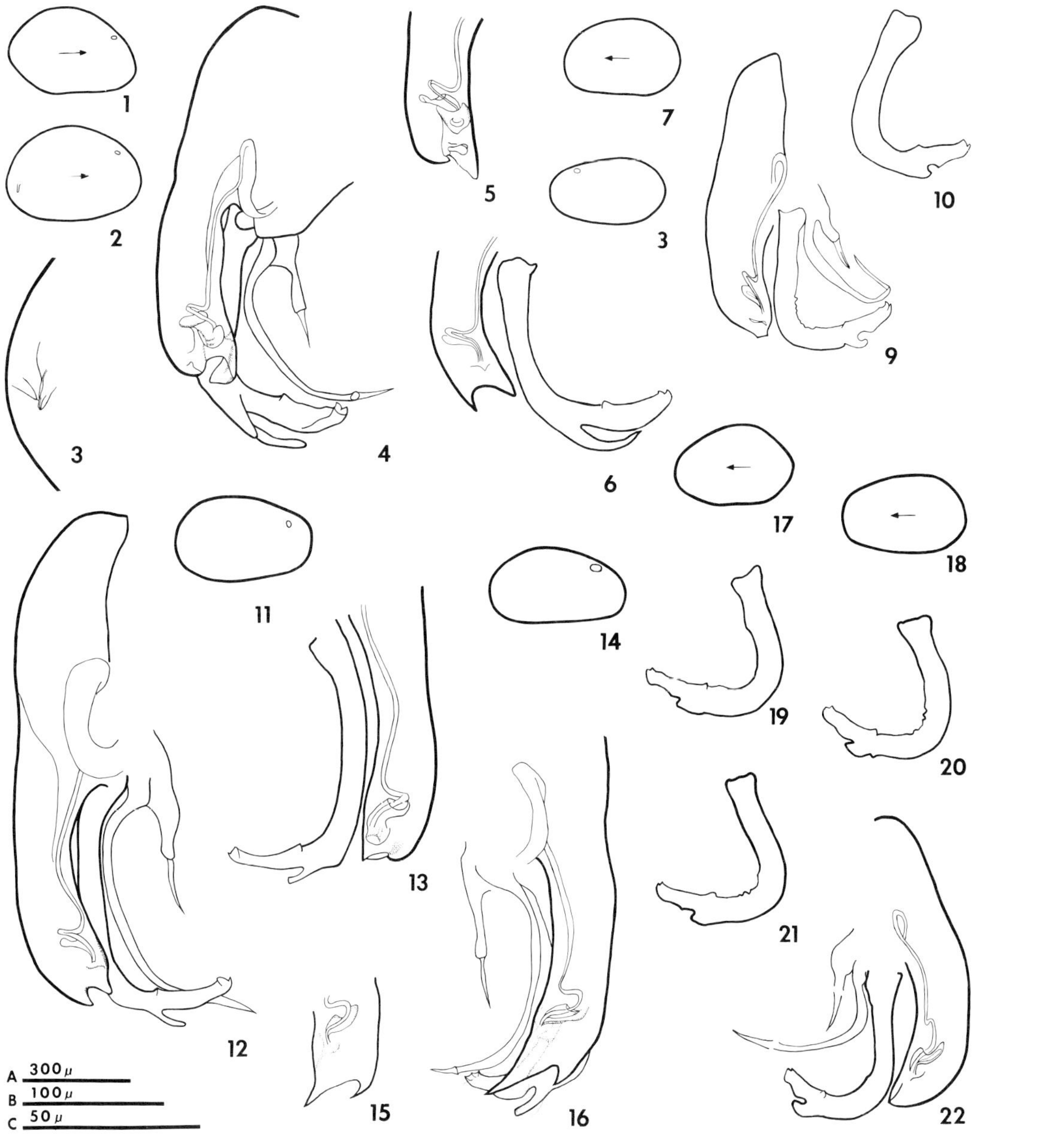

PLATE I

Figures 1-6, **Ankylocythere ancyla:** 1, Right valve of male holotype; 2, Right valve of female allotype; 3, Genital apparatus of allotype; 4, Copulatory complex of holotype; 5, Peniferum of holotype; 6, Copulatory complex of male (from Hobbs, 1966).

Figures 7-10, **Ankylocythere bidentata:** 7, Left valve of male; 8, Left valve of female; 9, Copulatory complex of male topotype; 10, Clasping apparatus.

Figures 11-13, **Ankylocythere burkeorum:** 11, Right valve of male holotype; 12, Copulatory complex of holotype; 13, Copulatory complex of male paratype.

Figures 14-16, **Ankylocythere copiosa:** 14, Right valve of male allotype; 15, Peniferum of allotype; 16, Copulatory complex of allotype.

Figures 17-22, **Ankylocythere cubensis:** 17, Left valve of male; 18, Left valve of female; 19, 20, 21, Variations in clasping apparatus; 22, Copulatory complex of male topotype. (Figures 17-22 from Hobbs, 1971.)

Figures 1, 2, 7, 8, 11, 14, 17, and 18 drawn to scale A; figure 3 drawn to scale B; figures 4-6, 9, 10, 12, 13, 15, 16, and 19-22 drawn to scale C.

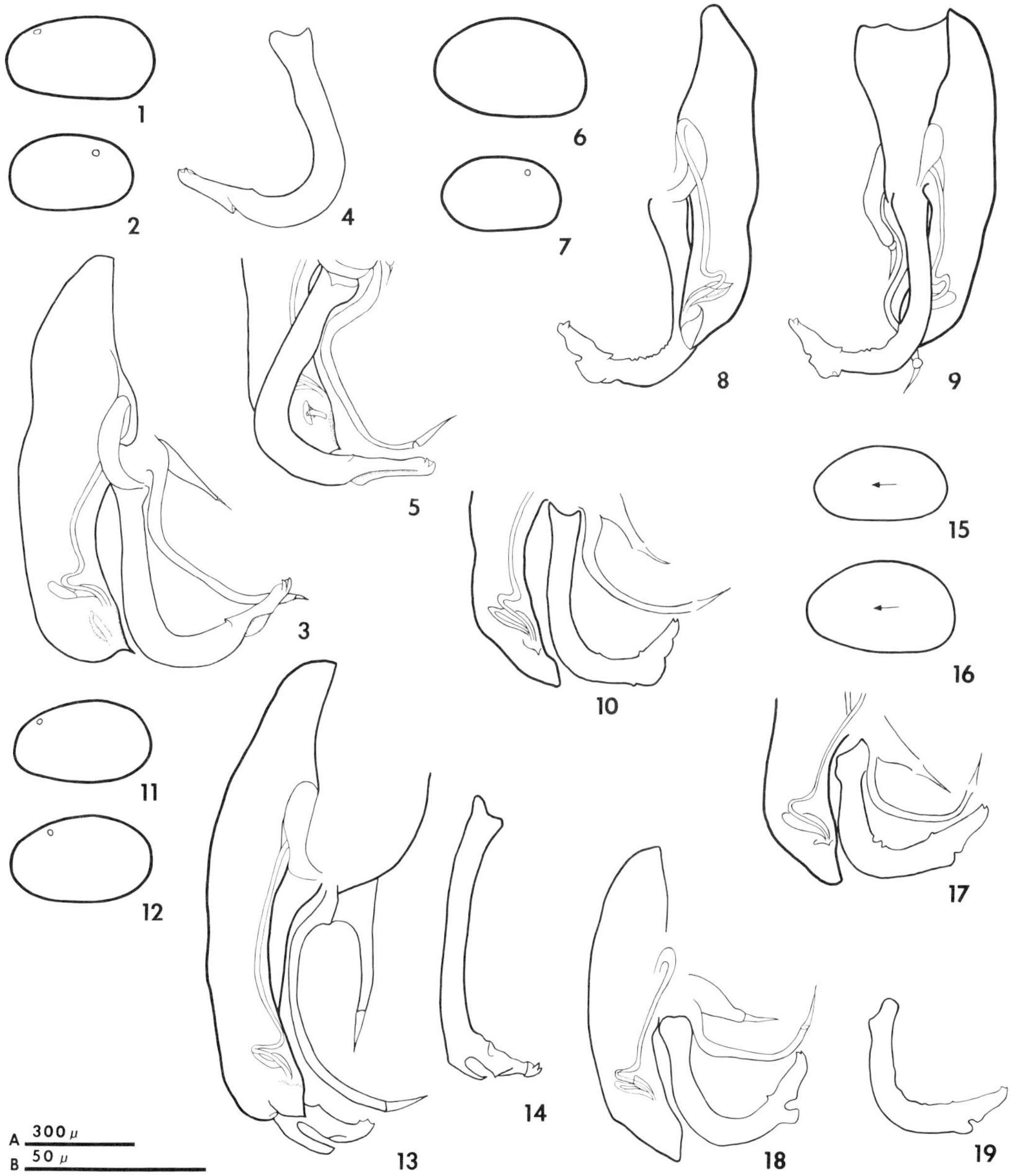

PLATE II

Figures 1-5, **Ankylocythere ephydra:** 1, Left valve of male holotype; 2, Right valve of biunguis female allotype; 3, Peniferum of holotype; 4, Other clasper of holotype; 5, Peniferum of paratype.

Figures 6-10, **Ankylocythere hamata:** 6, Left valve of female allotype; 7, Right valve of male holotype; 8, 9, Copulatory complexes of holotype; 10, Copulatory complex of male (drawing from Hobbs, 1966).

Figures 11-14, **Ankylocythere harmani:** 11, Left valve of male holotype; 12, Left valve of female allotype; 13, Copulatory complex of holotype; 14, Clasping apparatus of paratype.

Figures 15-19, **Ankylocythere heterodonta:** 15, Left valve of male; 16, Left valve of female; 17, Copulatory complex of male; 18, Copulatory complex of male; 19, Clasping apparatus. (Figures 15, 16, 18, 19 from Hobbs, 1971; Figure 17 from Hobbs, 1966.)

Figures 1, 2, 6, 7, 11, 12, 15, and 16 drawn to scale A; figures 3-5, 8-10, 13, 14, and 17-19 drawn to scale B.

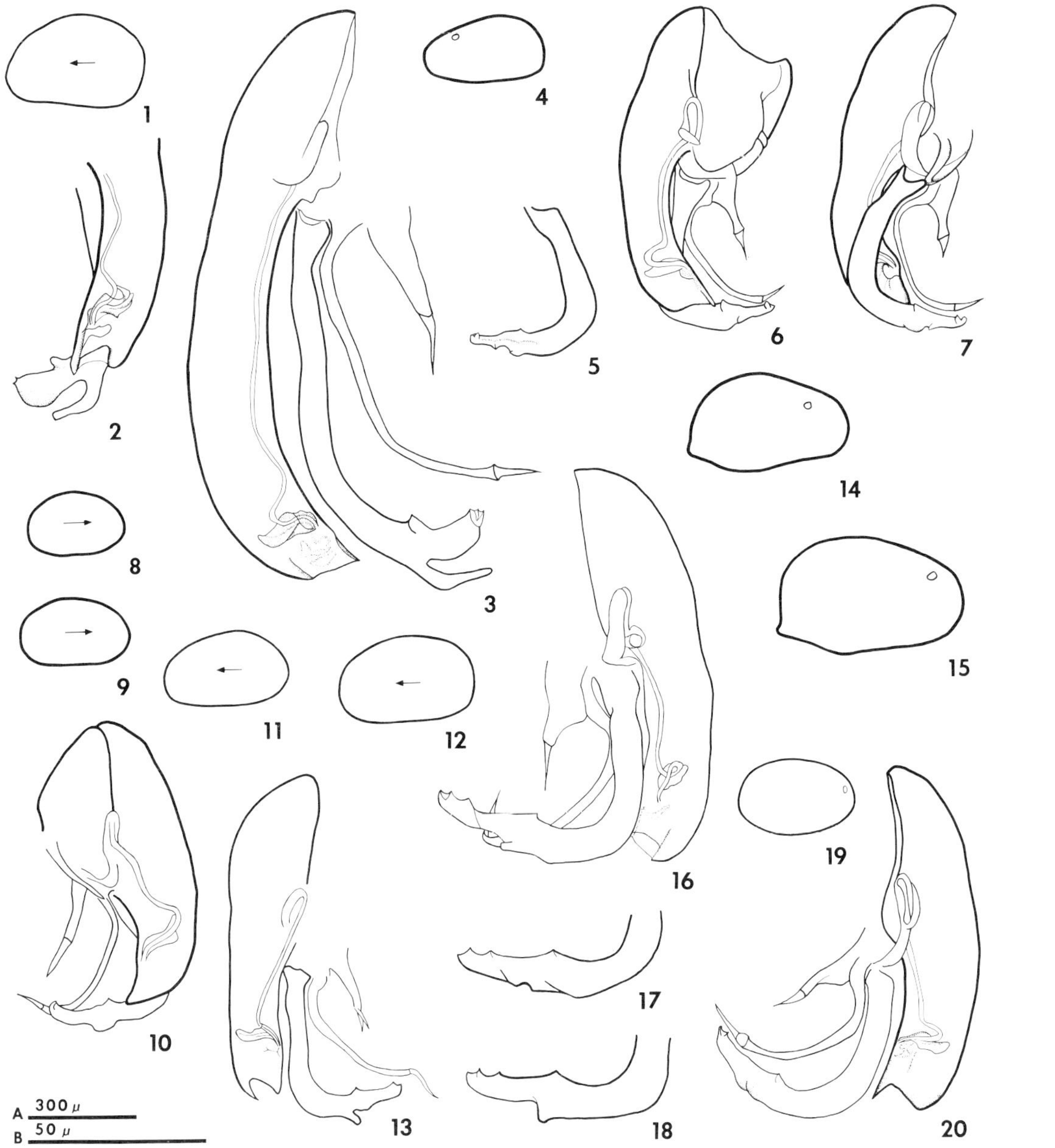

PLATE III

Figures 1-3, **Ankylocythere hobbsi:** 1, Left valve of male holotype; 2, Copulatory complex of male paratype; 3, Copulatory complex of holotype.

Figures 4-7, **Ankylocythere hyba:** 4, Left valve of male holotype; 5, Clasping apparatus of male paratype; 6, 7, Copulatory complexes of holotype.

Figures 8-10, **Ankylocythere maya:** 8, Right valve of male holotype; 9, Right valve of female allotype; 10, Copulatory complex of holotype.

Figures 11-13, **Ankylocythere sinuosa:** 11, Valve of male; 12, Valve of female; 13, Copulatory complex of male topotype. (Figures 11-13 from Hobbs, 1971.)

Figures 14-18, **Ankylocythere tallapoosa:** 14, Right valve of male holotype; 15, Right valve of female allotype; 16, Copulatory complex of holotype; 17, 18, Clasping apparatus of two male paratypes.

Figures 19-20, **Ankylocythere talulus:** 19, Right valve of male holotype; 20, Copulatory complex of holotype.

Figures 1, 4, 8, 9, 11, 12, 14, 15, and 19 drawn to scale A; figures 2, 3, 5-7, 10, 13, 16-18, and 20 drawn to scale B.

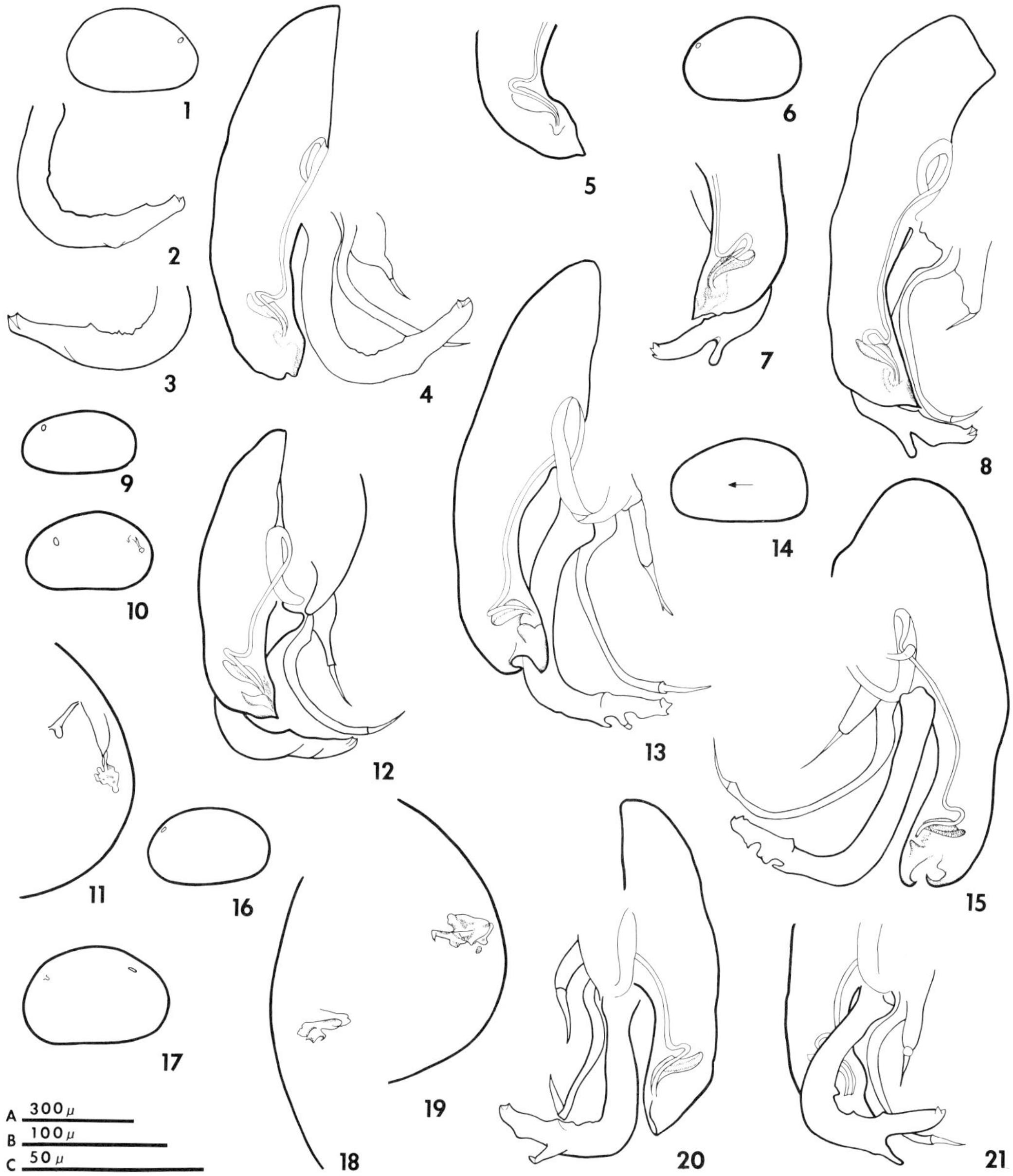

PLATE IV

Figures 1-5, **Ankylocythere telmoecea:** 1, Right valve of male holotype; 2, 3, Clasping apparatus of two male paratypes; 4, Copulatory complex of holotype; 5, Peniferum (from Hobbs, 1966).

Figures 6-8, **Ankylocythere tiphophila:** 6, Left valve of male holotype; 7, Copulatory complex of male paratype; 8, Copulatory complex of holotype.

Figures 9-12, **Ankylocythere toltecae:** 9, Left valve of male holotype; 10, Left valve of female allotype; 11, Genital apparatus of allotype; 12, Copulatory complex of holotype.

Figures 13-15, **Ankylocythere tridentata:** 13, Copulatory complex of holotype; 14, Right valve of male holotype; 15, Copulatory complex of male.

Figures 16-21, **Ankylocythere villalobosi:** 16, Left valve of male holotype; 17, Right valve of female allotype; 18, Genital apparatus of allotype; 19, Genital apparatus of female paratype; 20, Copulatory complex of holotype; 21, Copulatory complex of male paratype.

Figures 1, 6, 9, 10, 14, 16, and 17 drawn to scale A; figures 11, 18, and 19 drawn to scale B; figures 2-5, 7, 8, 12, 13, 15, 20, and 21 drawn to scale C.

PLATE V

Figures 1-5, **Ascetocythere asceta:** 1, Right valve of male holotype; 2, Right valve of female allotype; 3, Genital apparatus of allotype; 4, Peniferum; 5, Copulatory complex of holotype (drawing from Hobbs, 1966).

Figures 6-9, **Ascetocythere batchi:** 6, Left valve of male holotype; 7, Left valve of female allotype; 8, Genital apparatus of allotype; 9, Copulatory complex of holotype.

Figures 10-12, **Ascetocythere coryphodes:** 10, Right valve of male holotype; 11, Copulatory complex of holotype; 12, Peniferum of paratype.

Figures 1, 2, 6, 7, and 10 drawn to scale A; figures 3 and 8 drawn to scale B; figures 4, 5, 9, 11, and 12 drawn to scale C.

PLATE VI

Figures 1-3, **Ascetocythere cosmeta:** 1, Left valve of male holotype; 2, 3, Copulatory complex and peniferum, both halves of holotype.

Figures 4-6, **Ascetocythere didactylata:** 4, Copulatory complex of male (from Hobbs and Hart, 1966); 5, Copulatory complex of male; 6, Copulatory complex of male holotype.

Figures 7-10, **Ascetocythere hoffmani:** 7, Left valve of male holotype; 8, 9, Peniferum and copulatory complex, both halves of holotype; 10, Copulatory complex of male (from Hobbs and Hart, 1966).

Figures 1 and 7 drawn to scale A; figures 2-6 and 8-10 drawn to scale B.

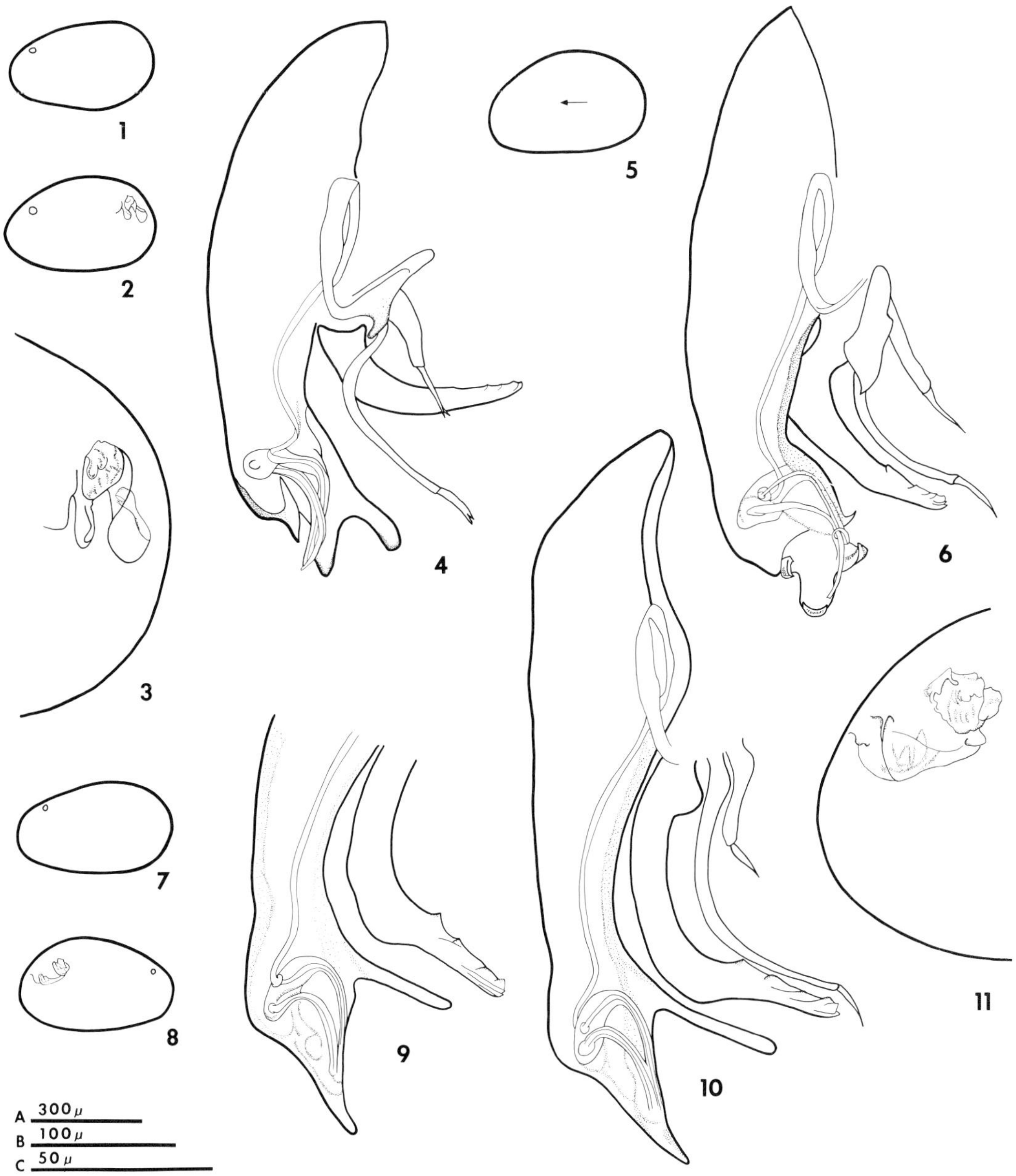

PLATE VII

Figures 1-4, **Ascetocythere holti:** 1, Left valve of male holotype; 2, Left valve of female allotype; 3, Genital apparatus of allotype; 4, Copulatory complex of holotype.

Figures 5-6, **Ascetocythere hyperoche:** 5, Left valve of male holotype; 6, Copulatory complex of holotype.

Figures 7-11, **Ascetocythere lita:** 7, Left valve of male holotype; 8, Right valve of female allotype; 9, 10, Copulatory complexes of holotype; 11, Genital apparatus of allotype.

Figures 1, 2, 5, 7, and 8 drawn to scale A; figures 3 and 11 drawn to scale B; figures 4, 6, 9, and 10 drawn to scale C.

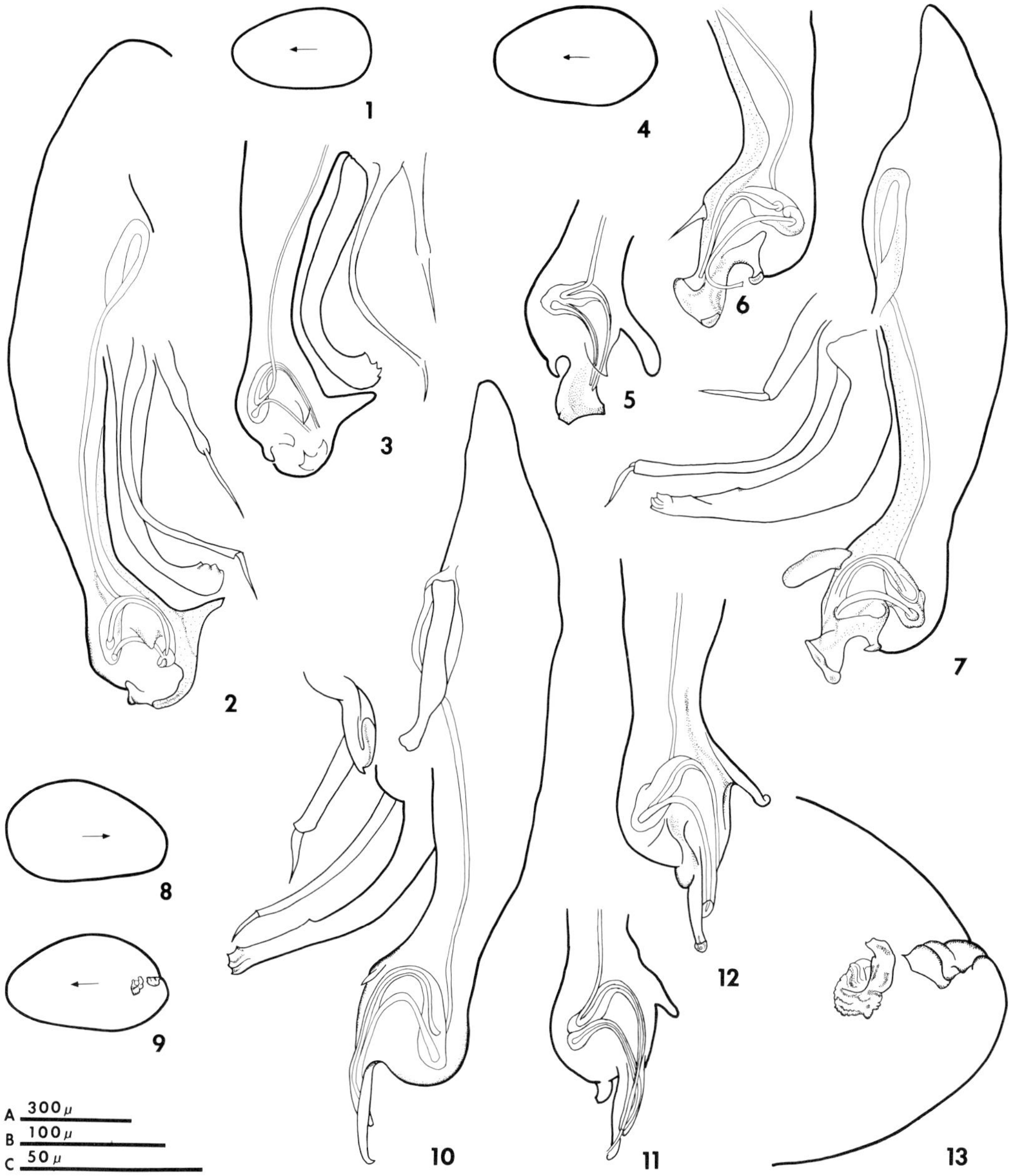

PLATE VIII

Figures 1-3, **Ascetocythere myxoides**: 1, Left valve of male holotype; 2, Copulatory complex of holotype; 3, Copulatory complex of male (from Hobbs and Hart, 1966).

Figures 4-7, **Ascetocythere ozalea**: 4, Left valve of male holotype; 5, Peniferum (from Hobbs and Hart, 1966); 6, Peniferum of paratype; 7, Copulatory complex of holotype.

Figures 8-13, **Ascetocythere sclera**: 8, Right valve of male holotype; 9, Left valve of female allotype; 10, Copulatory complex of holotype; 11, Peniferum (from Hobbs and Hart, 1966); 12, Peniferum of paratype; 13, Genital apparatus of allotype.

Figures 1, 4, 8, and 9 drawn to scale A; figure 13 drawn to scale B; figures 2, 3, 5-7, 10-12 drawn to scale C.

PLATE IX

Figures 1-6, **Cymocythere clavata:** 1, Right valve of male holotype; 2, Left valve of female allotype; 3, Genital apparatus of allotype; 4, Copulatory complex of male (from Hobbs and Hart, 1966); 5, 6, Peniferum and copulatory complex, both halves of holotype.

Figures 7-11, **Cymocythere cyma:** 7, Right valve of male holotype; 8, Copulatory complex of holotype; 9, Clasping apparatus of male; 10, Peniferum of holotype; 11, Copulatory complex of male (from Hobbs and Hart, 1966).

Figures 12-14, **Cymocythere gonia:** 12, Copulatory complex of male holotype; 13, Right valve of holotype; 14, Copulatory complex of male (from Hobbs and Hart, 1966).

Figures 1, 2, 7, and 13 drawn to scale A; figure 3 drawn to scale B; figures 4, 5, 6, 8-12, and 14 drawn to scale C.

PLATE X

Figures 1-4, **Dactylocythere amicula:** 1, Left valve of male holotype; 2, Right valve of female allotype; 3, Genital complex of allotype; 4, Copulatory complex of holotype.

Figures 5-10, **Dactylocythere amphiakis:** 5, Copulatory complex of male paratype; 6, Genital complex of female paratype; 7, Left valve of male holotype; 8, Left valve of female allotype; 9, Genital complex of allotype; 10, Copulatory complex of holotype.

Figures 1, 2, 7, and 8 drawn to scale A; figures 3, 6, and 9 drawn to scale B; figures 4, 5, and 10 drawn to scale C.

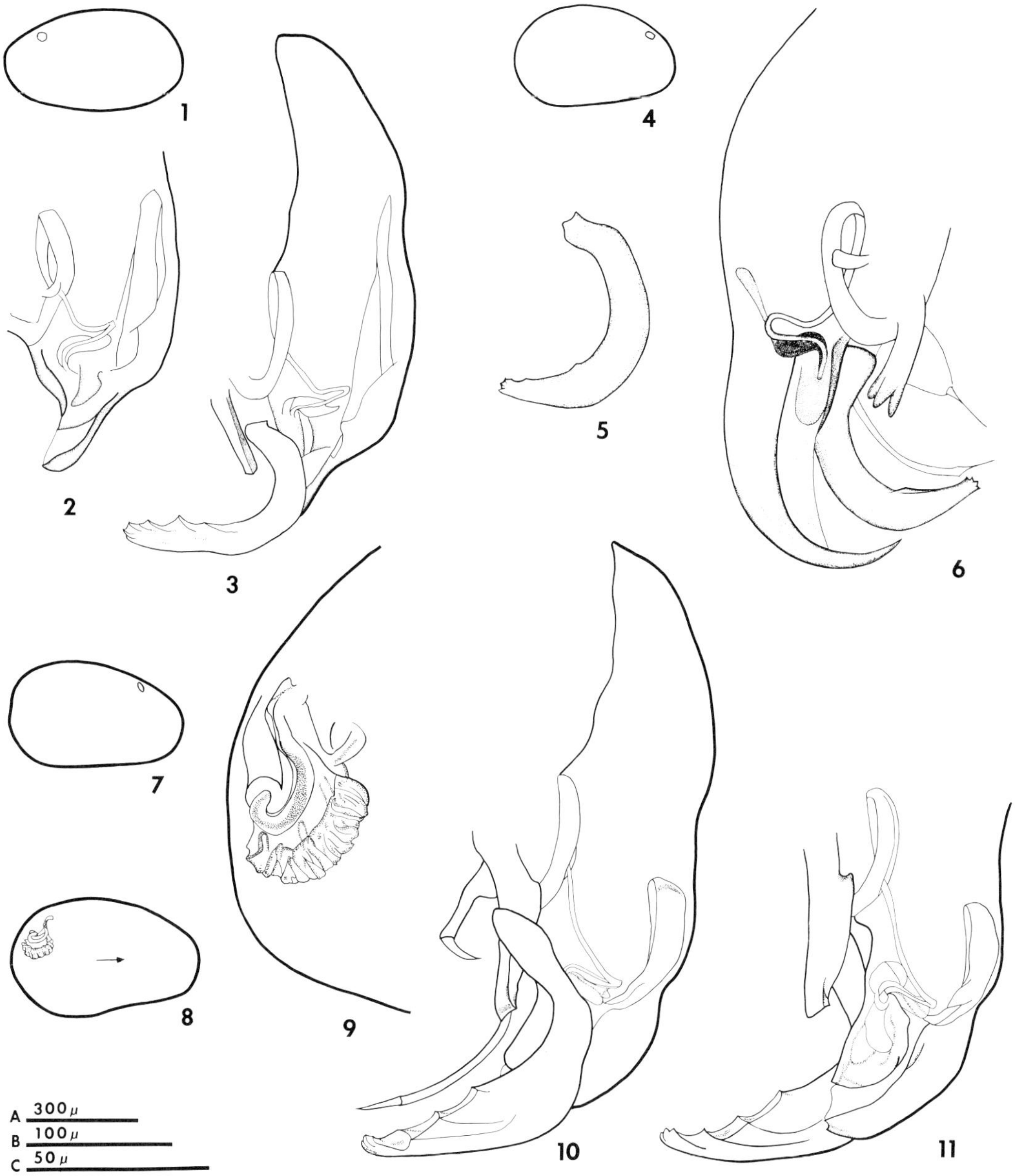

PLATE XI

Figures 1-3, **Dactylocythere arcuata:** 1, Left valve of male holotype; 2, 3, Peniferum and copulatory complex of holotype.

Figures 4-6, **Dactylocythere banana:** 4, Right valve of male holotype; 5, Clasping apparatus of male; 6, Copulatory complex of holotype.

Figures 7-11, **Dactylocythere brachystrix:** 7, Right valve of male holotype; 8, Right valve of female allotype; 9, Genital complex of allotype; 10, Copulatory complex of holotype; 11, Copulatory complex of paratype.

Figures 1, 4, 7, and 8 drawn to scale A; figure 9 drawn to scale B; figures 2, 3, 5, 6, 10, and 11 drawn to scale C.

PLATE XII

Figures 1-6, **Dactylocythere chalaza:** 1, Left valve of male holotype; 2, Left valve of female allotype; 3, Genital complex of allotype; 4, Genital complex of paratype; 5, Copulatory complex of holotype; 6, Copulatory complex of paratype.

Figures 7-10, **Dactylocythere charadra:** 7, Right valve of male holotype; 8, Genital complex of allotype; 9, Left valve of female allotype; 10, Copulatory complex of holotype.

Figures 11-15, **Dactylocythere chelomata:** 11, Left valve of male holotype; 12, Right valve of female allotype; 13, Genital complex of allotype; 14, Copulatory complex of holotype; 15, Clasping apparatus of holotype.

Figures 1, 2, 7, 9, 11, and 12 drawn to scale A; figures 3, 4, 8, and 13 drawn to scale B; figures 5, 6, 10, 14, and 15 drawn to scale C.

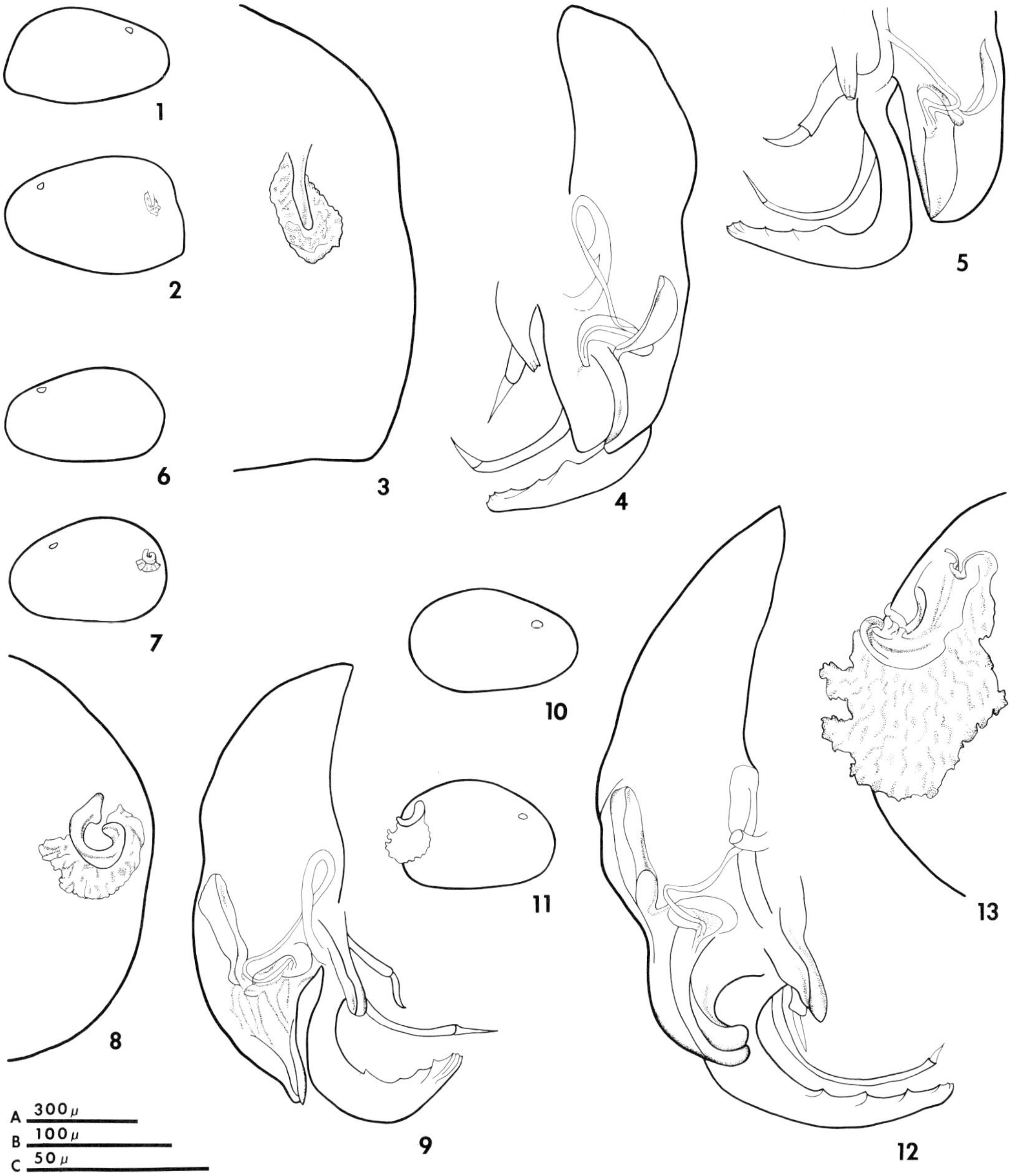

PLATE XIII

Figures 1-5, **Dactylocythere coloholca:** 1, Right valve of male holotype; 2, Left valve of female allotype; 3, Genital complex of allotype; 4, 5, Copulatory complexes of both halves of holotype.

Figures 6-9, **Dactylocythere cooperorum:** 6, Left valve of male holotype; 7, Left valve of female allotype; 8, Genital complex of allotype; 9, Copulatory complex of holotype.

Figures 10-13, **Dactylocythere crawfordi:** 10, Right valve of male holotype; 11, Right valve of female allotype; 12, Copulatory complex of holotype; 13, Genital complex of allotype.

Figures 1, 2, 6, 7, 10 and 11 drawn to scale A; figures 3, 8, and 13 drawn to scale B; figures 4, 5, 9, and 12 drawn to scale C.

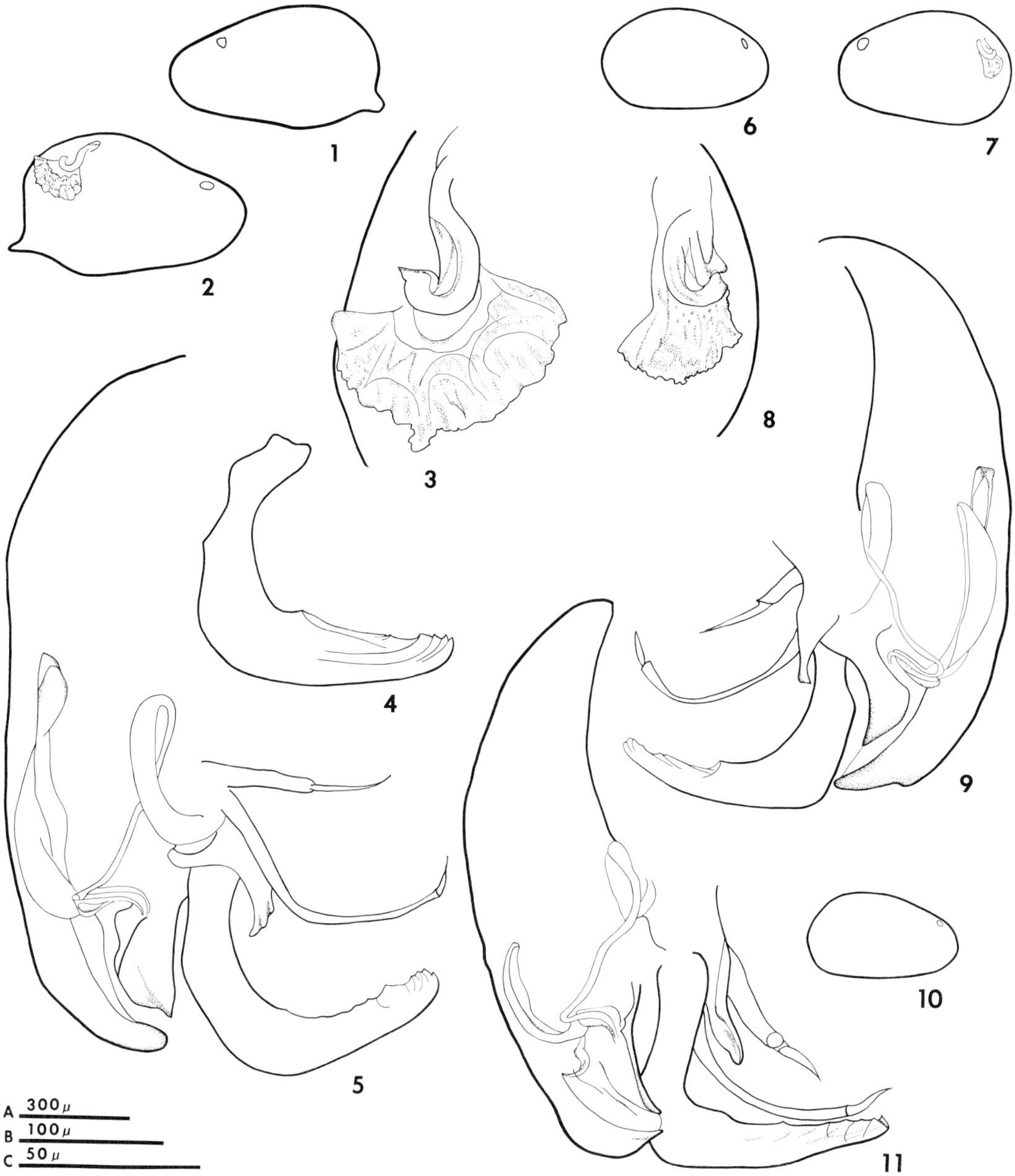

PLATE XIV

Figures 1-5, **Dactylocythere daphnioides:** 1, Left valve of male holotype; 2, Right valve of female allotype; 3, Genital complex of allotype; 4, Clasping apparatus of paratype; 5, Copulatory complex of holotype.

Figures 6-9, **Dactylocythere enoploholca:** 6, Right valve of male holotype; 7, Left valve of female allotype; 8, Genital complex of allotype; 9, Copulatory complex of holotype.

Figures 10-11, **Dactylocythere exoura:** 10, Right valve of male holotype; 11, Copulatory complex of holotype.

Figures 1, 2, 6, 7, and 10 drawn to scale A; figures 3 and 8 drawn to scale B; figures 4, 5, 9, and 10 drawn to scale C.

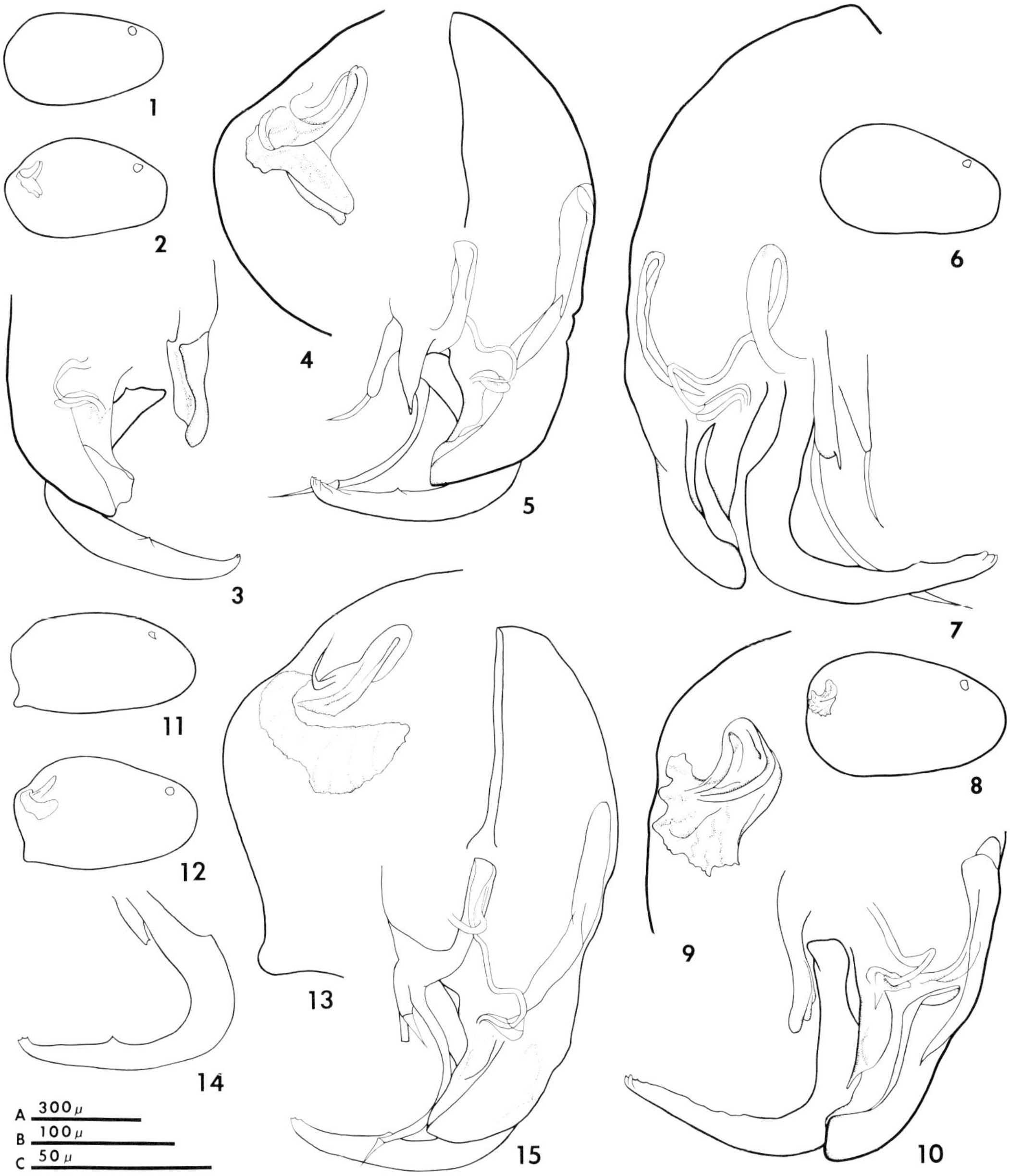

PLATE XV

Figures 1-5, **Dactylocythere falcata:** 1, Right valve of male holotype; 2, Right valve of female allotype; 3, Copulatory complex of paratype; 4, Genital complex of allotype; 5, Copulatory complex of holotype.

Figures 6-10, **Dactylocythere jeanae:** 6, Right valve of male holotype; 7, Copulatory complex of holotype; 8, Right valve of female allotype; 9, Genital complex of allotype; 10, Copulatory complex of paratype.

Figures 11-15, **Dactylocythere koloura:** 11, Right valve of male holotype; 12, Right valve of female allotype; 13, Genital complex of allotype; 14, Clasping apparatus of paratype; 15, Copulatory complex of holotype.

Figures 1, 2, 6, 8, 11, and 12 drawn to scale A; figures 4, 9, and 13 drawn to scale B; figures 3, 5, 7, 10, 14, and 15 drawn to scale C.

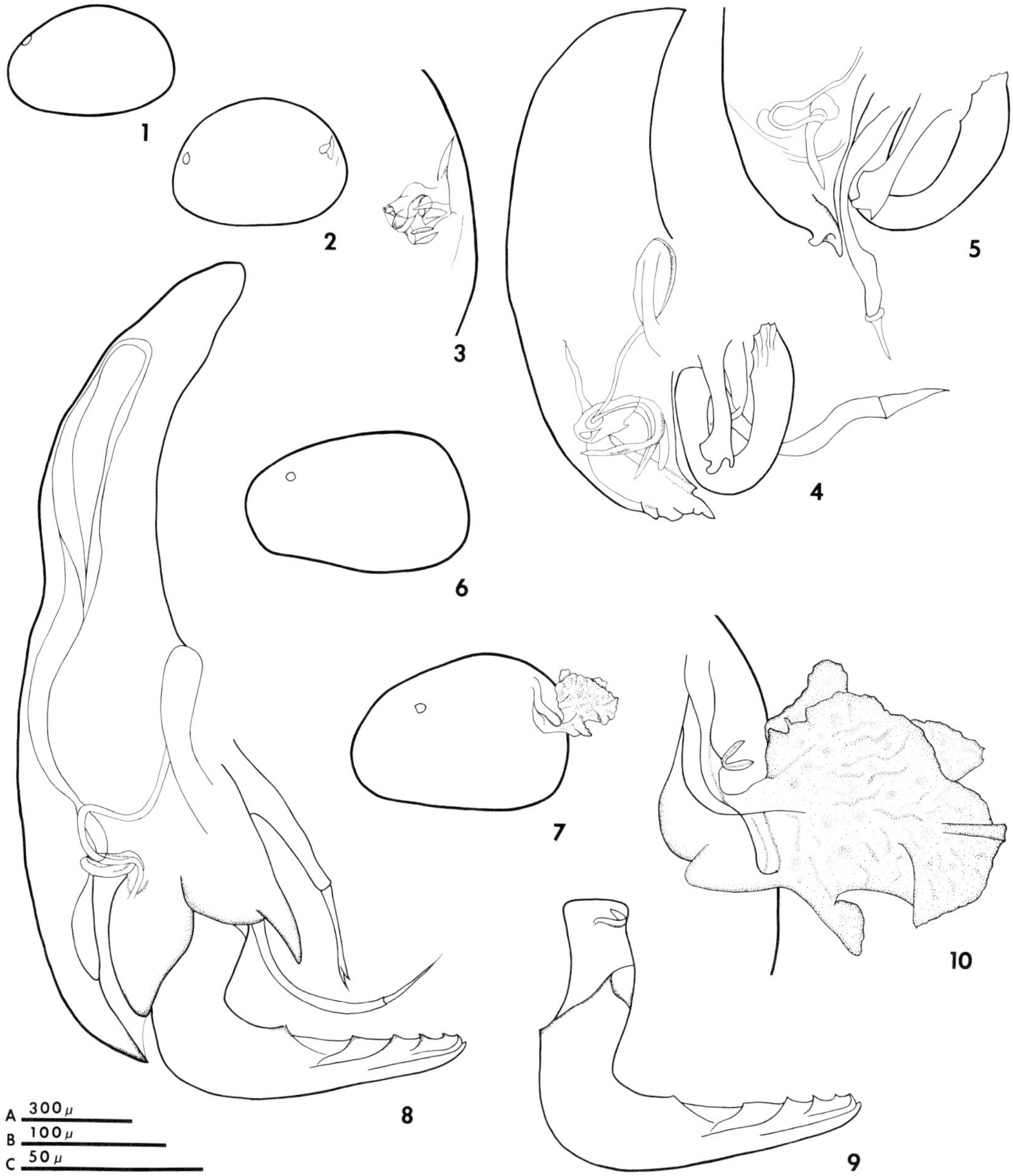

PLATE XVI

Figures 1-5, **Dactylocythere leptophylax:** 1, Left valve of male holotype; 2, Left valve of female allotype; 3, Genital complex of allotype; 4, Copulatory complex of paratype; 5, Copulatory complex of holotype.

Figures 6-10, **Dactylocythere macroholca:** 6, Left valve of male holotype; 7, Left valve of female allotype; 8, 9, Copulatory complex and clasping apparatus of holotype; 10, Genital complex of allotype.

Figures 1, 2, 6, and 7 drawn to scale A; figures 3 and 10 drawn to scale B; figures 4, 5, 8, and 9 drawn to scale C.

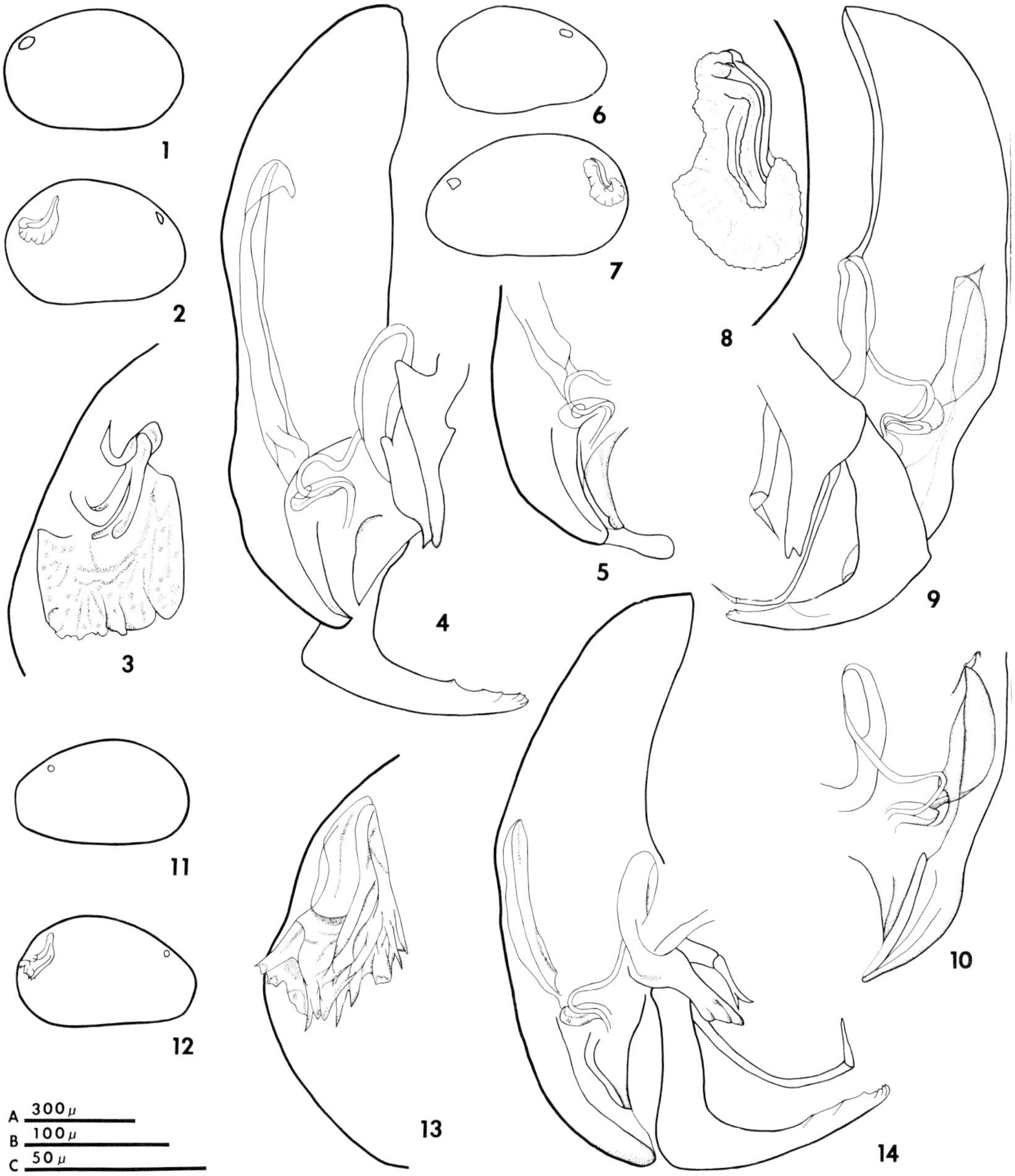

PLATE XVII

Figures 1-5, **Dactylocythere mecoscapha:** 1, Left valve of male holotype; 2, Right valve of female allotype; 3, Genital complex of allotype; 4, 5, Copulatory complex and peniferum of holotype.

Figures 6-10, **Dactylocythere megadactylus:** 6, Right valve of male holotype; 7, Left valve of female allotype; 8, Genital complex of allotype; 9, Copulatory complex of holotype; 10, Peniferum of paratype.

Figures 11-14, **Dactylocythere myura:** 11, Left valve of male holotype; 12, Right valve of female allotype; 13, Genital complex of allotype; 14, Copulatory complex of holotype.

Figures 1, 2, 6, 7, 11, and 12 drawn to scale A; figures 3, 8, and 13 drawn to scale B; figures 4, 5, 9, 10, and 14 drawn to scale C.

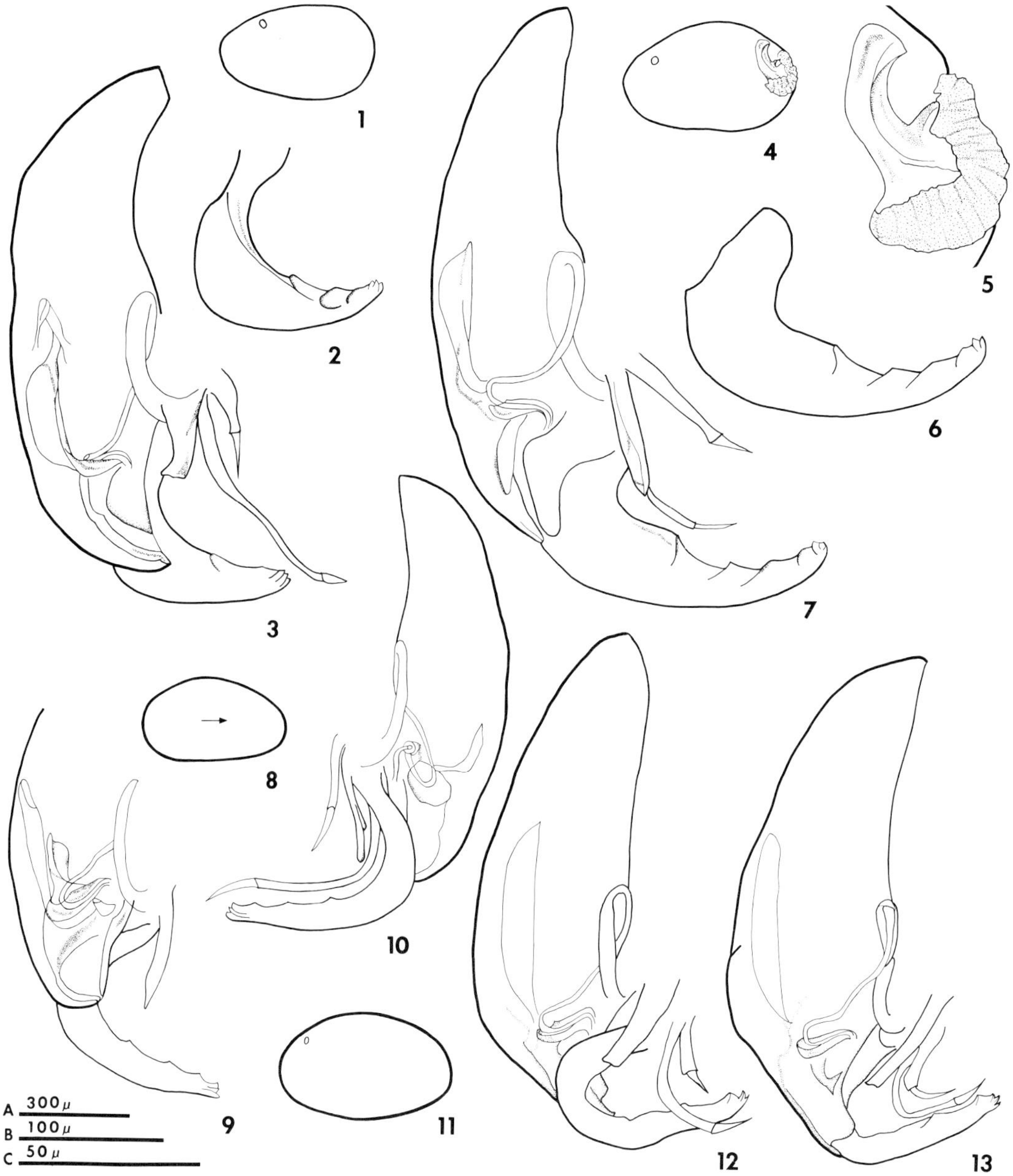

PLATE XVIII

Figures 1-3, **Dactylocythere pachysphyrata:** 1, Left valve of male holotype; 2, Clasping apparatus of paratype; 3, Copulatory complex of holotype.

Figures 4-7, **Dactylocythere phoxa:** 4, Left valve of female allotype; 5, Genital apparatus of allotype; 6, 7, Clasping apparatus and copulatory complex of holotype.

Figures 8-10, **Dactylocythere prinsi:** 8, Right valve of male holotype; 9, Copulatory complex of paratype; 10, Copulatory complex of holotype.

Figures 11-13, **Dactylocythere prionata:** 11, Left valve of male holotype; 12, 13, Copulatory complexes of both halves of holotype.

Figures 1, 4, 8, and 11 drawn to scale A; figure 5 drawn to scale B; figures 3, 6, 7, 9, 10, 12, and 13 drawn to scale C.

PLATE XIX

Figures 1-5, **Dactylocythere pughae:** 1, Left valve of female allotype; 2, Left valve of male holotype; 3, Genital complex of allotype; 4, 5, Clasping apparatus and copulatory complex of holotype.

Figures 6-9, **Dactylocythere runki:** 6, Right valve of male holotype; 7, Left valve of allotype; 8, Genital complex of allotype; 9, Copulatory complex of holotype.

Figures 10-15, **Dactylocythere sandbergi:** 10, Right valve of male holotype; 11, Right valve of female allotype; 12, Genital complex of allotype; 13, 14, Copulatory complexes of males; 15, Copulatory complex of holotype.

Figures 1, 2, 6, 7, 10, and 11 drawn to scale A; figures 3, 8, and 12 drawn to scale B; figures 4, 5, 9, 13, 14, and 15 drawn to scale C.

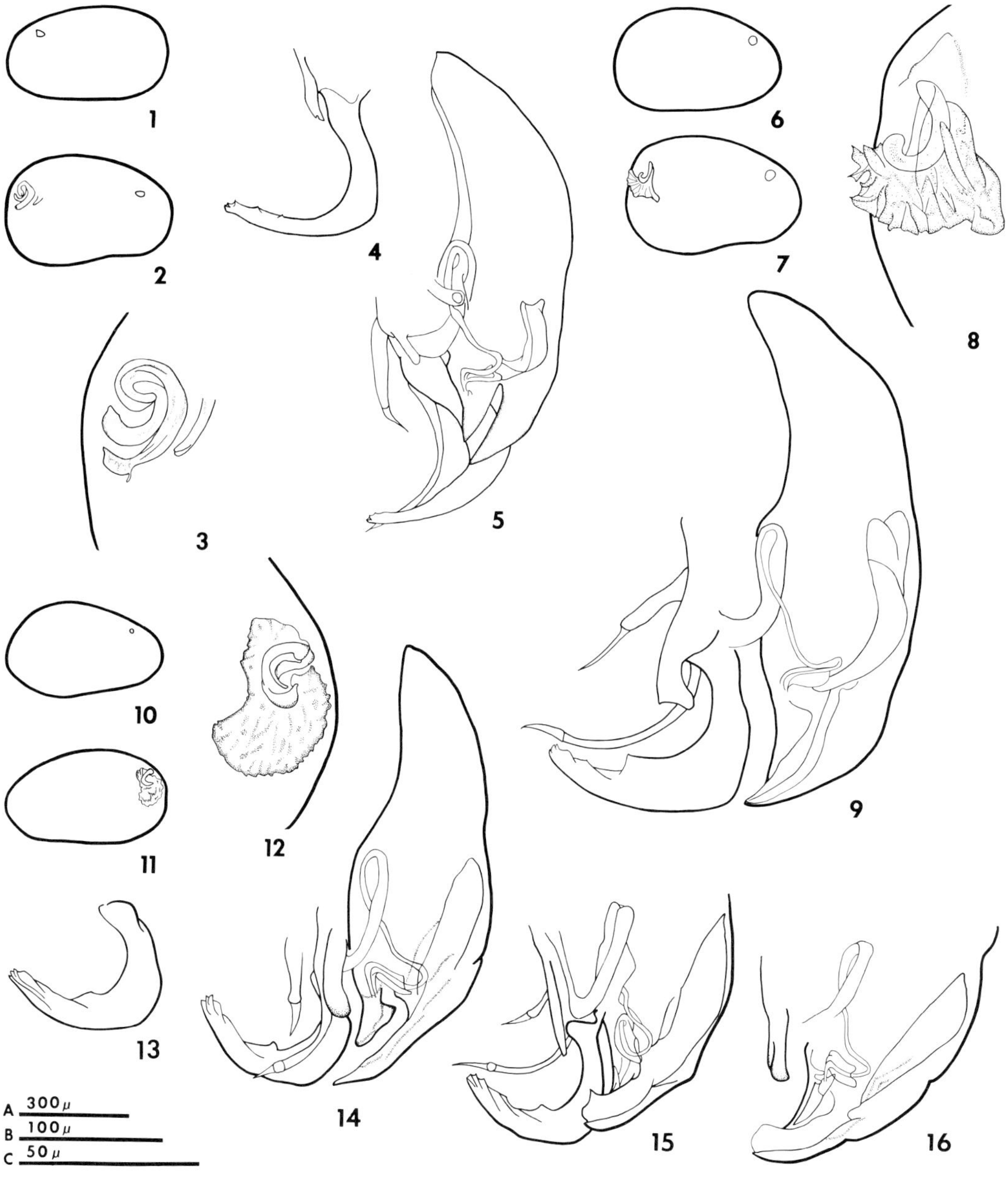

PLATE XX

Figures 1-5, **Dactylocythere speira:** 1, Left valve of male holotype; 2, Right valve of female allotype; 3, Genital complex of allotype; 4, Clasping apparatus of paratype; 5, Copulatory complex of holotype.

Figures 6-9, **Dactylocythere spinata:** 6, Right valve of male holotype; 7, Right valve of female allotype; 8, Genital complex of allotype; 9, Copulatory complex of holotype.

Figures 10-16, **Dactylocythere steevesi:** 10, Right valve of male topotype; 11, Left valve of female topotype; 12, Genital complex of topotype; 13, 14, 15, 16, Clasping apparatus and copulatory complexes of topotypes.

Figures 1, 2, 6, 7, 10, and 11 drawn to scale A; figures 3, 8, and 12 drawn to scale B; figures 4, 5, 9, 13, 14, 15, and 16 drawn to scale C.

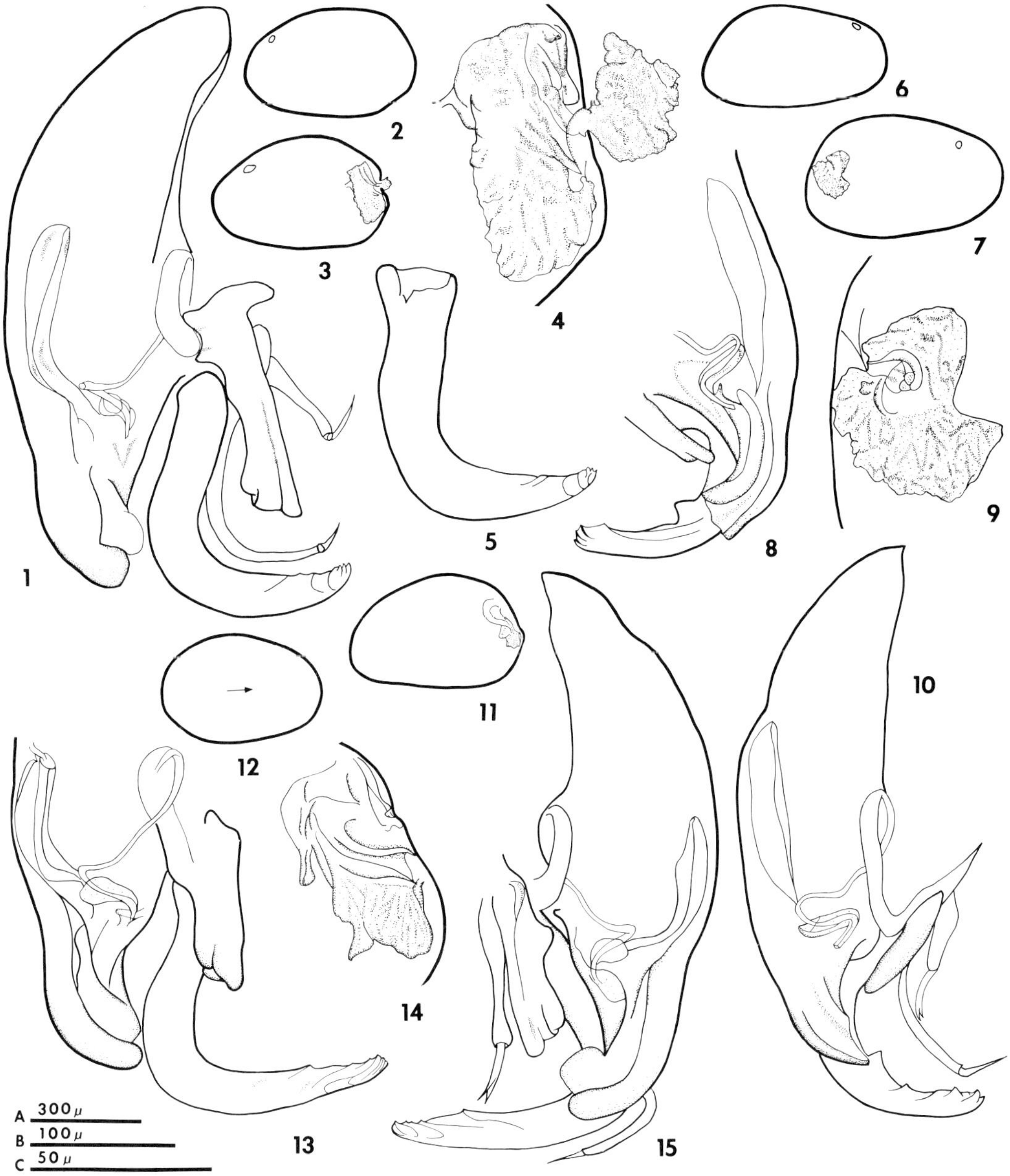

PLATE XXI

Figures 1-5, **Dactylocythere striophylax:** 1, Copulatory complex of holotype; 2, Left valve of male holotype; 3, Left valve of female allotype; 4, Genital complex of allotype; 5, Clasping apparatus of paratype.

Figures 6-10, **Dactylocythere susanae:** 6, Right valve of male holotype; 7, Right valve of female allotype; 8, Copulatory complex of paratype; 9, Genital complex of allotype; 10, Copulatory complex of holotype.

Figures 11-15, **Dactylocythere suteri:** 11, Left valve of female allotype; 12, Right valve of male holotype; 13, Copulatory complex of paratype; 14, Genital complex of allotype; 15, Copulatory complex of holotype.

Figures 2, 3, 6, 7, 11, and 12 drawn to scale A; figures 4, 9, and 14 drawn to scale B; figures 1, 5, 8, 10, 13, and 15 drawn to scale C.

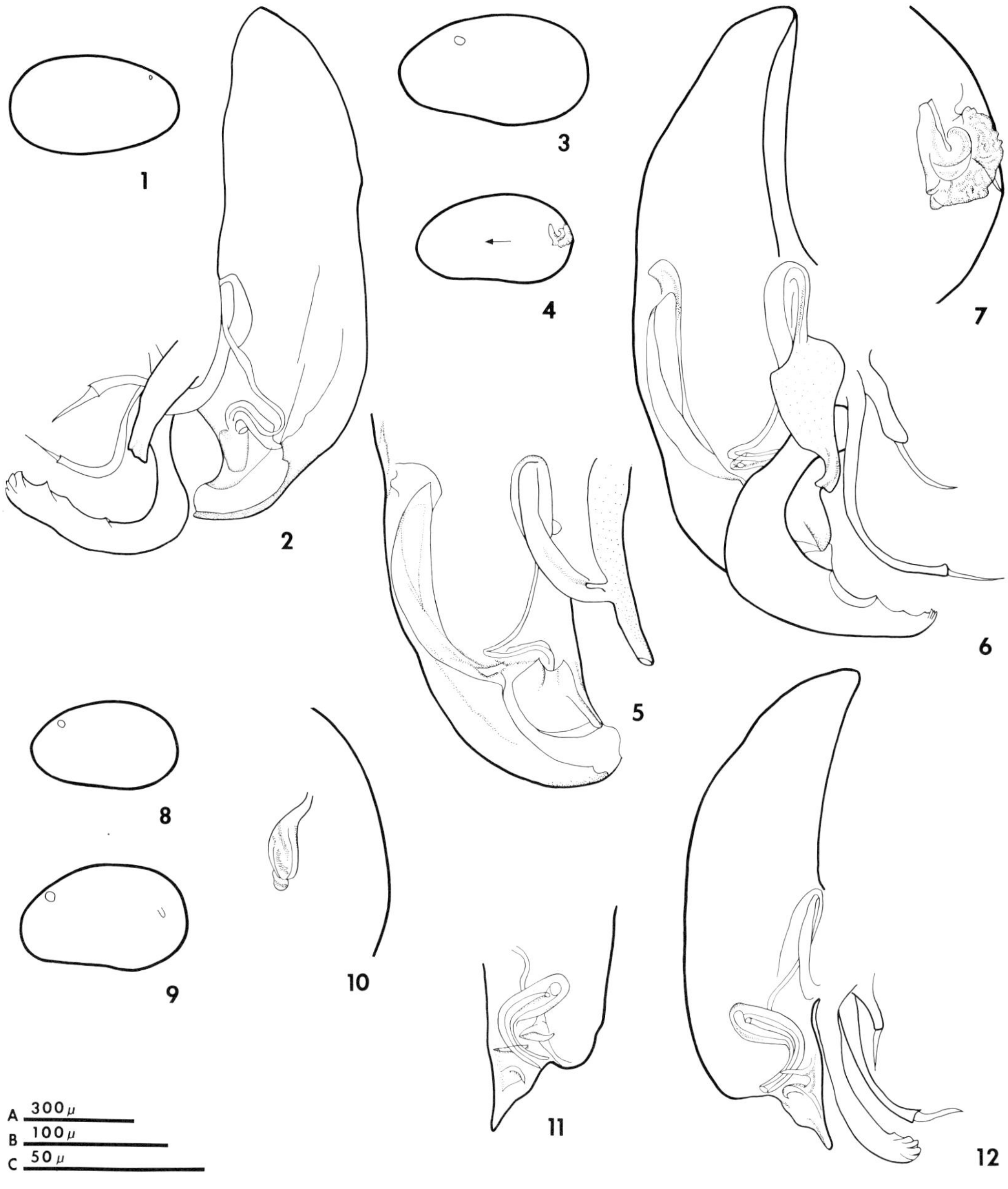

PLATE XXII

Figures 1-2, **Dactylocythere ungulata:** 1, Right valve of male holotype; 2, Copulatory complex of holotype.

Figures 3-7, **Dactylocythere xystroides:** 3, Left valve of male holotype; 4, Left valve of female allotype; 5, Peniferum of paratype; 6, Copulatory complex of holotype; 7, Genital complex of allotype.

Figures 8-12, **Donnaldsoncythere ardis:** 8, Left valve of male holotype; 9, Left valve of female allotype; 10, Genital complex of allotype; 11, Peniferum of paratype; 12, Copulatory complex of holotype.

Figures 1, 3, 4, 8, and 9 drawn to scale A; figures 7 and 10 drawn to scale B; figures 2, 5, 6, 11, and 12 drawn to scale C.

PLATE XXIII

Figures 1-5, **Donnaldsoncythere cayugaensis**: 1, Right valve of male holotype; 2, Right valve of female allotype; 3, Genital complex of allotype; 4, Peniferum of paratype; 5, Copulatory complex of holotype.

Figure 6, **Donnaldsoncythere donnaldsonensis**: 6, Copulatory complex of male (from Klie, 1931).

Figures 7-10, **Donnaldsoncythere hiwasseensis**: Right valve of male holotype; 8, Right valve of female allotype; 9, Peniferum of paratype; 10, Copulatory complex of holotype.

Figures 11-14, **Donnaldsoncythere humesi**: 11, Right valve of female paratype; 12, Left valve of male allotype; 13, 14, Peniferum and copulatory complex of allotype.

Figures 1, 2, 7, 8, 11, and 12 drawn to scale A; figure 3 drawn to scale B; figures 4, 5, 9, 10, 13, and 14 drawn to scale C. Scale of figure 6 unknown.

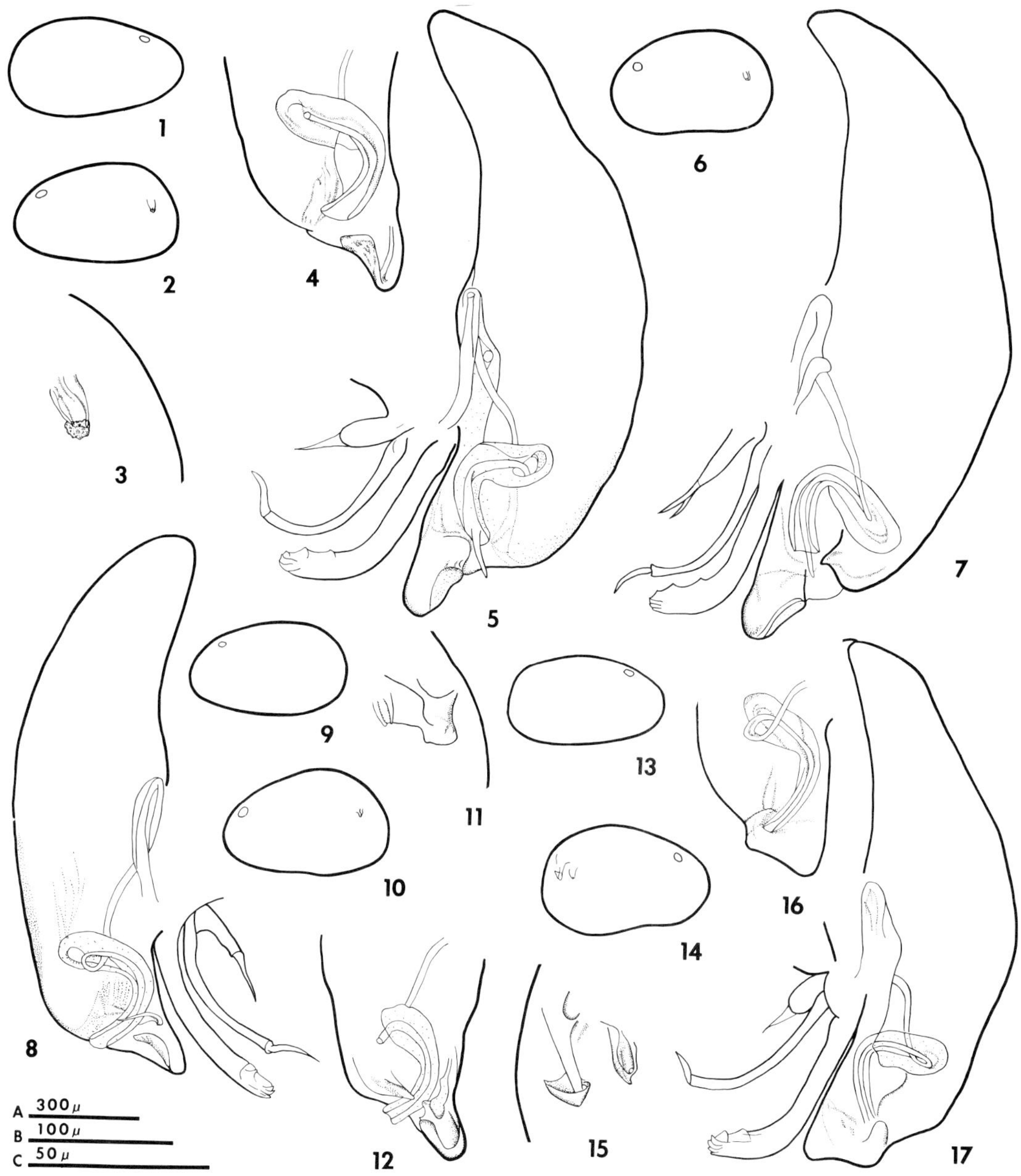

PLATE XXIV

Figures 1-5, **Donnaldsoncythere ileata:** 1, Right valve of male holotype; 2, Left valve of female allotype; 3, Genital complex of allotype; 4, Peniferum of paratype; 5, Copulatory complex of holotype.

Figures 6-7, **Donnaldsoncythere pennsylvanica:** 6, Left valve of female allotype; 7, Copulatory complex of paratype.

Figures 8-12, **Donnaldsoncythere scalis:** 8, Copulatory complex of holotype; 9, Left valve of male holotype; 10, Left valve of female allotype; 11, Genital complex of allotype; 12, Peniferum of paratype.

Figures 13-17, **Donnaldsoncythere truncata:** 13, Right valve of male holotype; 14, Right valve of female allotype; 15, Genital complex of allotype; 16, Peniferum of paratype; 17, Copulatory complex of holotype.

Figures 1, 2, 6, 9, 10, 13, and 14 drawn to scale A; figures 3, 11, and 15 drawn to scale B; figures 4, 5, 7, 8, 12, 16, and 17 drawn to scale C.

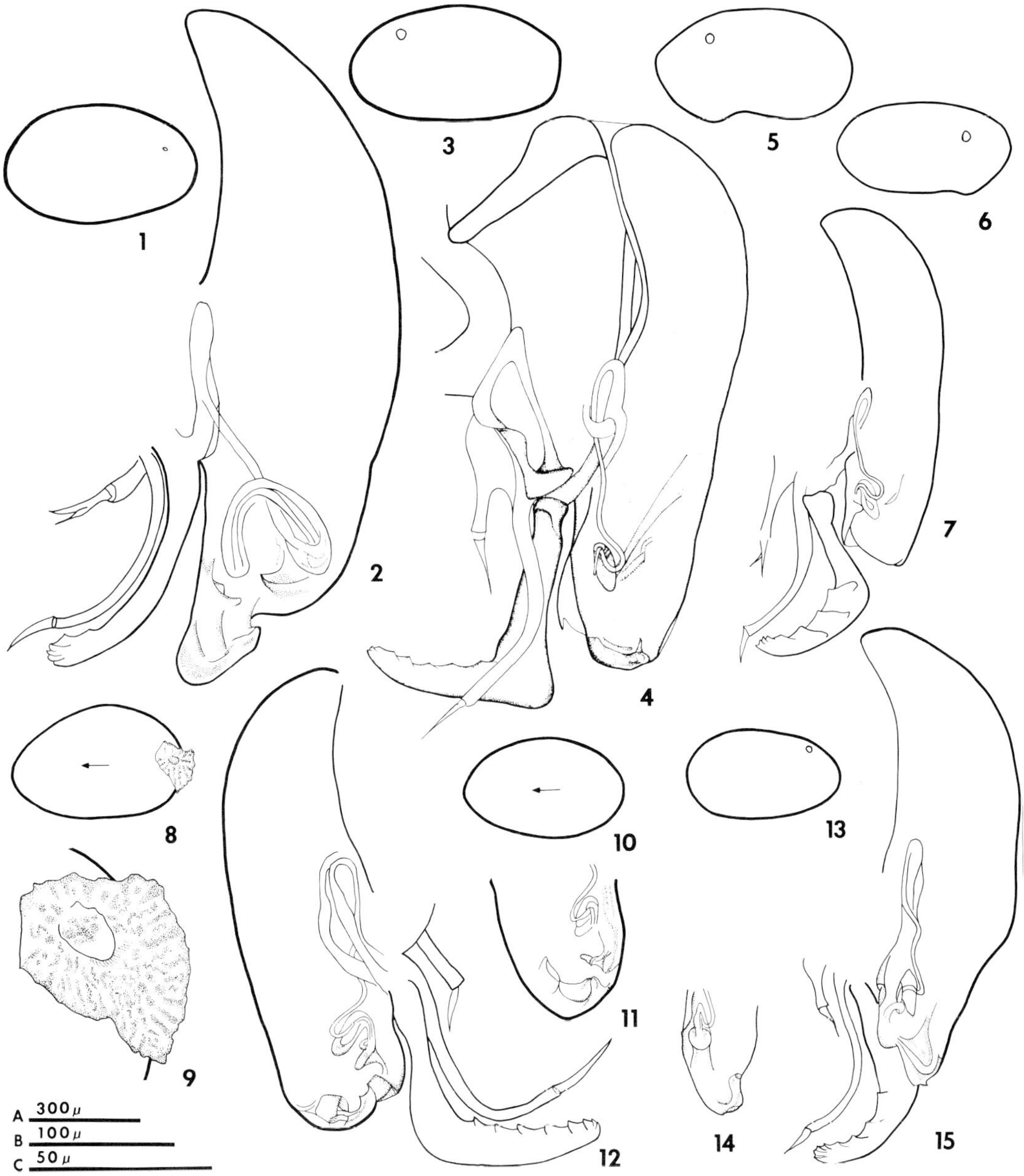

PLATE XXV

Figures 1-2, **Donnaldsoncythere tuberosa:** 1, Right valve of male holotype; 2, Copulatory complex of holotype.

Figures 3-4, **Entocythere cambaria;** 3, Left valve of male; 4, Copulatory complex of male.

Figures 5-7, **Entocythere claytonhoffi:** 5, Left valve of male; 6, Right valve of female; 7, Copulatory complex of male. (Figures 5-7 from Hobbs, 1971.)

Figures 8-12, **Entocythere dentata:** 8, Left valve of female allotype; 9, Genital complex of allotype; 10, Left valve of male holotype; 11, 12, Peniferum and copulatory complex of holotype.

Figures 13-15, **Entocythere dorsorotunda:** 13, Right valve of male holotype; 14, 15, Peniferum and copulatory complex of holotype.

Figures 1, 3, 5, 6, 8, 10, and 13 drawn to scale A; figure 9 drawn to scale B; figures 2, 4, 7, 11, 12, 14, and 15 drawn to scale C.

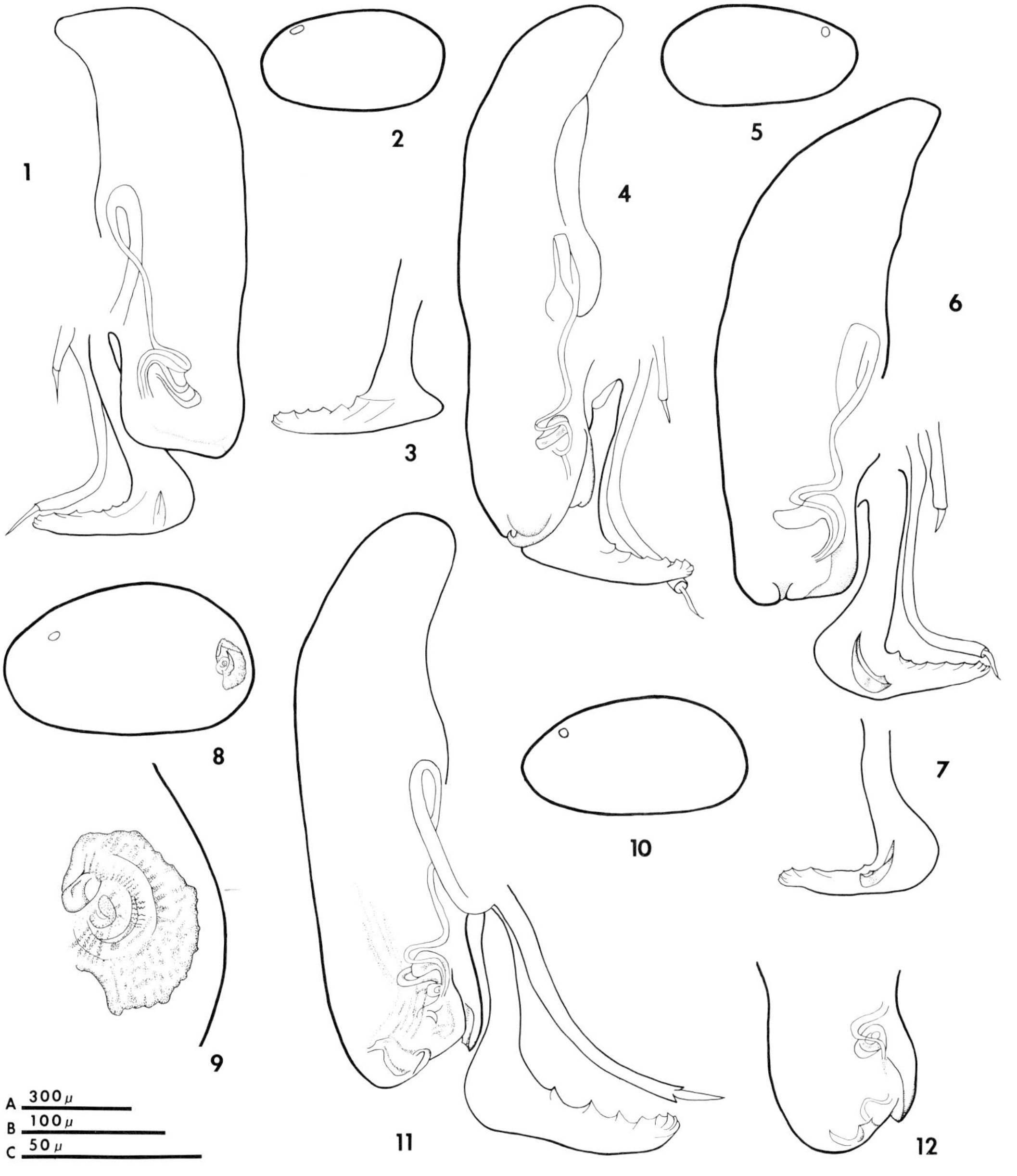

PLATE XXVI

Figures 1-2, **Entocythere elliptica:** 1, Copulatory complex of holotype; 2, Left valve of male holotype.

Figures 3-4, **Entocythere illinoisensis:** 3, Clasping apparatus; 4, Copulatory complex of male allotype.

Figures 5-7, **Entocythere internotalus:** 5, Right valve of male holotype; 6, Copulatory complex of paratype; 7, Clasping apparatus of holotype.

Figures 8-12, **Entocythere kanawhaensis:** 8, Left valve of female allotype; 9, Genital complex of allotype; 10, Left valve of male holotype; 11, Copulatory complex of holotype; 12, Peniferum of holotype.

Figures 2, 5, 8, and 10 drawn to scale A; figure 9 drawn to scale B; figures 1, 3, 4, 6, 7, 11, and 12 drawn to scale C.

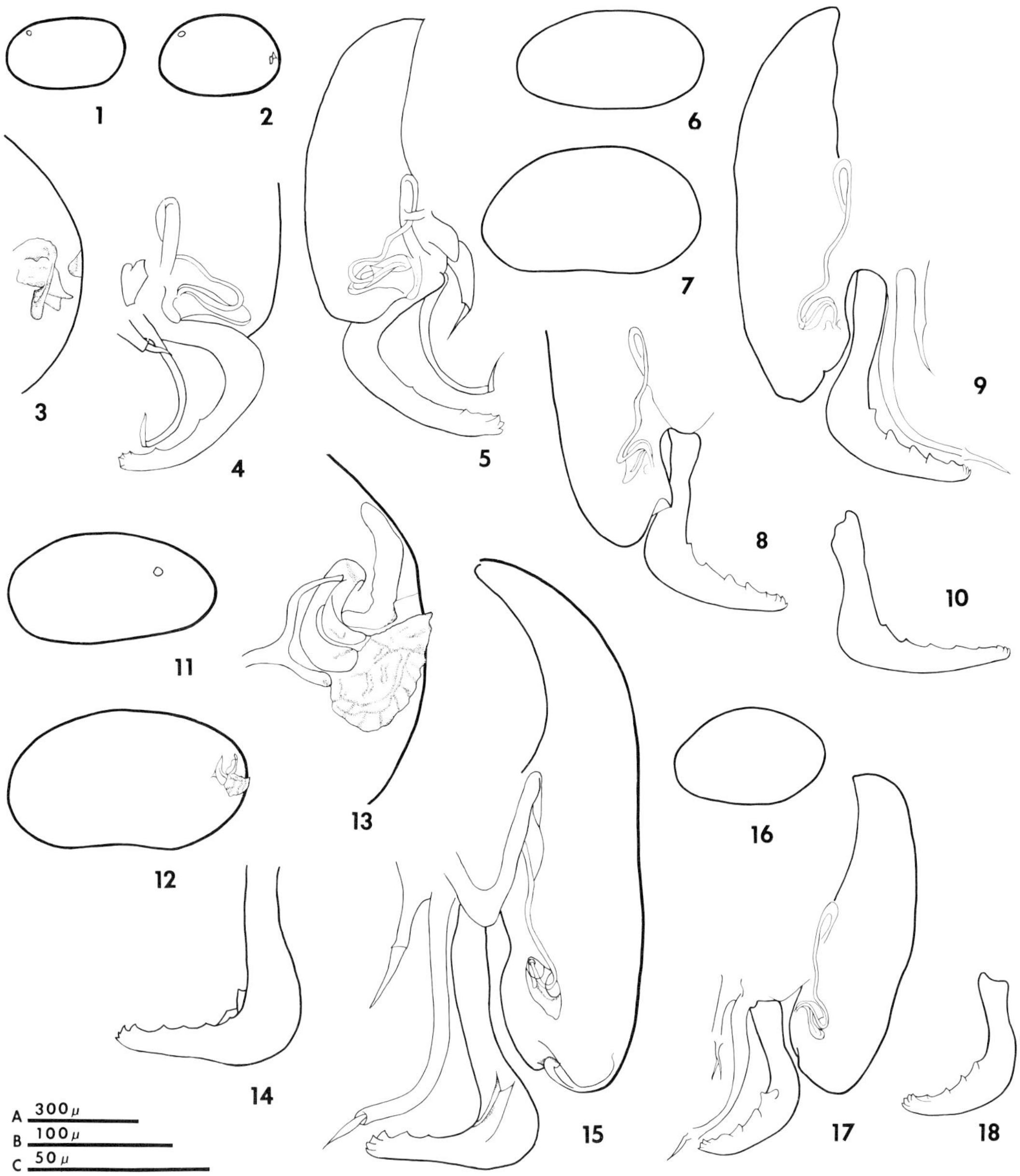

PLATE XXVII

Figures 1-5, **Entocythere lepta:** 1, Left valve of male holotype; 2, Left valve of female allotype; 3, Genital complex of allotype; 4, Copulatory complex of paratype; 5, Copulatory complex of holotype.

Figures 6-10, **Entocythere mexicana:** 6, Right valve of male; 7, Right valve of female; 8, 9, Compulatory complex of males; 10, Clasping apparatus. (Figures 6-10 from Hobbs, 1971.)

Figures 11-15, **Entocythere redelli:** 11, Right valve of male holotype; 12, Left valve of female allotype; 13, Genital complex of allotype; 14, Clasping apparatus of holotype; 15, Copulatory complex of other holotype.

Figures 16-18, **Entocythere ruibali:** 16, Left valve of male; 17, Copulatory complex of male; 18, Clasping apparatus of male. (Figures 16-18 from Hobbs, 1971.)

Figures 1, 2, 6, 7, 11, 12, and 16 drawn to scale A; figures 3 and 13 drawn to scale B; figures 4, 5, 8, 9, 10, 14, 15, 17, and 18 drawn to scale C.

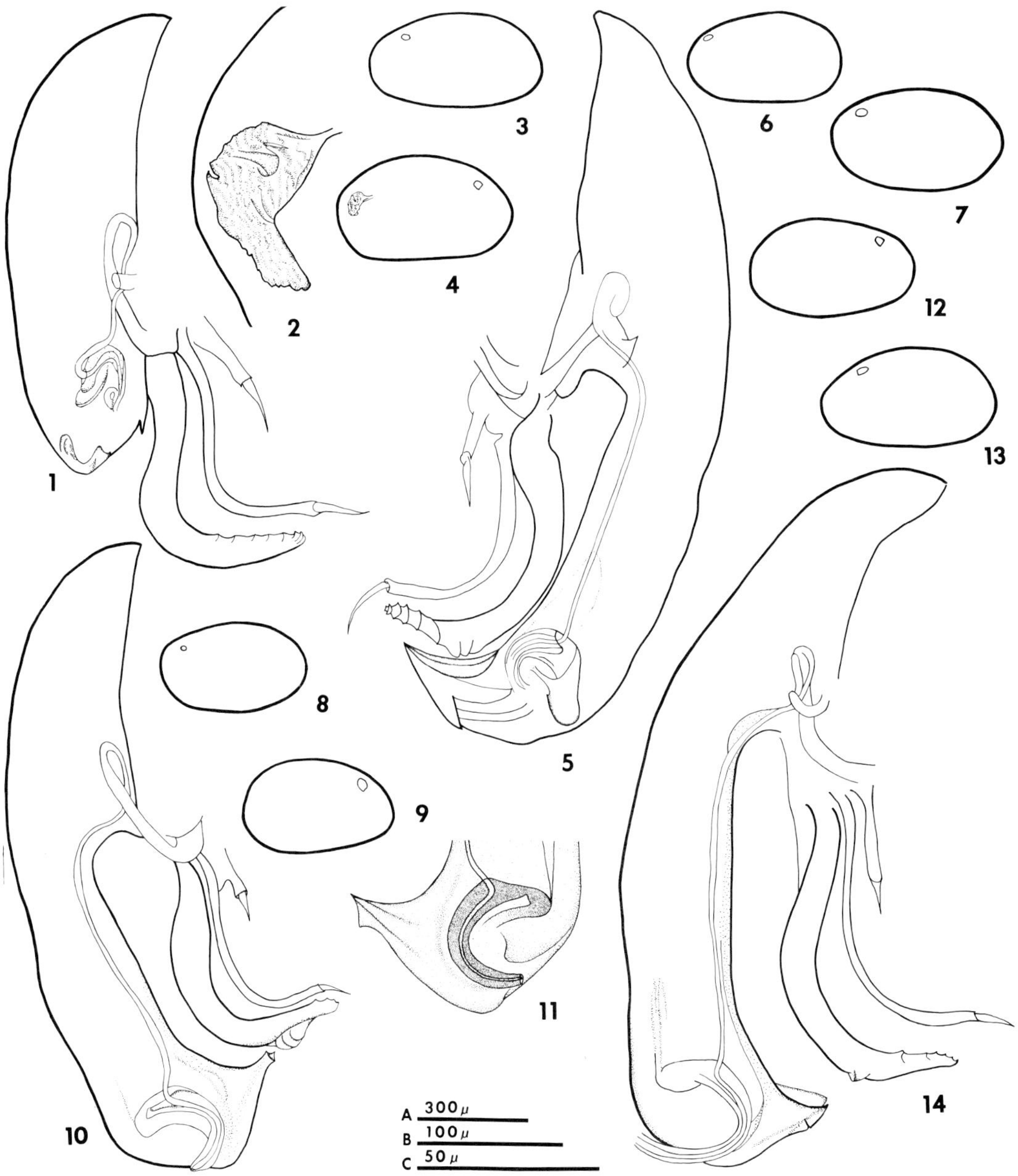

PLATE XXVIII

Figures 1-4, **Entocythere tyttha:** 1, Copulatory complex of holotype; 2, Genital complex of allotype; 3, Left valve of holotype; 4, Right valve of allotype.

Figures 5-7, **Geocythere acuta:** 5, Copulatory complex of holotype; 6, Left valve of female allotype; 7, Left valve of male holotype.

Figures 8-11, **Geocythere geophila:** 8, Left valve of female allotype; 9, Right valve of male holotype; 10, Copulatory complex of holotype; 11, Detail of peniferum.

Figures 12-14, **Geocythere gyralea:** 12, Right valve of female allotype; 13, Left valve of male holotype; 14, Copulatory complex of holotype.

Figures 3, 4, 6-9, 12, and 13 drawn to scale A; figure 2 drawn to scale B; figures 1, 5, 10, and 14 drawn to scale C; figure 11 not drawn to scale.

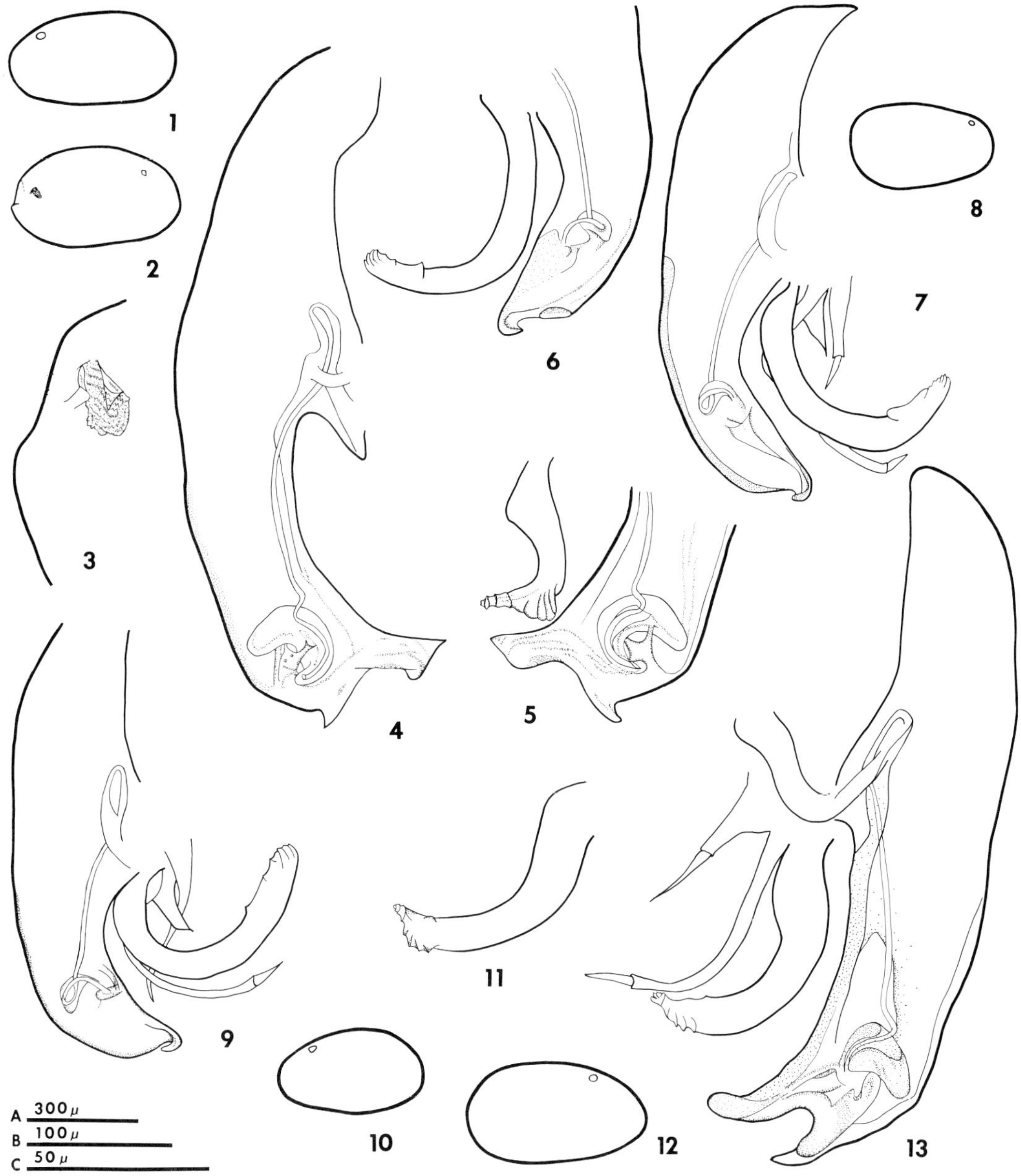

PLATE XXIX

Figures 1-5, **Geocythere nessoides:** 1, Left valve of male holotype; 2, Right valve of female allotype; 3, Genital complex of allotype; 4, Copulatory complex of holotype; 5, Copulatory complex of paratype.

Figures 6-8, **Harpagocythere georgiae:** 6, Copulatory complex of paratype; 7, Copulatory complex of holotype; 8, Right valve of male holotype.

Figures 9-10, **Harpagocythere tertius:** 9, Copulatory complex of holotype; 10, Left valve of male holotype.

Figures 11-13, **Hartocythere torreya:** 11, Clasping apparatus of paratype; 12, Right valve of male holotype; 13, Copulatory complex of holotype.

Figures 1, 2, 8, 10, 12, drawn to scale A; figure 3 drawn to scale B; figures 4, 5, 6, 7, 9, 11, 13 drawn to scale C.

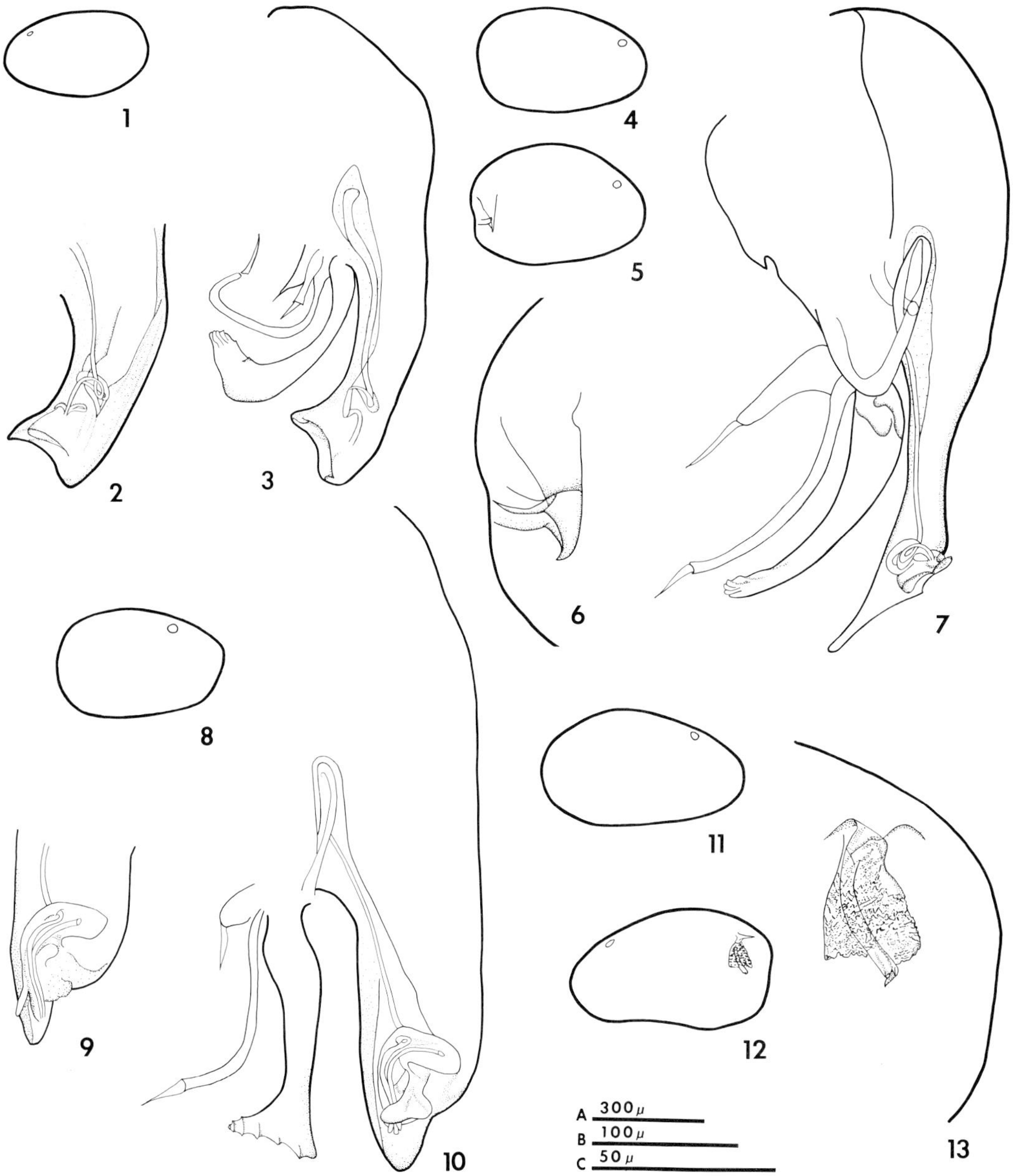

PLATE XXX

Figures 1-3, **Litocythere lucileae:** 1, Left valve of male holotype; 2, Peniferum of paratype; 3, Copulatory complex of holotype.

Figures 4-7, **Lordocythere petersi:** 4, Right valve of male holotype; 5, Right valve of female allotype; 6, Genital complex of allotype; 7, Copulatory complex of holotype.

Figures 8-10, **Okriocythere cheia:** 8, Right valve of male holotype; 9, Peniferum of paratype; 10, Copulatory complex of holotype.

Figures 11-13, **Ornithocythere aetodes:** 11, Right valve of male holotype; 12, Left valve of female allotype; 13, Genital complex of allotype.

Figures 1, 4, 5, 8, 11, 12 drawn to scale A; figures 6 and 13 drawn to scale B; figures 2, 3, 7, 9, and 10 drawn to scale C.

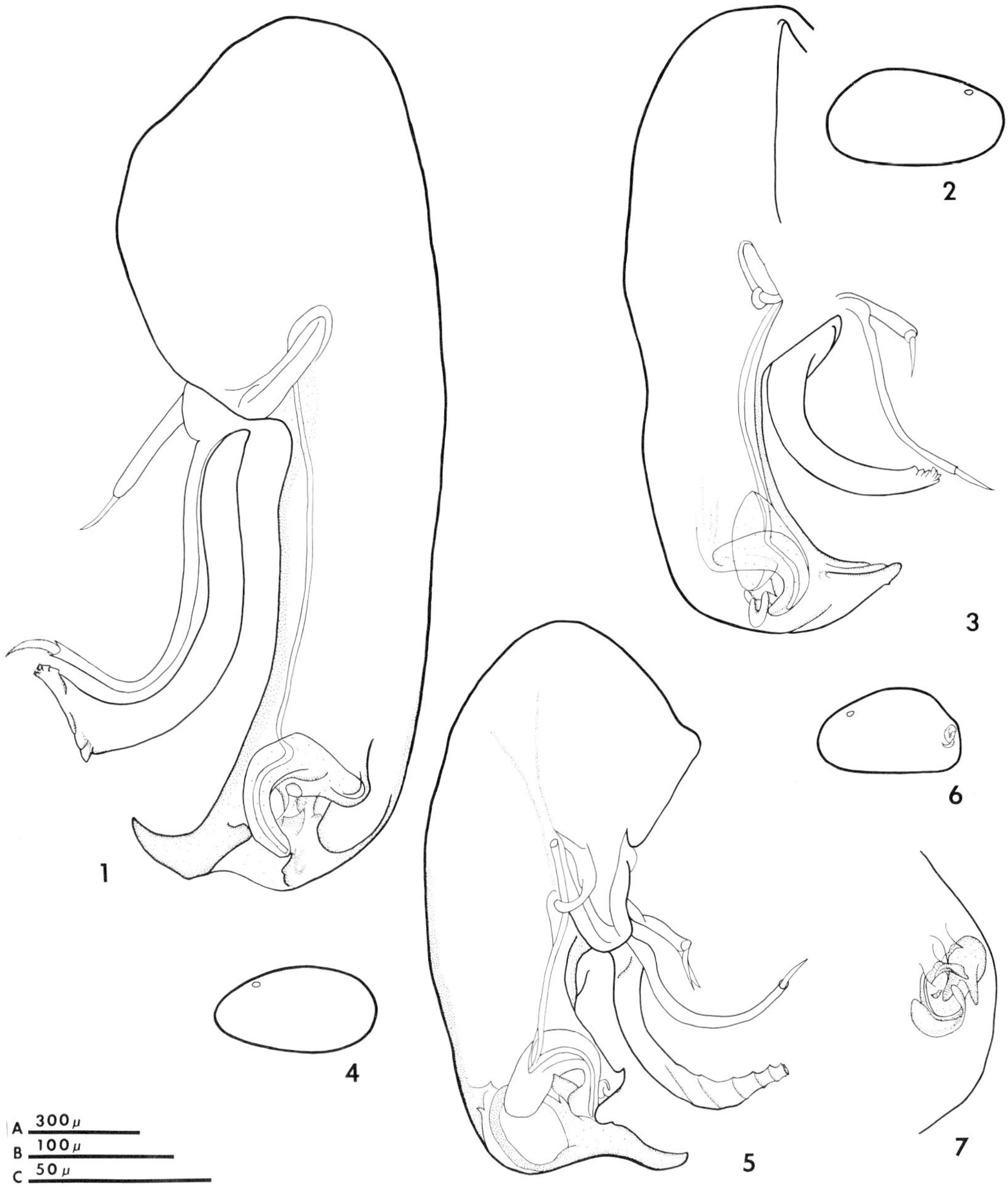

PLATE XXXI

Figure 1, **Ornithocythere aetodes** (continued): 1, Copulatory complex of holotype.

Figures 2-3, **Ornithocythere gypodes:** 2, Right valve of male holotype; 3, Copulatory complex of holotype.

Figures 4-7, **Ornithocythere popi:** 4, Left valve of male holotype; 5, Copulatory complex of holotype; 6, Left valve of female allotype; 7, Genital complex of allotype.

Figures 2, 4, and 6 drawn to scale A; figure 7 drawn to scale B; figures 1, 3, and 5 drawn to scale C.

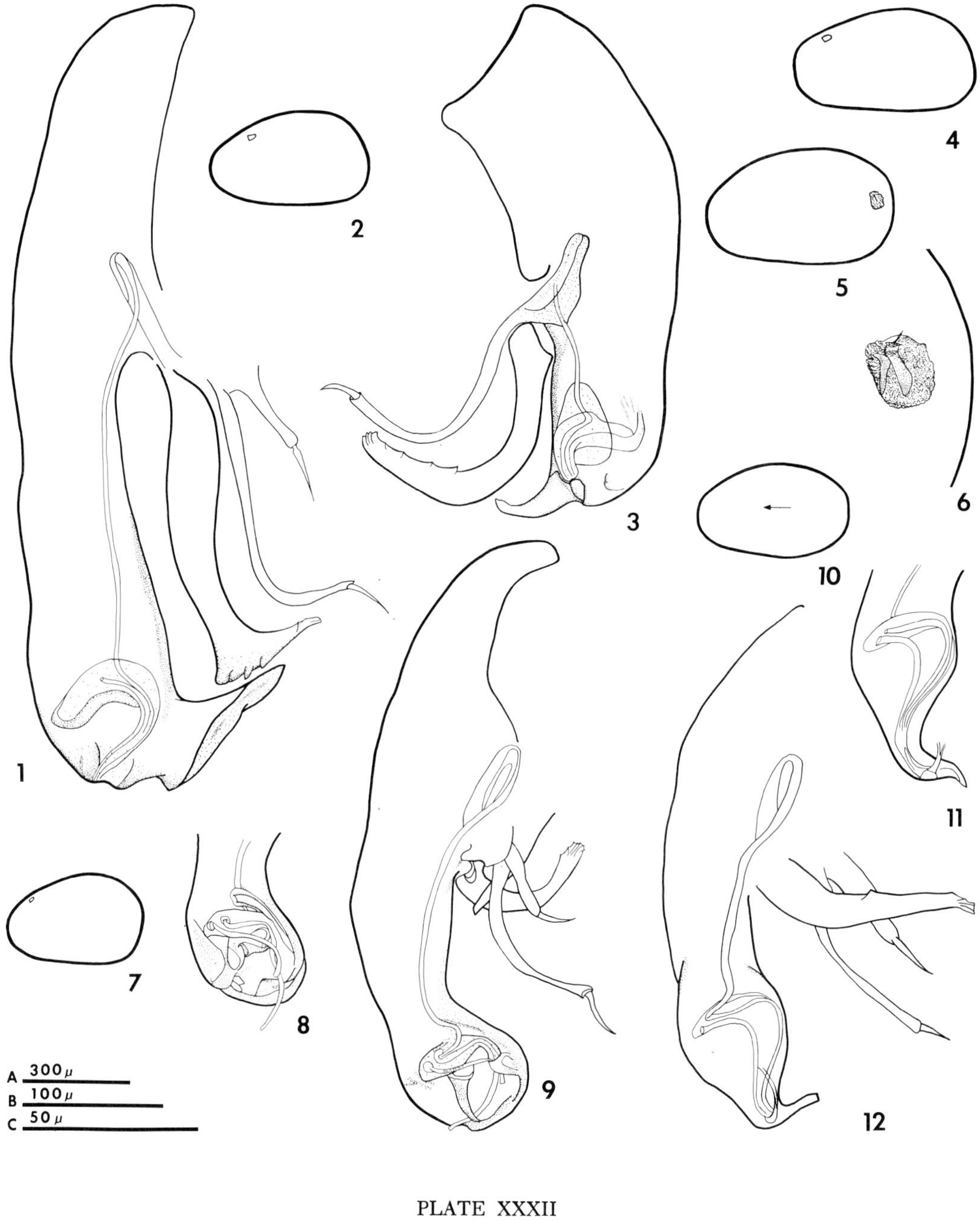

PLATE XXXII

Figures 1-2, **Ornithocythere rhea:** 1, Copulatory complex of holotype; 2, Left valve of holotype.

Figures 3-6, **Ornithocythere waltonae:** 3, Copulatory complex of holotype; 4, Left valve of male holotype; 5, Left valve of female allotype; 6, Genital complex of allotype.

Figures 7-9, **Phymocythere phyma:** 7, Left valve of male holotype; 8, Peniferum of holotype; 9, Copulatory complex of holotype.

Figures 10-12, **Plectocythere crotophis:** 10, Left valve of male holotype; 11, Peniferum of holotype; 12, Copulatory complex of holotype.

Figures 2, 4, 5, 7, and 10 drawn to scale A; figure 6 drawn to scale B; figures 1, 3, 8, 9, 11, and 12 drawn to scale C.

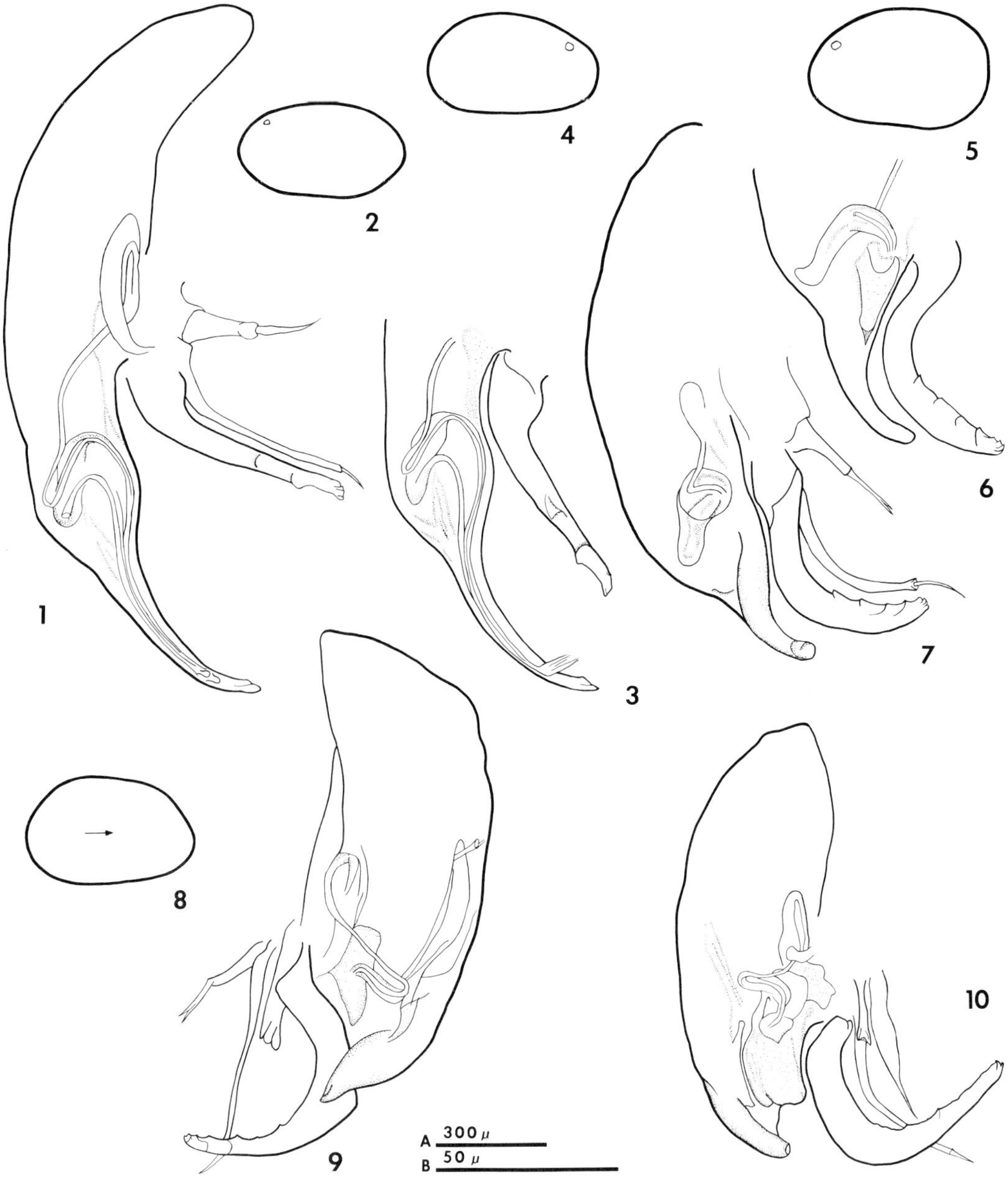

PLATE XXXIII

Figures 1-3, **Plectocythere johnsonae:** 1, Copulatory complex of holotype; 2, Left valve of male holotype; 3, Peniferum of holotype.

Figures 4-7, **Rhadinocythere serrata:** 4, Right valve of male holotype; 5, Left valve of female allotype; 6, Copulatory complex of paratype; 7, Copulatory complex of holotype.

Figures 8-9, **Sagittocythere barri:** 8, Right valve of male holotype; 9, Copulatory complex of holotype.

Figure 10, **Sagittocythere stygia:** 10, Copulatory complex of holotype.

Figures 2, 4, 5, and 8 drawn to scale A; figures 1, 3, 6, 7, 9, and 10 drawn to scale B.

PLATE XXXIV

Figures 1-2, **Saurocythere rhipis:** 1, Copulatory complex of holotype; 2, Left valve of male holotype.

Figures 3-5, **Thermastrocythere riojai:** 3, Right valve of male allotype; 4, Right valve of female holotype; 5, Copulatory complex of allotype.

Figures 6-8, **Uncinocythere allenae:** 6, Left valve of male holotype; 7, 8, Both halves of copulatory complex of holotype.

Figures 9-11, **Uncinocythere ambophora:** 9, Right valve of male holotype; 10, Left valve of female allotype; 11, Copulatory complex of holotype.

Figures 12-15, **Uncinocythere bicuspide:** 12, Left valve of male; 13, Left valve of female; 14, Copulatory complex; 15, Clasping apparatus. (Drawings 12-15 from Hobbs, 1971.)

Figures 16-19, **Uncinocythere cassiensis:** 16, Left valve of male holotype; 17, Left valve of female allotype; 18, Copulatory complex of paratype; 19, Copulatory complex of holotype.

Figures 2, 3, 4, 6, 9, 10, 12, 13, 16, and 17 drawn to scale A; figures 1, 5, 7, 8, 11, 14, 15, 18, and 19 drawn to scale B.

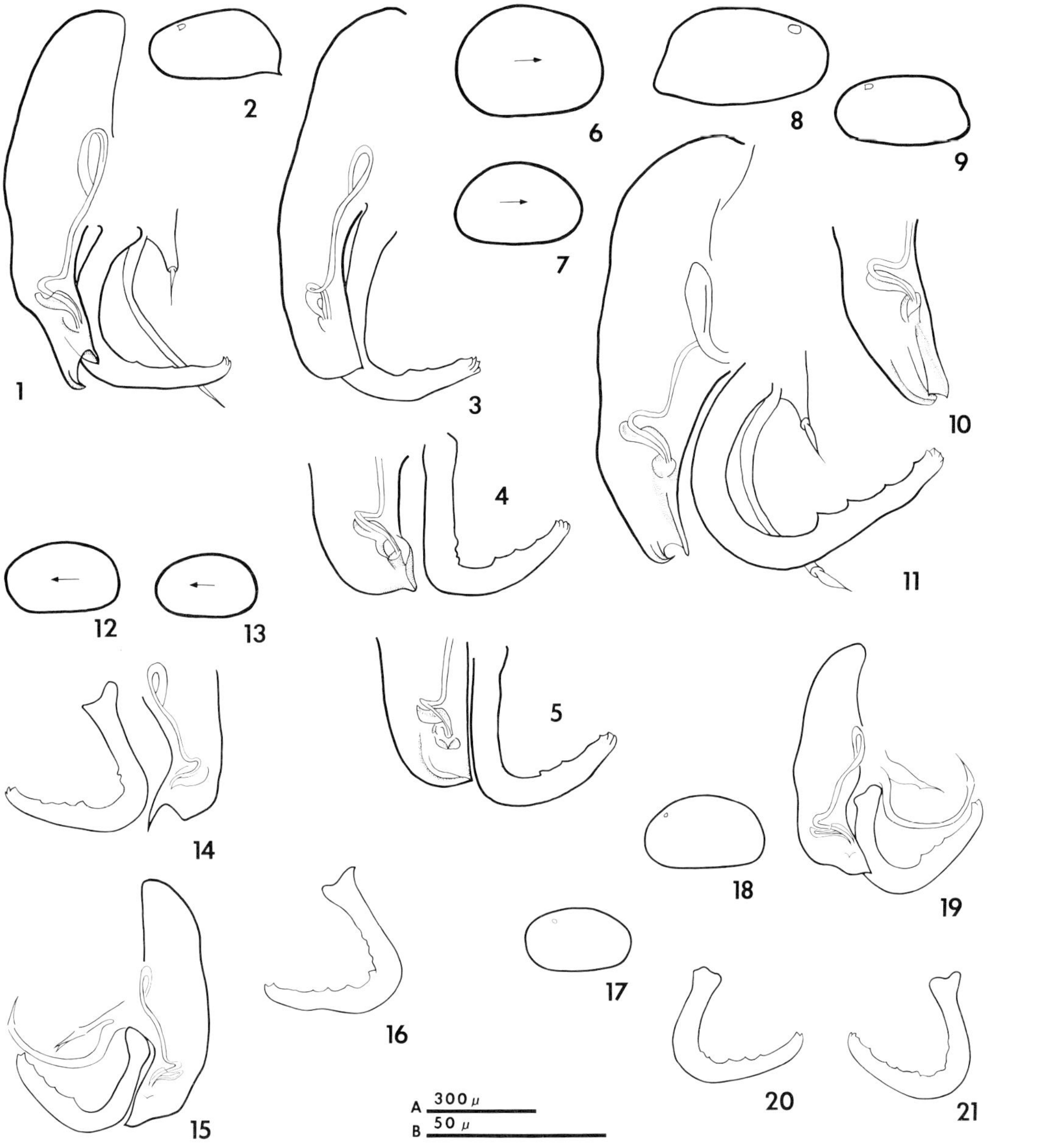

PLATE XXXV

Figures 1-2, **Uncinocythere caudata:** 1, Copulatory complex of holotype; 2, Left valve of male holotype.

Figures 3-7, **Uncinocythere clemsonella:** 3, Copulatory complex of holotype; 4, 5, Both halves of copulatory complex of paratype; 6, Right valve of female allotype; 7, Right valve of male holotype.

Figures 8-11, **Uncinocythere columbia:** 8, Right valve of male; 9, Left valve of female; 10, Copulatory complex; 11, Peniferum.

Figures 12-16, **Uncinocythere cuadricuspide**: 12, Left valve of female; 13, Left valve of male; 14, 15, Copulatory complexes; 16, Clasping apparatus. (Drawings from Hobbs, 1971.)

Figures 17-21, **Uncinocythere dobbinae:** 17, Left valve of male; 18, Left valve of female; 19, Copulatory complex; 20, 21, Clasping apparatuses. (Drawings 17-21 from Hobbs, 1971.)

Figures 2, 6, 7, 8, 9, 12, 13, 17, and 18 drawn to scale A; figures 1, 3, 4, 5, 10, 11, 14, 15, 16, 19, 20, and 21 drawn to scale B.

PLATE XXXVI

Figures 1-3, **Uncinocythere equicurva:** 1, Left valve of male holotype; 2, Left valve of female allotype; 3, Copulatory complex of holotype.

Figures 4-5, **Uncinocythere ericksoni:** 4, Left valve of holotype; 5, Copulatory complex of holotype.

Figures 6-8, **Uncinocythere holti:** 6, Left valve of male holotype; 7, Left valve of female allotype; 8, Copulatory complex of holotype.

Figures 9-12, **Uncinocythere lucifuga:** 9, Right valve of male holotype; 10, Right valve of female allotype; 11, Copulatory complex of holotype; 12, Copulatory complex of paratype.

Figures 13-15, **Uncinocythere neglecta:** 13, Right valve of male paratype; 14, Left valve of female paratype; 15, Copulatory complex of paratype.

Figures 16-19, **Uncinocythere occidentalis:** 16, Left valve of male holotype; 17, Right valve of female allotype; 18, Peniferum of paratype; 19, Copulatory complex of holotype.

Figures 1, 2, 4, 6, 7, 9, 10, 13, 14, 16, and 17 drawn to scale A; figures 3, 5, 8, 11, 12, 15, 18, and 19 drawn to scale B.

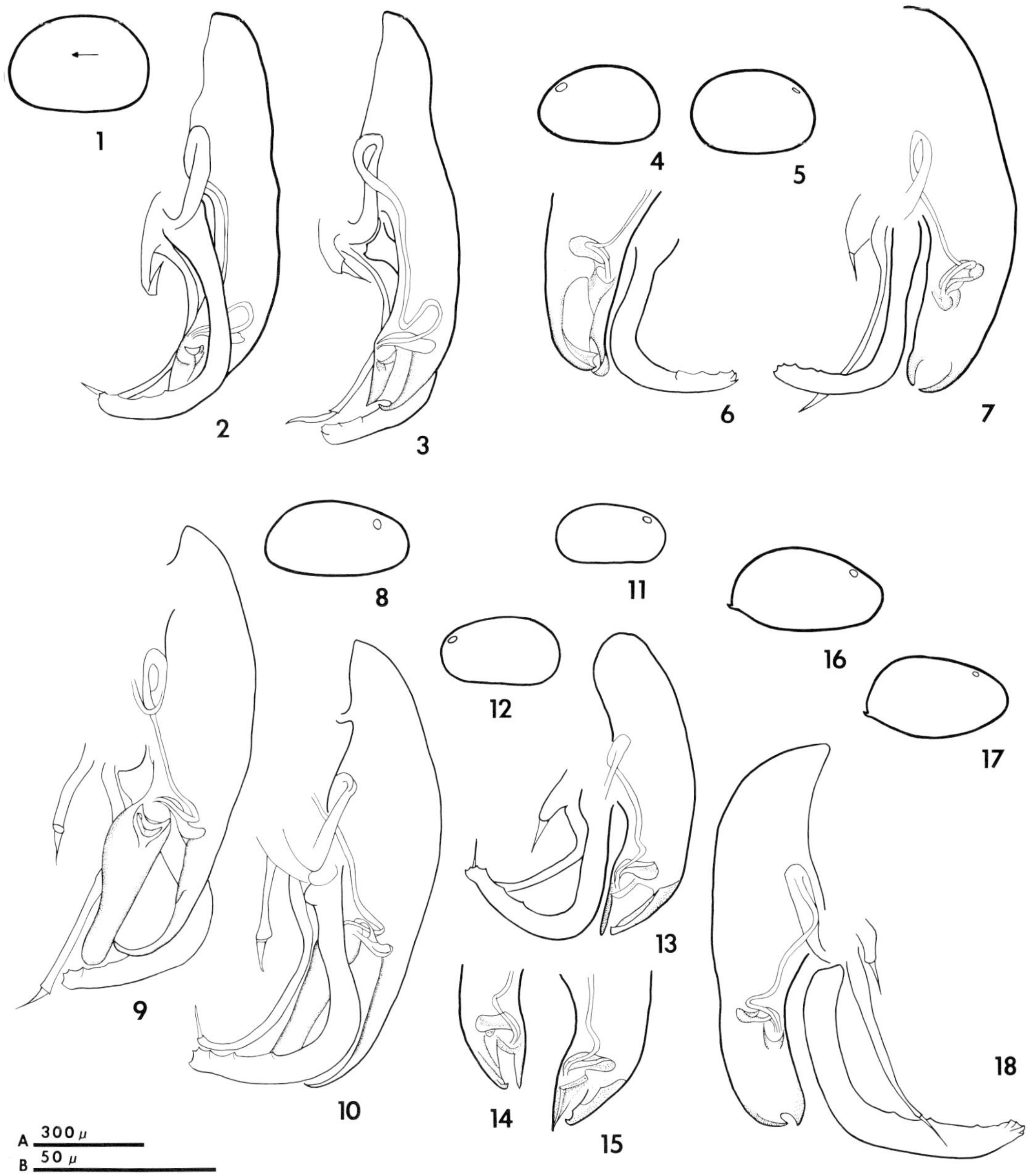

PLATE XXXVII

Figures 1-3, **Uncinocythere pholetera:** 1, Left valve of female allotype; 2, 3, Both halves of copulatory complex of holotype.

Figures 4-7, **Uncinocythere simondsi:** 4, Left valve of male holotype; 5, Right valve of female allotype; 6, Copulatory complex of paratype; 7, Copulatory complex of holotype.

Figures 8-10, **Uncinocythere spathe:** 8, Right valve of male holotype; 9, 10, Both halves of copulatory complex of holotype.

Figures 11-15, **Uncinocythere stubbsi:** 11, Right valve of male holotype; 12, Left valve of female allotype; 13, Copulatory complex of holotype; 14, Peniferum of paratype; 15, Peniferum of holotype.

Figures 16-18, **Uncinocythere thektura:** 16, Right valve of male holotype; 17, Right valve of female allotype; 18, Copulatory complex of holotype.

Figures 1, 4, 5, 8, 11, 12, 16, and 17 drawn to scale A; figures 2, 3, 6, 7, 9, 10, 13, 14, 15, and 18 drawn to scale B.

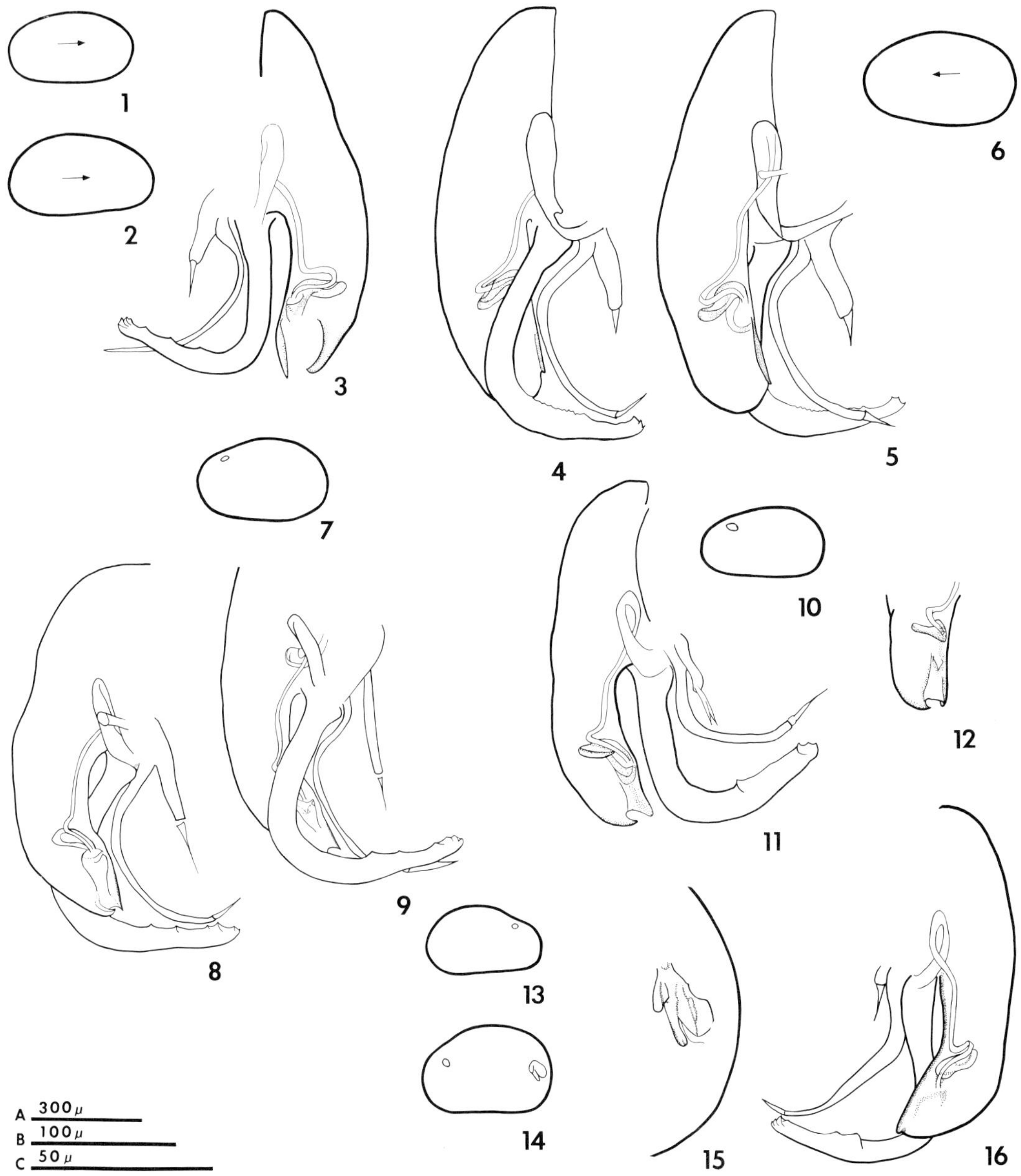

PLATE XXXVIII

Figures 1-3, **Uncinocythere warreni:** 1, Right valve of male holotype; 2, Right valve of female allotype; 3, Copulatory complex of holotype.

Figures 4-6, **Uncinocythere xania:** 4, 5, Both halves of copulatory complex of holotype; 6, Left valve of male holotype.

Figures 7-9, **Uncinocythere xena:** 7, Left valve of male holotype; 8, 9, Both halves of copulatory complex of holotype.

Figures 10-12, **Uncinocythere zancla:** 10, Left valve of male holotype; 11, Copulatory complex of holotype; 12, Peniferum of holotype.

Figures 13-16, **Uncinocythere zaruri:** 13, Right valve of male holotype; 14, Left valve of female allotype; 15, Genital complex of allotype; 16, Copulatory complex of holotype.

Figures 1, 2, 6, 7, 10, 13, and 14 drawn to scale A; figure 15 drawn to scale B; figures 3, 4, 5, 8, 9, 11, 12, and 16 drawn to scale C.

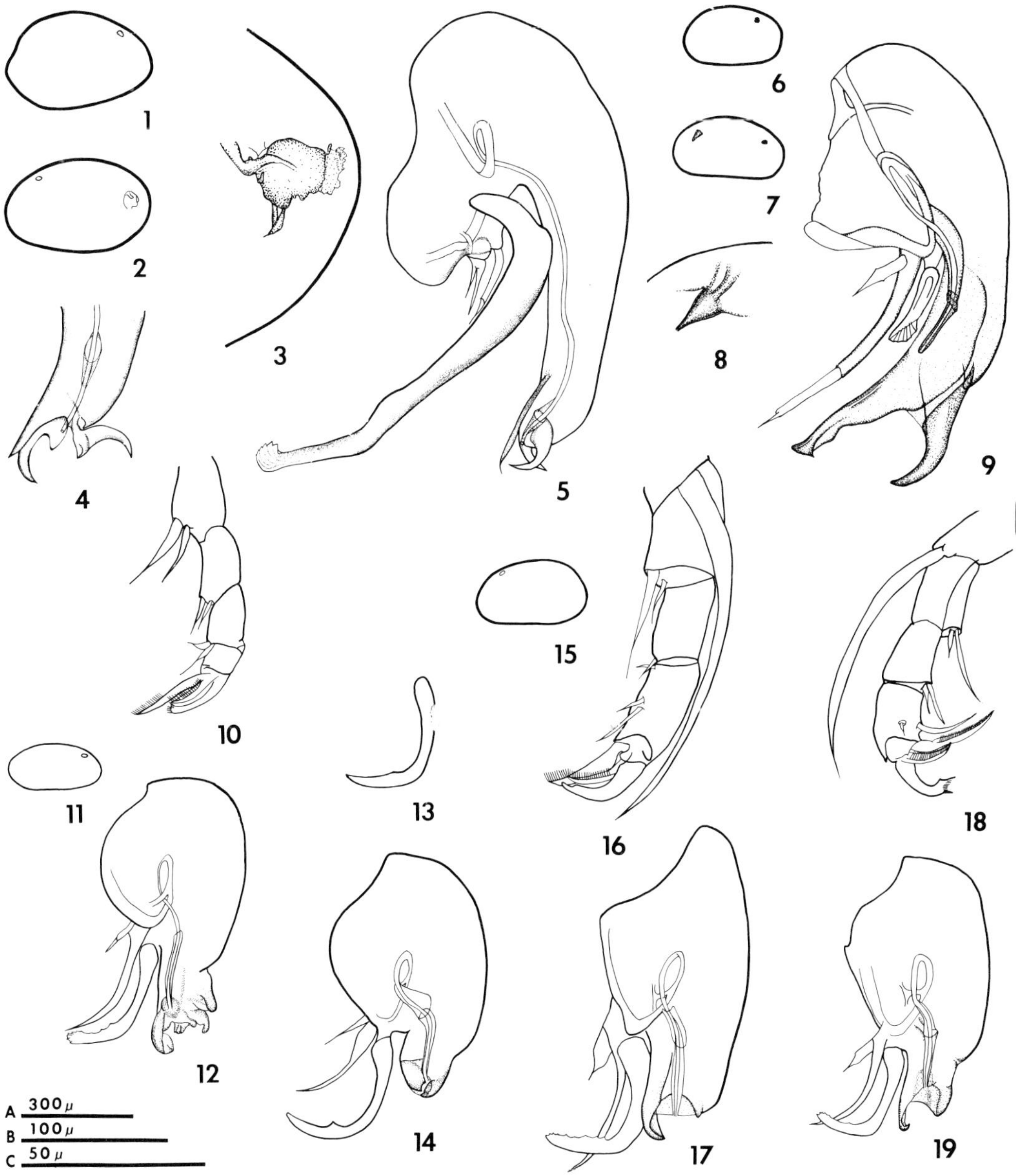

PLATE XXXIX

Figures 1-5, **Microsyssitria indica:** 1, Right valve of male; 2, Left valve of female; 3, Genital complex of female; 4, Peniferum; 5, Copulatory complex.

Figures 6-9, **Elachistocythere merista:** 6, Right valve of male; 7, Right valve of female; 8, Genital complex; 9, Copulatory complex.

Figures 10-12, **Chelocythere kalganensis:** 10, Antenna; 11, Right valve of male; 12, Copulatory complex.

Figures 13-14, **Herpetocythere acanthoides:** 13, Clasping apparatus; 14, Copulatory complex.

Figures 15-17, **Herpetocythere australensis:** 15, Left valve of male; 16, Antenna of male; 17, Copulatory complex.

Figures 18-19, **Herpetocythere bendora:** 18, Antenna: 19, Copulatory complex.

Figures 1, 2, 6, 7, 11, and 15 drawn to scale A; figures 3 and 8 drawn to scale B; figures 5, 9, 10, 12, 13, 14, 16, 17, 18, and 19 drawn to scale C; figure 4 enlarged detail not drawn to scale.

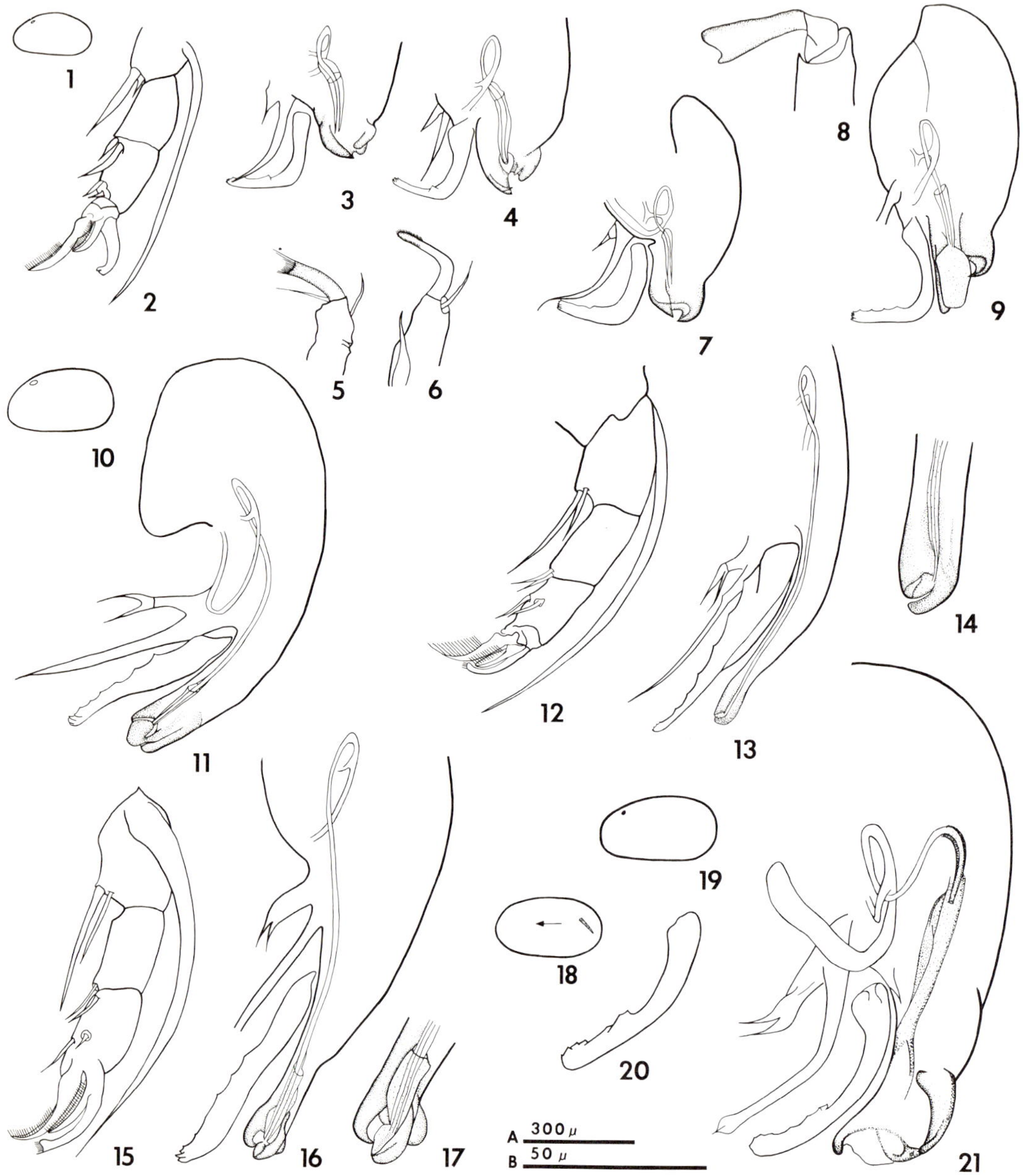

PLATE XL

Figures 1-6, **Herpetocythere gnoma:** 1, Left valve of male; 2, Antenna; 3, Copulatory complex of paratype; 4, Copulatory complex of holotype; 5, Terminal position of mandibular palp; 6, Maxilla.

Figure 7, **Herpetocythere labidoides:** 7, Copulatory complex.

Figures 8-9, **Herpetocythere mackenziei:** 8, Terminal portion of mandibular palp; 9, Copulatory complex.

Figures 10-11, **Hesperocythere klasteroides:** 10, Left valve of male; 11, Copulatory complex.

Figures 12-14, **Hesperocythere tallanalla:** 12, Antenna; 13, Copulatory complex; 14, Detail of peniferum.

Figures 15-17, **Hesperocythere xiphoides:** 15, Antenna; 16, Copulatory complex; 17, Detail of peniferum.

Figures 18-21, **Laccocythere aotearoa:** 18, Left valve of female; 19, Left valve of male; 20, Clasping apparatus; 21, Copulatory complex.

Figures 1, 10, 18, and 19 drawn to scale A; figures 2, 3, 4, 5, 6, 7, 8, 9, 11, 12, 13, 15, 16, 20, and 21 drawn to scale B; figures 14 and 17 not drawn to scale.

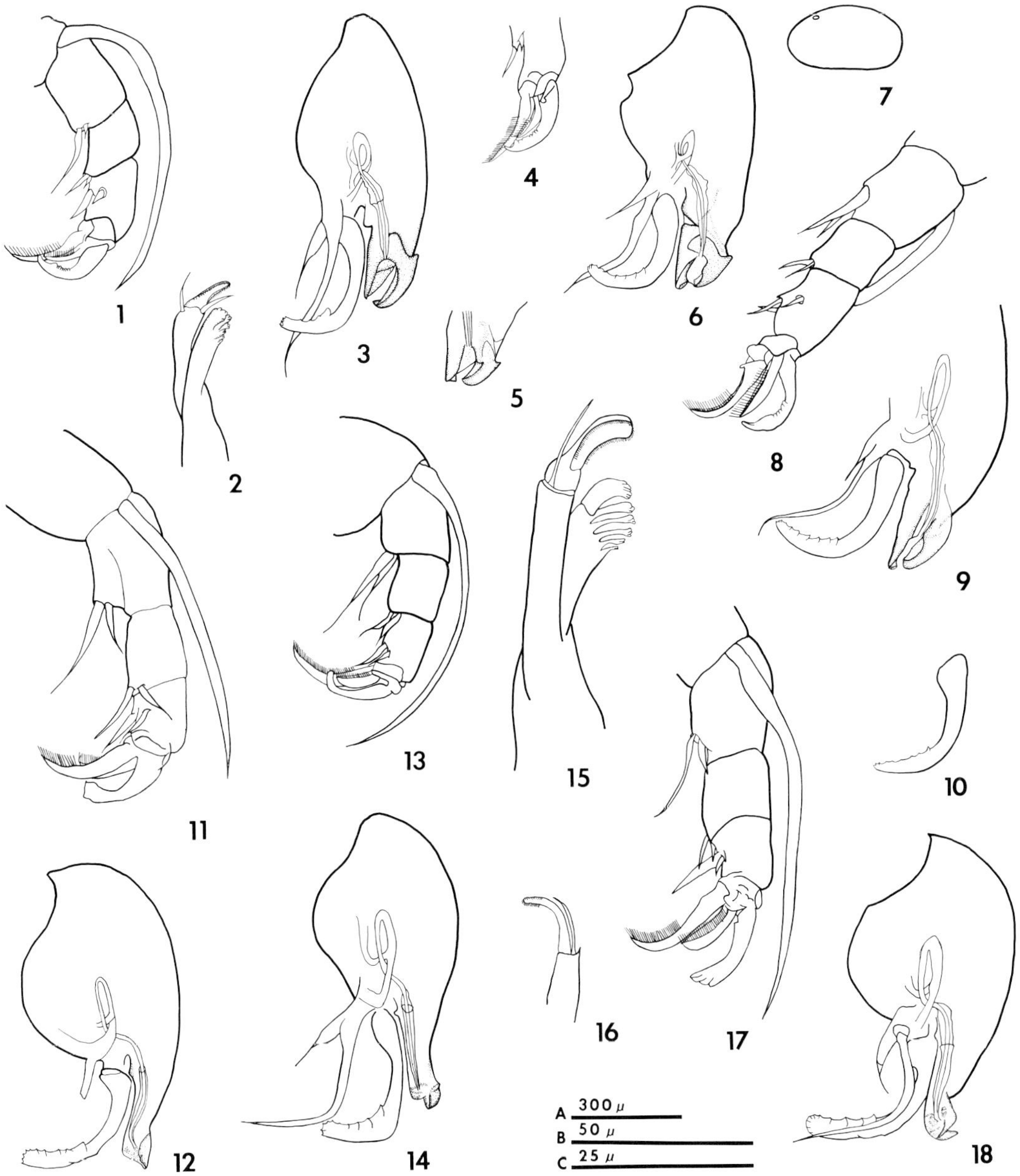

PLATE XLI

Figures 1-3, **Lichnocythere synethes:** 1, Antenna; 2, Mandible; 3, Copulatory complex.

Figures 4-6, **Lichnocythere tubrabucca:** 4, Distal portion of antenna; 5, Tip of peniferum; 6, Copulatory complex.

Figures 7-10, **Lichnocythere victoria:** 7, Left valve of male; 8, Antenna; 9, Copulatory complex; 10, Clasping apparatus.

Figures 11-12, **Notocythere antichthon:** 11, Antenna; 12, Copulatory complex.

Figures 13-16, **Notocythere blundelli:** 13, Antenna; 14, Copulatory complex; 15, Mandible; 16, Maxilla.

Figures 17-18, **Notocythere erica:** 17, Antenna; 18, Copulatory complex.

Figure 7 drawn to scale A; figures 1, 2, 3, 4, 5, 6, 8, 9, 10, 11, 12, 13, 14, 17, and 18 drawn to scale B; figures 15 and 16 drawn to scale C.

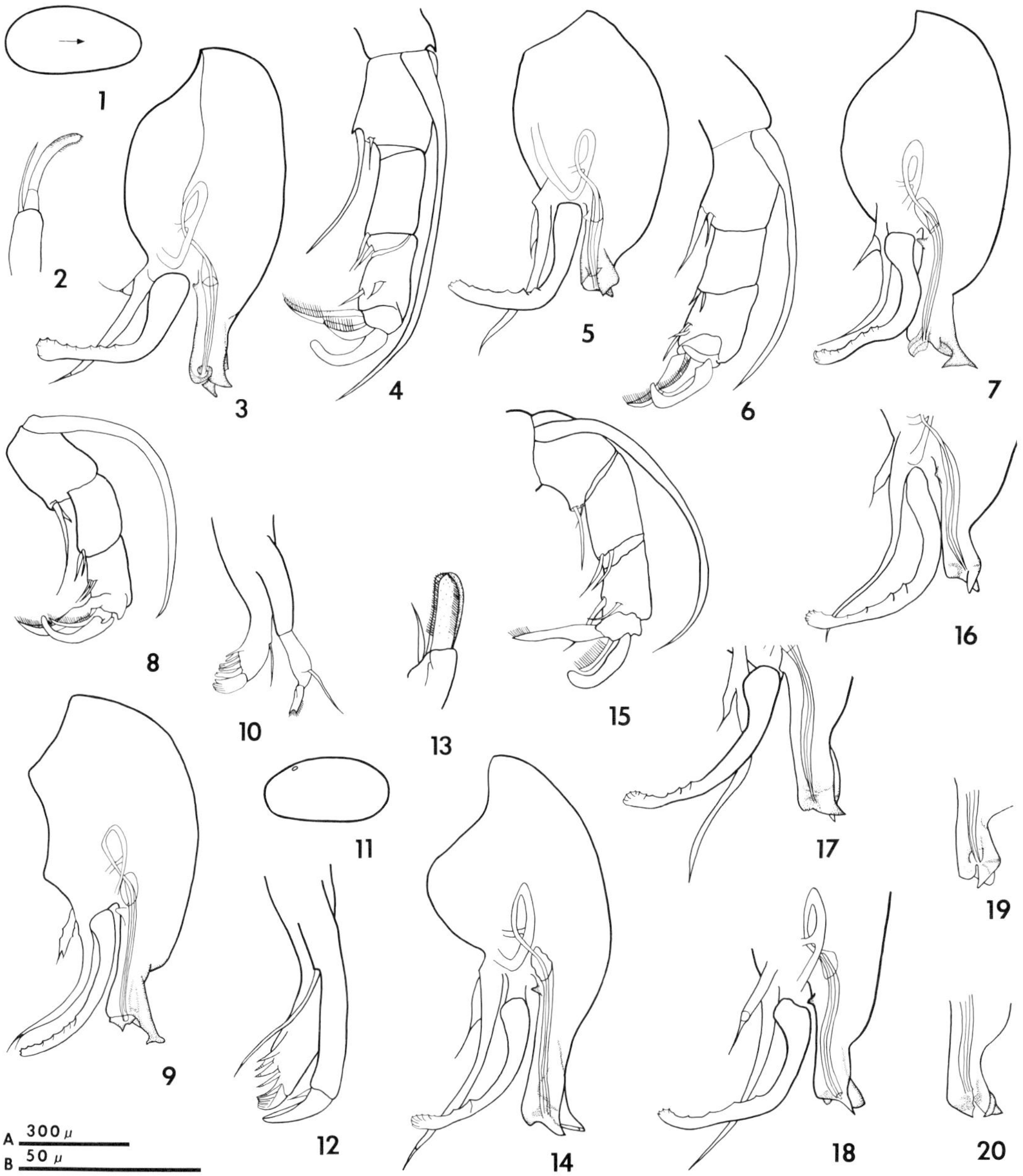

PLATE XLII

Figures 1-3, **Notocythere grampians:** 1, Right valve of male; 2, Maxilla; 3, Copulatory complex.

Figures 4-5, **Notocythere mirranatwa:** 4, Antenna; 5, Copulatory complex.

Figures 6-7, **Notocythere rieki:** 6, Antenna; 7, Copulatory complex.

Figures 8-10, **Notocythere synomodites:** 8, Antenna; 9, Copulatory complex; 10, Mandible.

Figures 11-20, **Notocythere syssitos:** 11, Left valve of male; 12, Mandible; 13, Terminal portion of mandibular palp; 14, Copulatory complex of holotype; 15, Antenna; 16, Copulatory complex of paratype; 17, Copulatory complex of paratype; 18, Copulatory complex of paratype; 19, 20, Distal portions of peniferums, same specimen as in figure 16.

Figures 1 and 11 drawn to scale A; figures 3, 4, 5, 6, 7, 8, 9, 10, 12, 13, 14, 15, 16, 17, 18, 19, and 20 drawn to scale B.

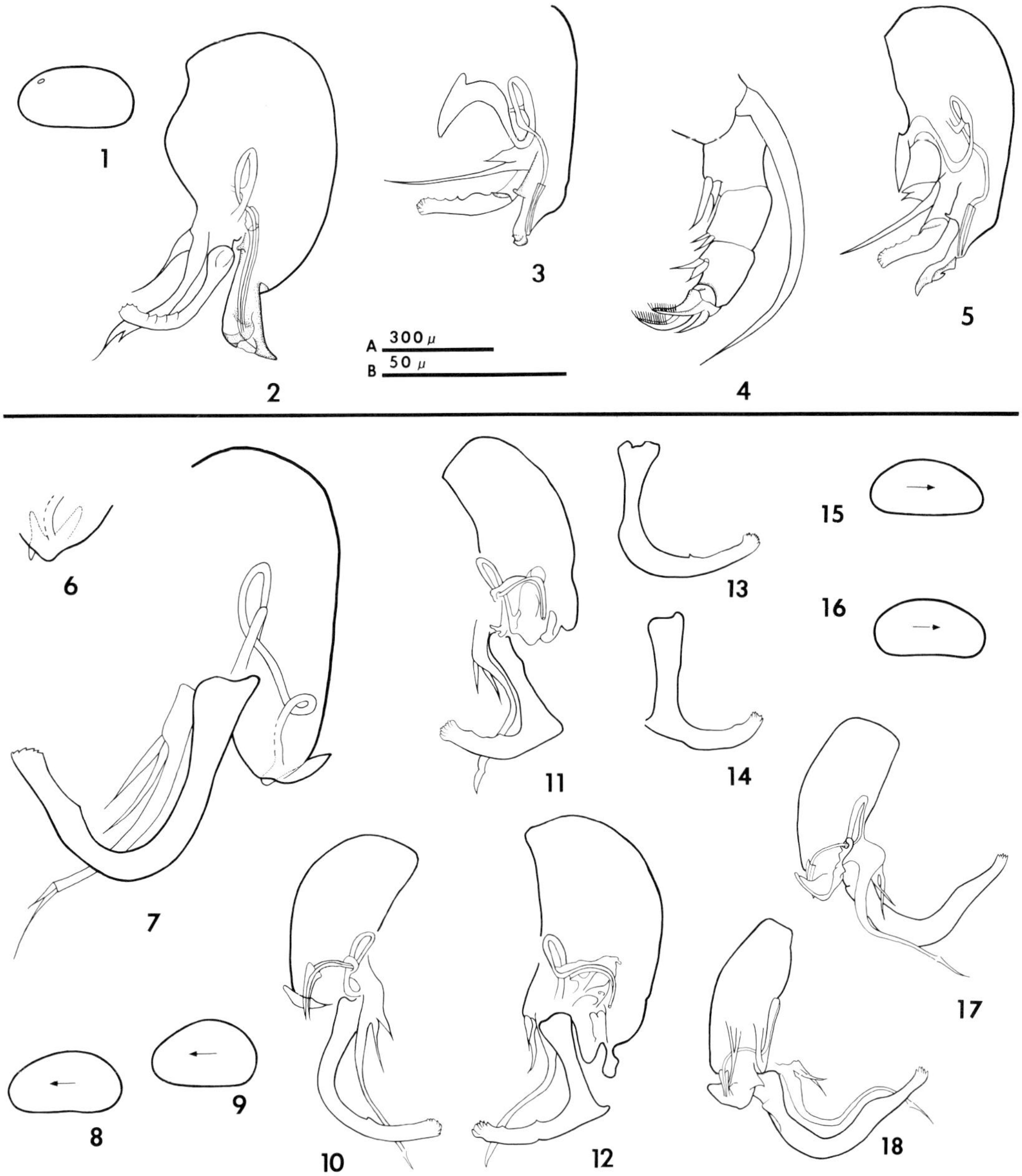

PLATE XLIII

Figures 1-2, **Notocythere tasmanica:** 1, Left valve of male; 2, Copulatory complex.

Figure 3, **Rickocythere cherax:** 3, Copulatory complex.

Figures 4-5, **Riekocythere xenika:** 4, Antenna; 5, Copulatory complex.

Figures 6-7, **Sphaeromicola cebennica:** 6, Tip of peniferum; 7, Copulatory complex.

Figures 8-14, **Sphaeromicola cirolanae:** 8, Left valve of male; 9, Left valve of female; 10, 11, 12, Variations in copulatory complexes; 13, 14, Variations in clasping apparatuses. (Figures 8-14 from Hobbs, 1971.)

Figures 15-18, **Sphaeromicola coahuiltecae:** 15, Right valve of male; 16, Right valve of female; 17, Copulatory complex of holotype; 18, Copulatory complex of paratype. (Figures 15-18 from Hobbs and Hobbs, 1973.)

Figures 1, 8, 9, 15, and 16 drawn to scale A; figures 2-5, 10-14, and 17-18 drawn to scale B; scale for figures 6 and 7 unknown.

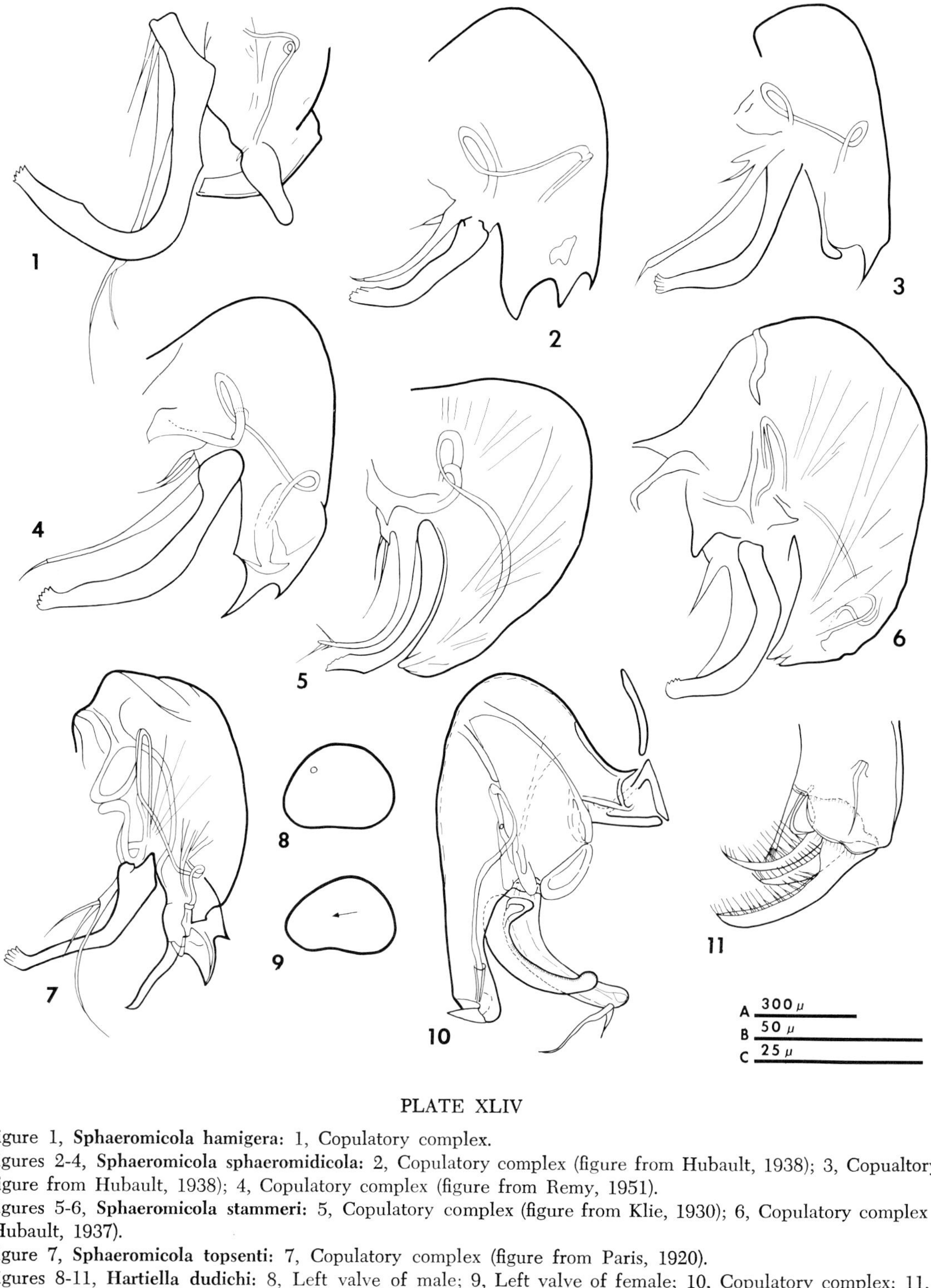

PLATE XLIV

Figure 1, **Sphaeromicola hamigera:** 1, Copulatory complex.

Figures 2-4, **Sphaeromicola sphaeromidicola:** 2, Copulatory complex (figure from Hubault, 1938); 3, Copualtory complex (figure from Hubault, 1938); 4, Copulatory complex (figure from Remy, 1951).

Figures 5-6, **Sphaeromicola stammeri:** 5, Copulatory complex (figure from Klie, 1930); 6, Copulatory complex (figure from Hubault, 1937).

Figure 7, **Sphaeromicola topsenti:** 7, Copulatory complex (figure from Paris, 1920).

Figures 8-11, **Hartiella dudichi:** 8, Left valve of male; 9, Left valve of female; 10, Copulatory complex; 11, Distal portion of antenna. (Figures from Danielopol, 1971b.)

Figures 8 and 9 drawn to scale A; figure 10 drawn to scale B; figure 11 drawn to scale C; scales for figures 1-7 unknown.

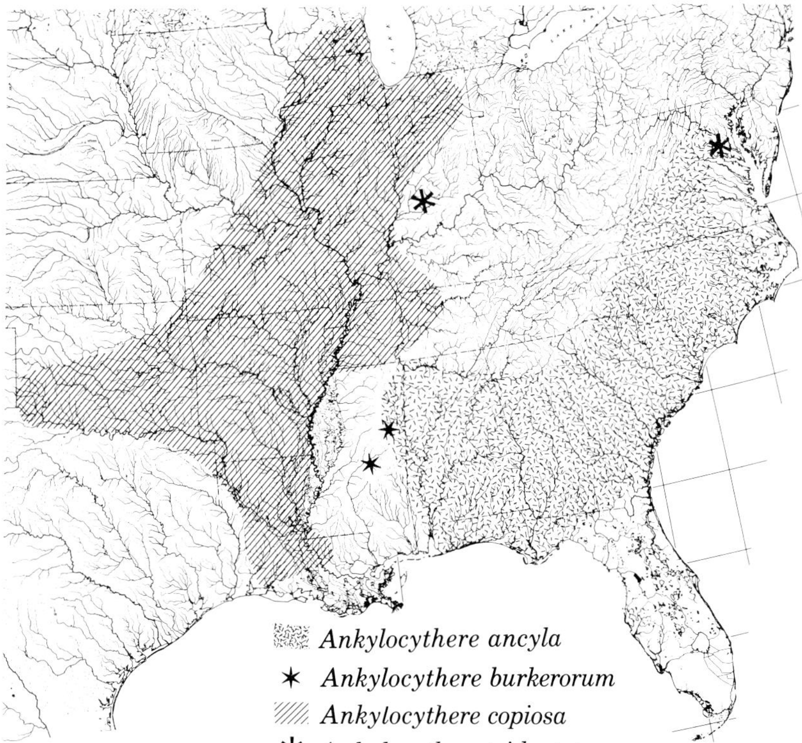

Ankylocythere ancyla
Ankylocythere burkerorum
Ankylocythere copiosa
Ankylocythere tridentata

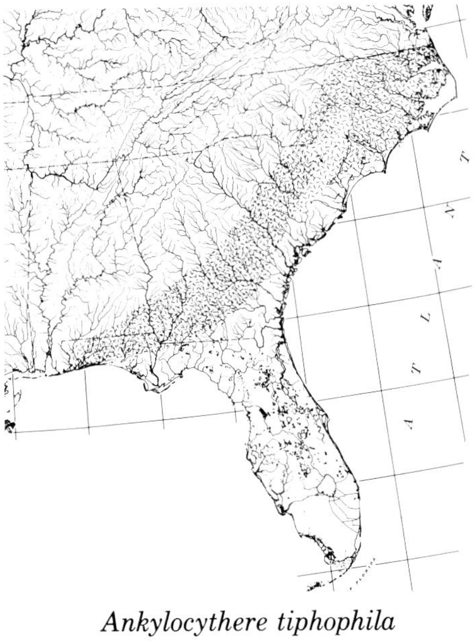

Ankylocythere tiphophila

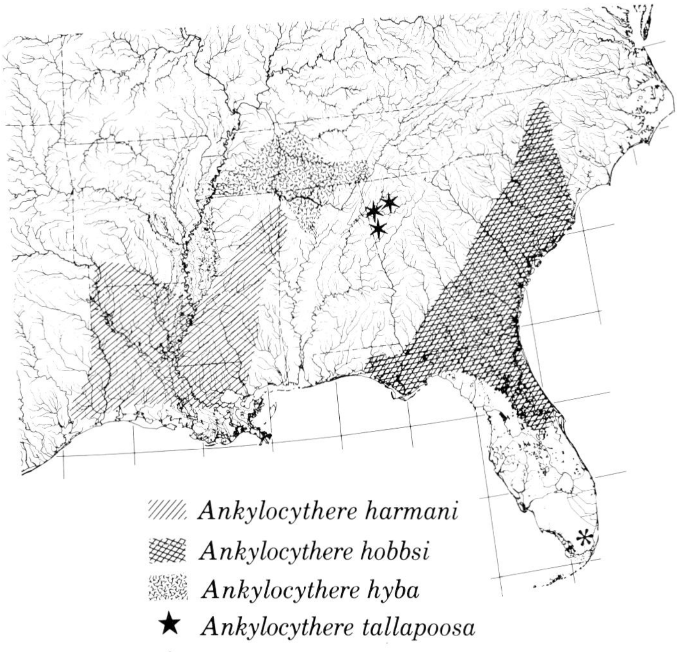

Ankylocythere harmani
Ankylocythere hobbsi
Ankylocythere hyba
Ankylocythere tallapoosa
Ankylocythere talulus

PLATE XLV

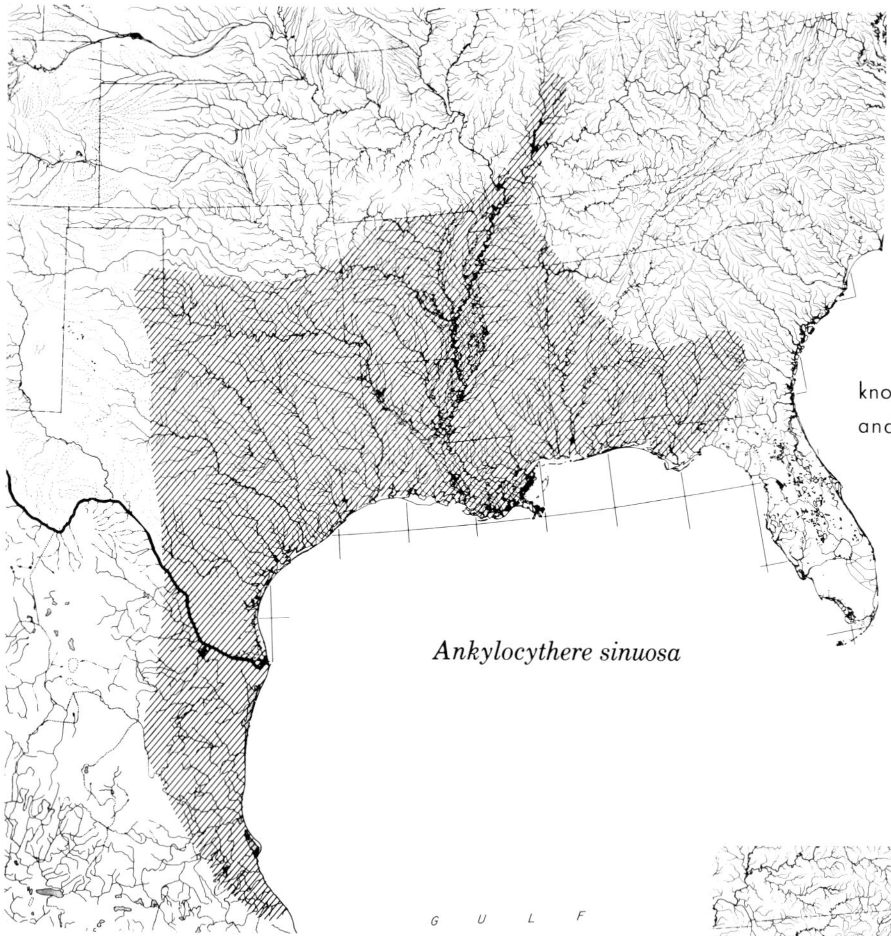

In addition, A. sinuosa is known from Lake Co. California and Oahu, Hawaii

Ankylocythere sinuosa

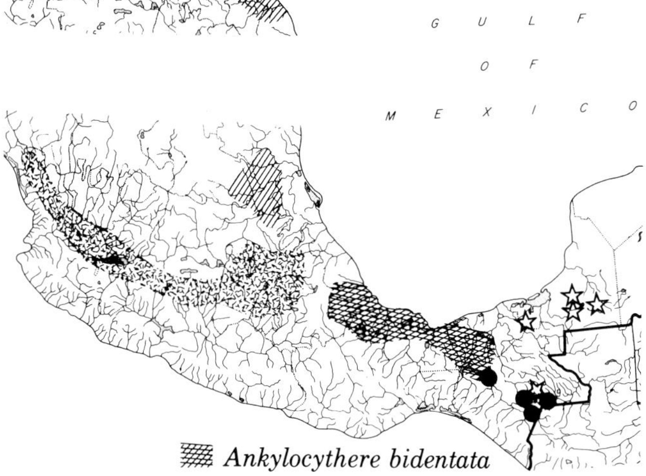

▦ *Ankylocythere bidentata*
░ *Ankylocythere heterodonta*
● *Ankylocythere maya*
//// *Ankylocythere toltecae*
☆ *Ankylocythere villalobosi*

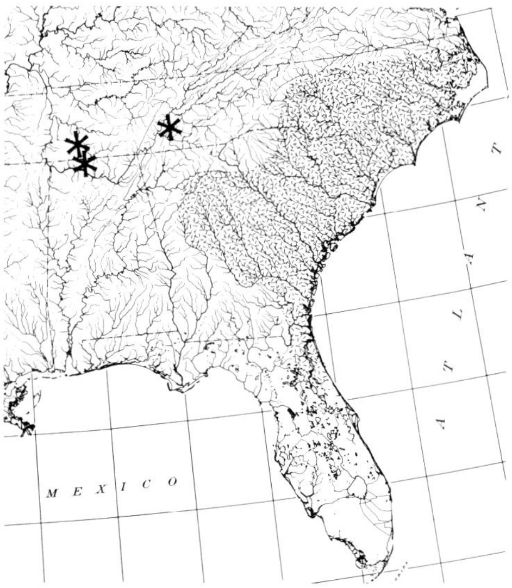

* *Ankylocythere ephydra*
░ *Ankylocythere telmoecea*

PLATE XLVI

Ankylocythere cubensis

Ankylocythere hamata

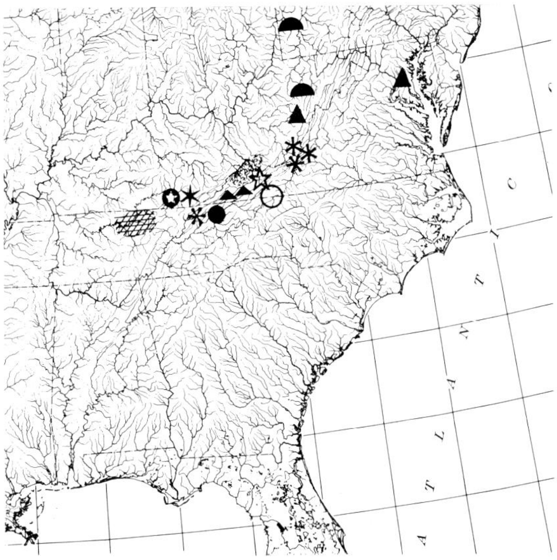

- ✱ *Ascetocythere asceta*
- ✶ *Ascetocythere batchi*
- ◓ *Ascetocythere coryphodes*
- ○ *Ascetocythere cosmeta*
- ✳ *Ascetocythere didactylata*
- ☆ *Ascetocythere hoffmani*
- ▦ *Ascetocythere holti*
- ● *Ascetocythere hyperoche*
- ✪ *Ascetocythere lita*
- ▲ *Ascetocythere myxoides*
- ▲ *Ascetocythere ozalea*
- ░ *Ascetocythere sclera*

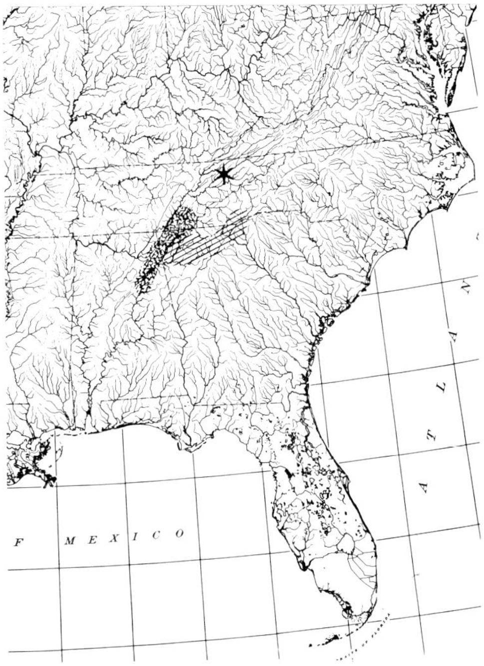

- ▨ *Cymocythere clavata*
- ░ *Cymocythere cyma*
- ✶ *Cymocythere gonia*

PLATE XLVII

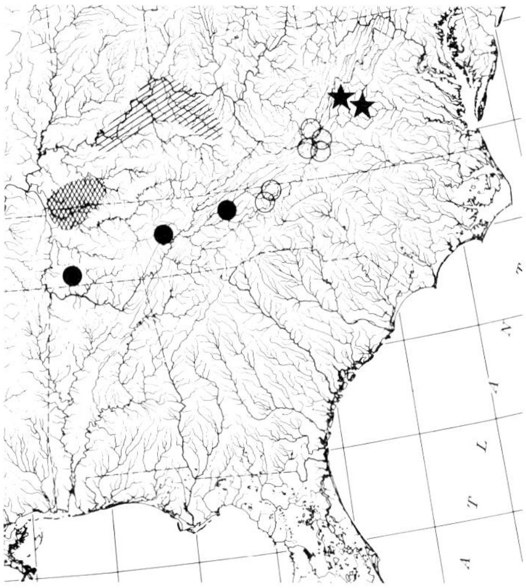

Dactylocythere amicula
Dactylocythere amphiakis
★ Dactylocythere banana
● Dactylocythere brachystrix
○ Dactylocythere chalaza

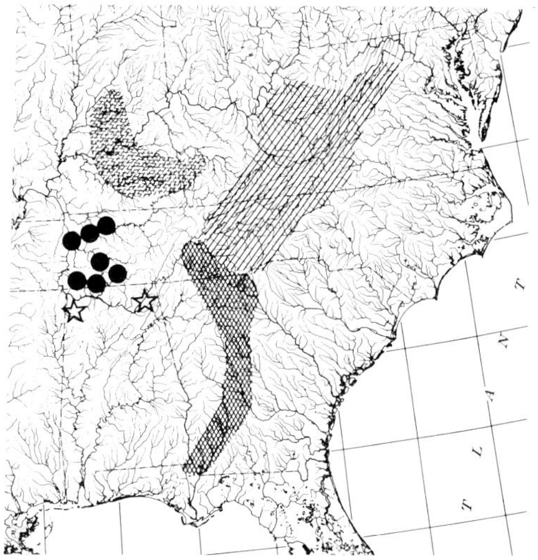

☆ Dactylocythere arcuata
Dactylocythere daphnioides
Dactylocythere leptophylax
Dactylocythere susanae
● Dactylocythere xystroides

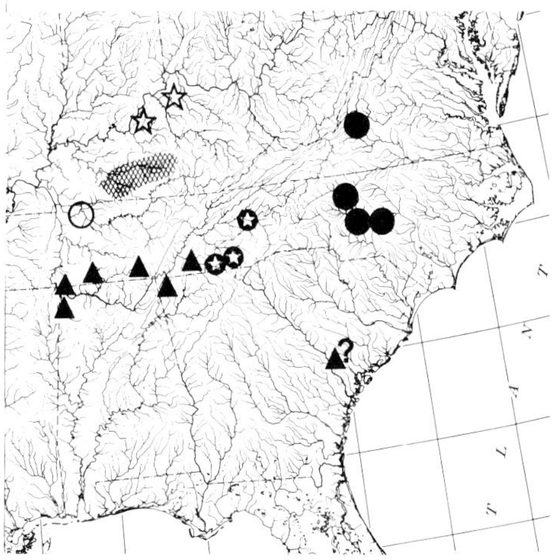

Dactylocythere chelomata
Dactylocythere cooperorum
☆ Dactylocythere exoura
○ Dactylocythere koloura
▲ Dactylocythere mecoscapha
● Dactylocythere megadactylus

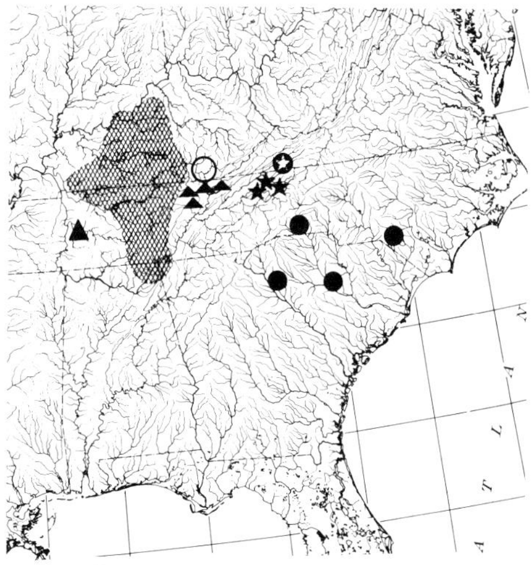

★ Dactylocythere charadra
○ Dactylocythere coloholca
Dactylocythere enoploholca
▲ Dactylocythere sandbergi
▲ Dactylocythere spinata
Dactylocythere steevesi
● Dactylocythere striophylax

PLATE XLVIII

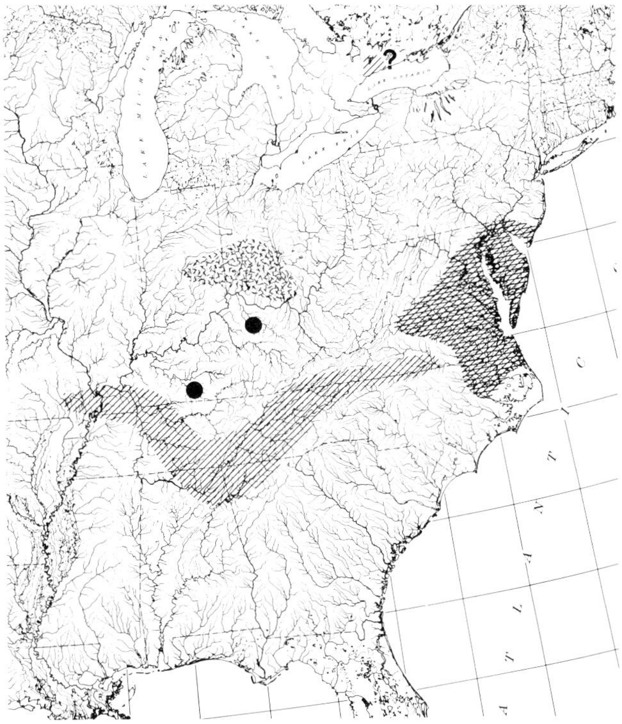

Dactylocythere crawfordi
Dactylocythere falcata
Dactylocythere jeanae
● Dactylocythere macroholca

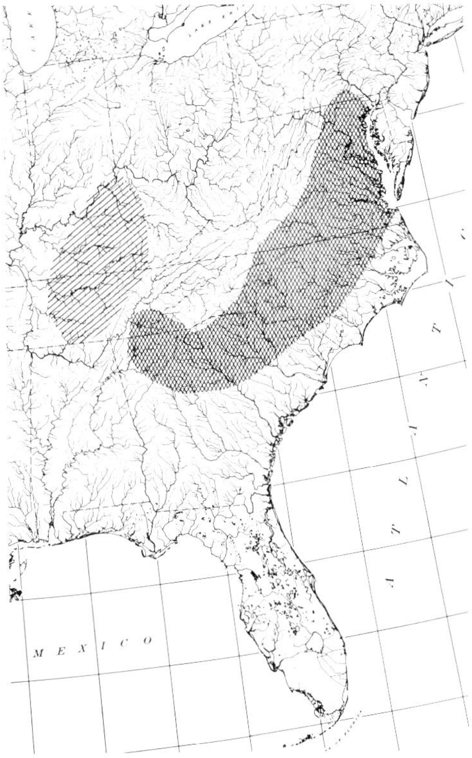

Dactylocythere suteri
Dactylocythere ungulata

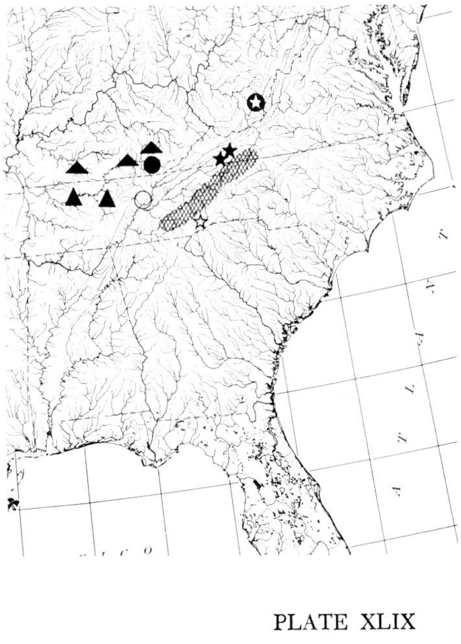

★ Dactylocythere myura
○ Dactylocythere pachysphyrata
✪ Dactylocythere phoxa
☆ Dactylocythere prinsi
▲ Dactylocythere prionata
● Dactylocythere pughae
Dactylocythere runki
▲ Dactylocythere speira

PLATE XLIX

Donnaldsoncythere ardis

Donnaldsoncythere ileata

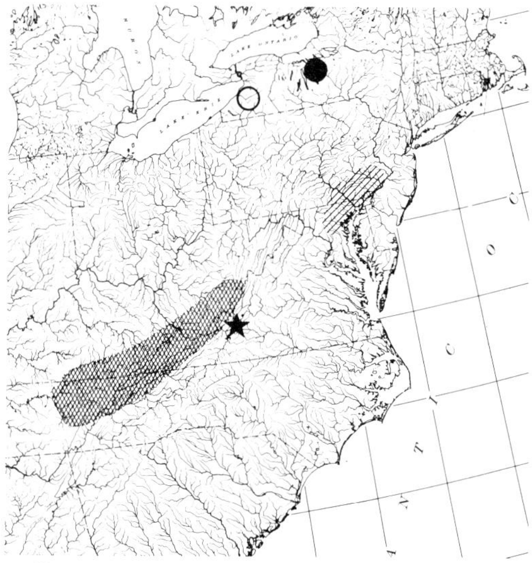

● *Donnaldsoncythere cayugaensis*
○ *Donnaldsoncythere humesi*
▨ *Donnaldsoncythere pennsylvanica*
★ *Donnaldsoncythere truncata*
▩ *Donnaldsoncythere tuberosa*

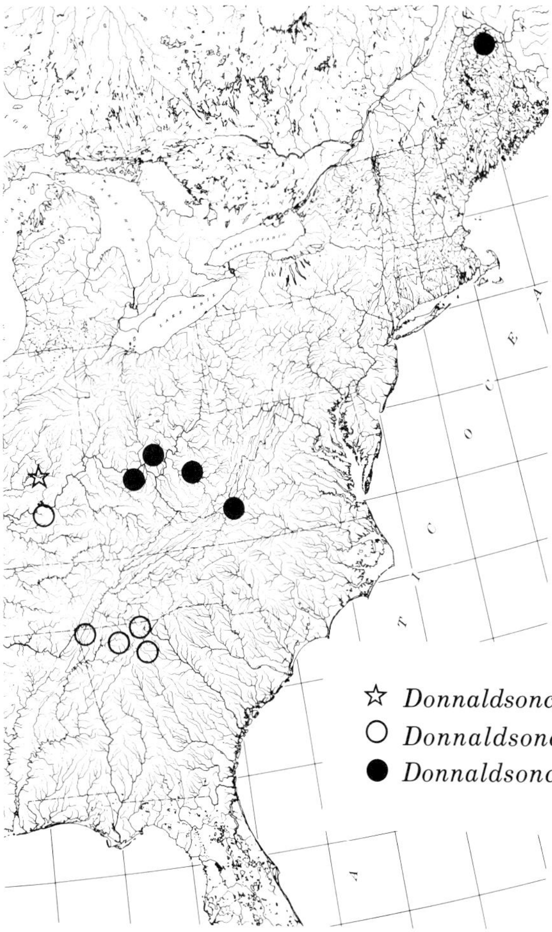

☆ *Donnaldsoncythere donnaldsonensis*
○ *Donnaldsoncythere hiwasseensis*
● *Donnaldsoncythere scalis*

PLATE L

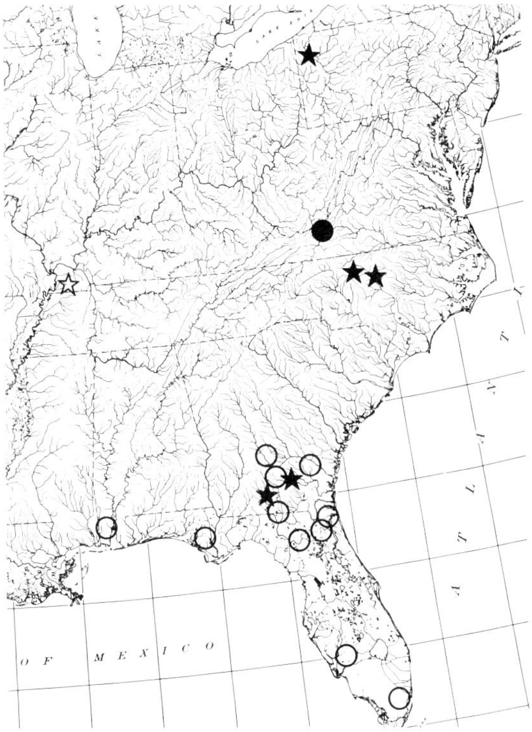

★ *Entocythere dentata*
○ *Entocythere dorsorotunda*
● *Entocythere kanawhaensis*
☆ *Entocythere tyttha*

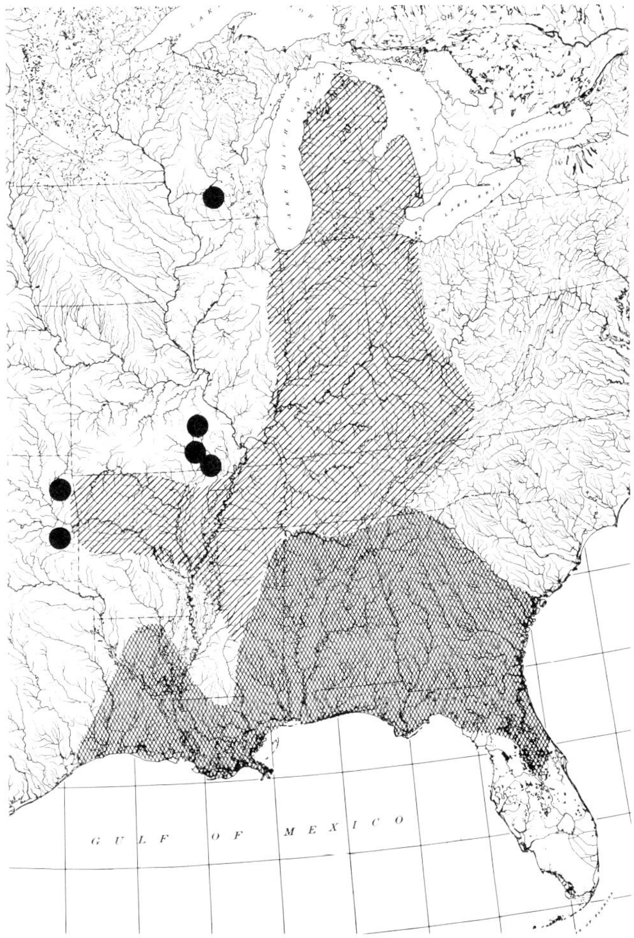

● *Entocythere cambaria*
▒ *Entocythere elliptica*
▨ *Entocythere illinoisensis*

Entocythere ruibali

PLATE LI

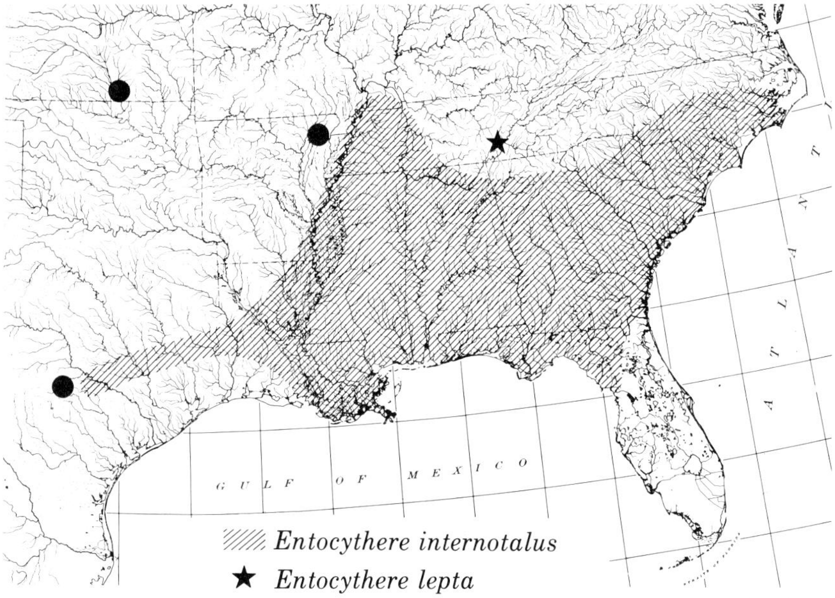

▨ *Entocythere internotalus*
★ *Entocythere lepta*
● *Entocythere reddelli*

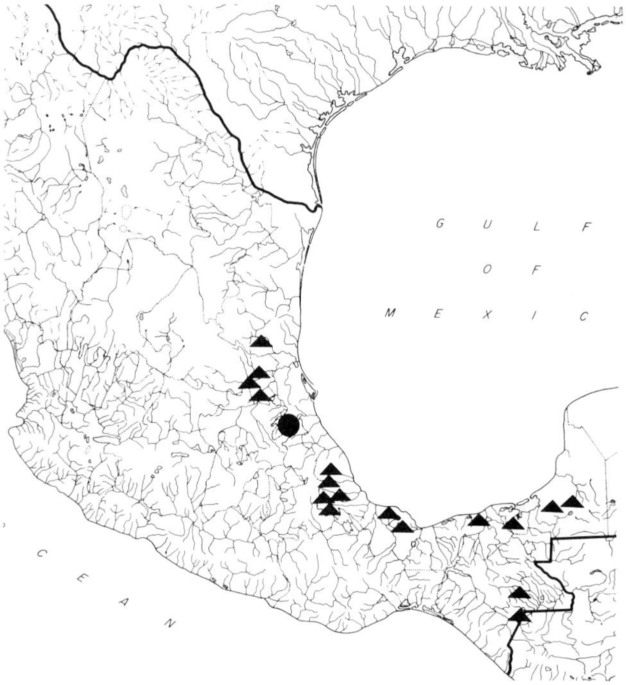

▲ *Entocythere claytonhoffi*
● *Entocythere mexicana*

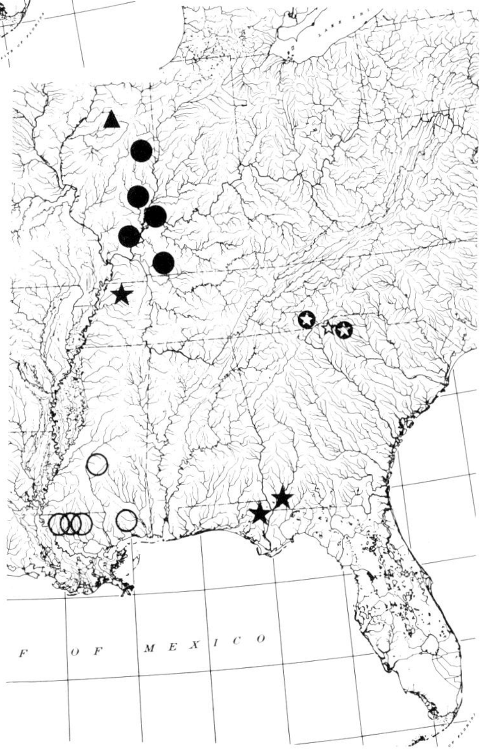

▲ *Geocythere acuta*
★ *Geocythere geophila*
● *Geocythere gyralea*
○ *Geocythere nessoides*
☆ *Harpagocýthere georgiae*
✪ *Harpagocythere tertius*

PLATE LII

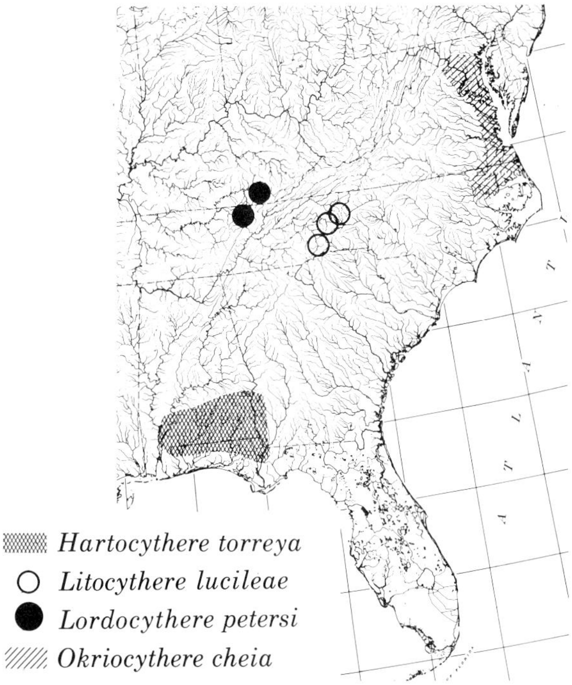
Hartocythere torreya
Litocythere lucileae
Lordocythere petersi
Okriocythere cheia

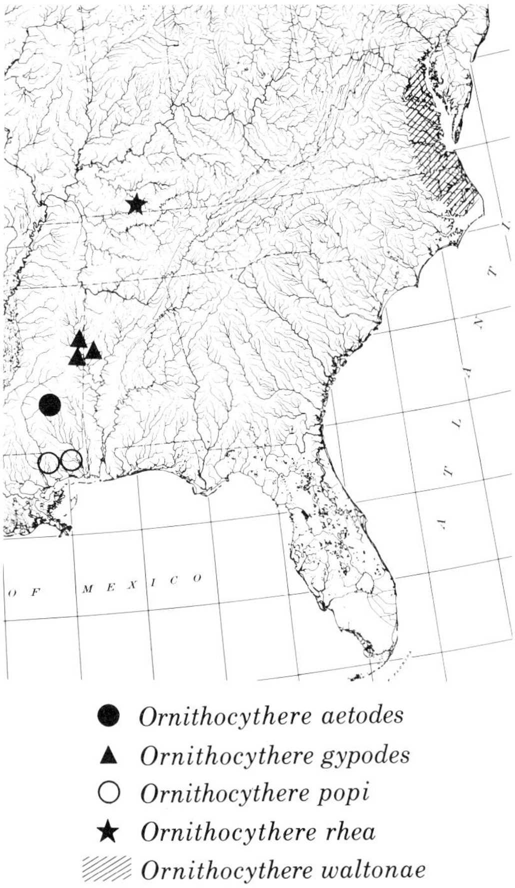
Ornithocythere aetodes
Ornithocythere gypodes
Ornithocythere popi
Ornithocythere rhea
Ornithocythere waltonae

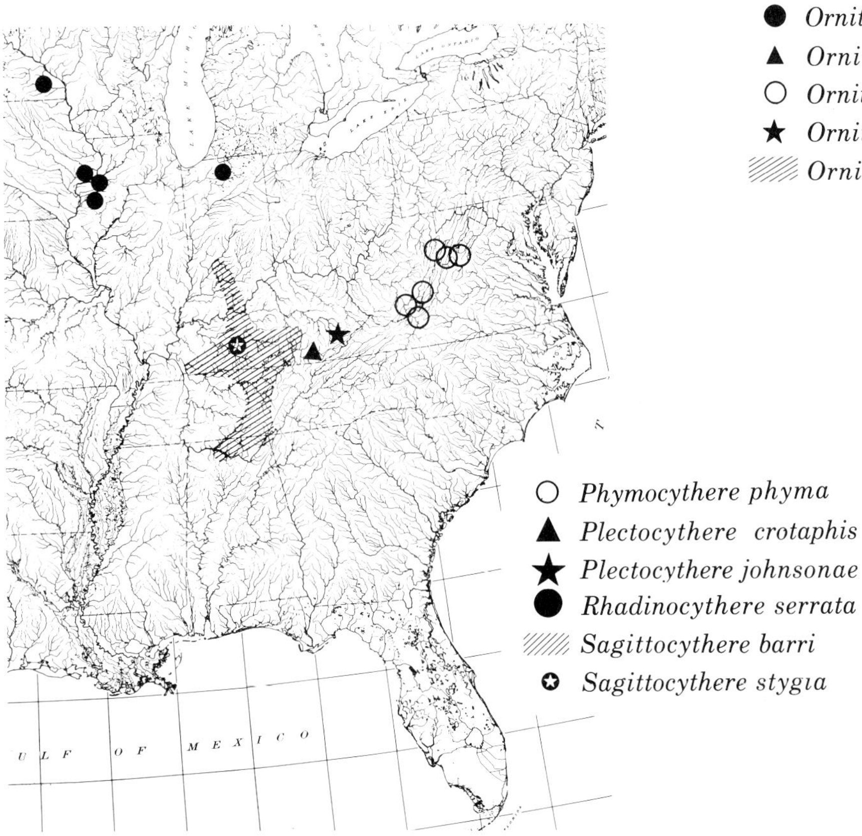
Phymocythere phyma
Plectocythere crotaphis
Plectocythere johnsonae
Rhadinocythere serrata
Sagittocythere barri
Sagittocythere stygia

PLATE LIII

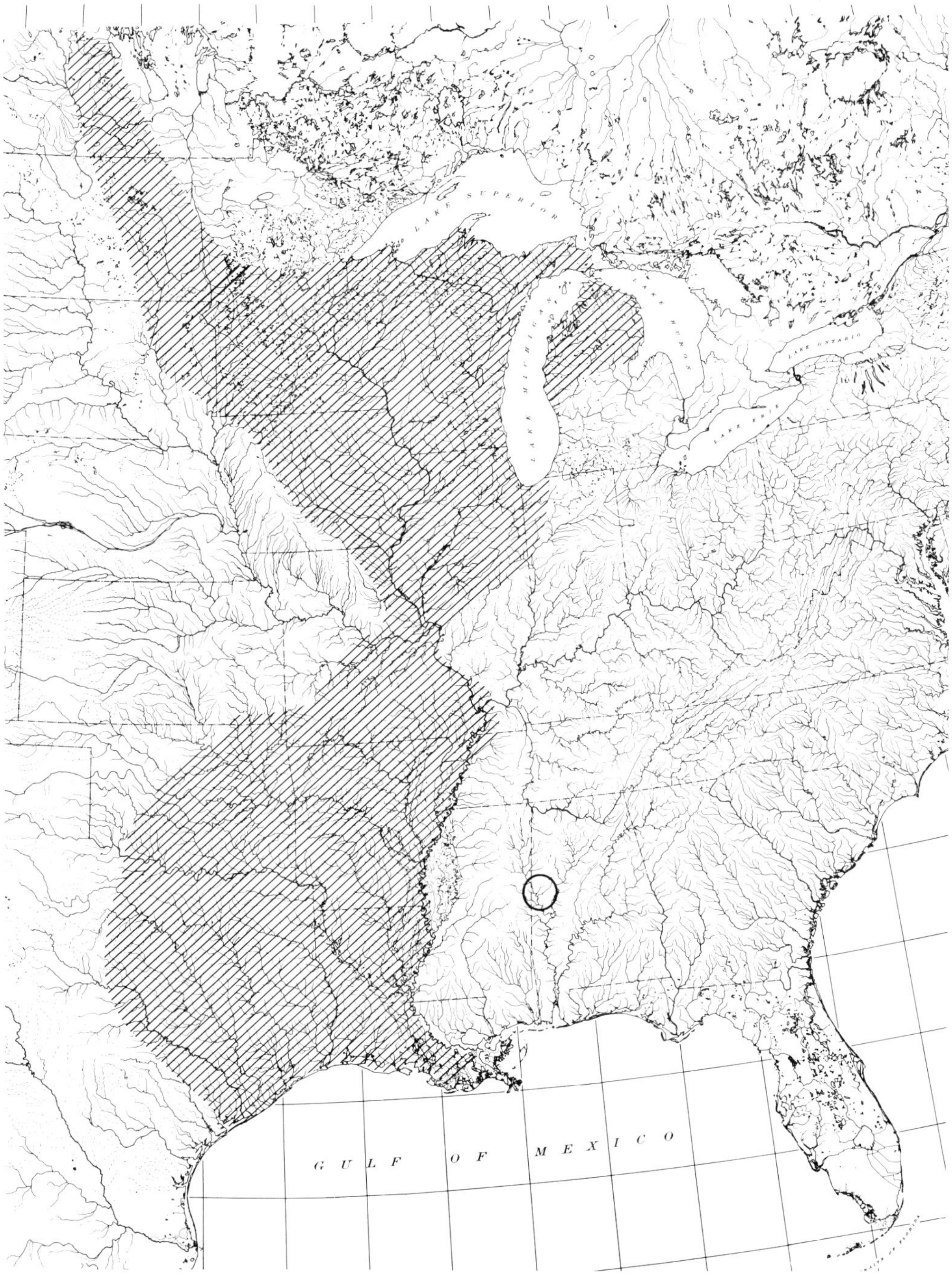

○ *Saurocythere rhipis*

▨ *Thermastrocythere riojai*

PLATE LIV

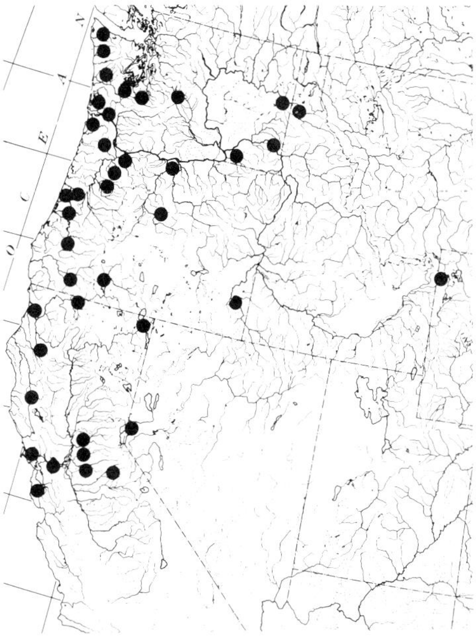

Uncinocythere occidentalis

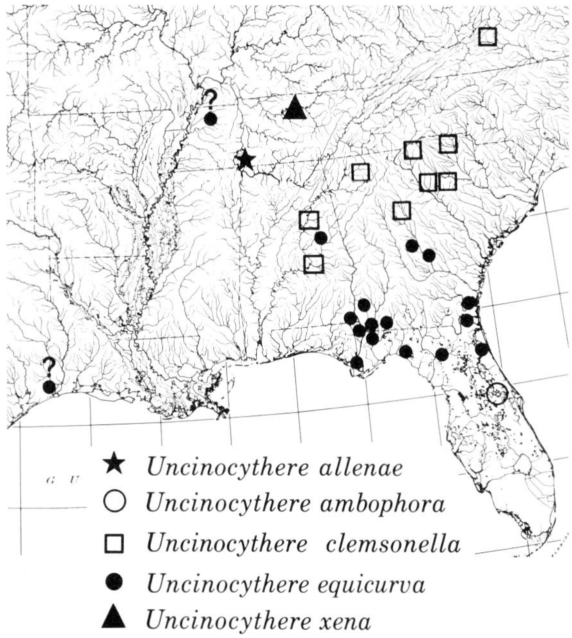

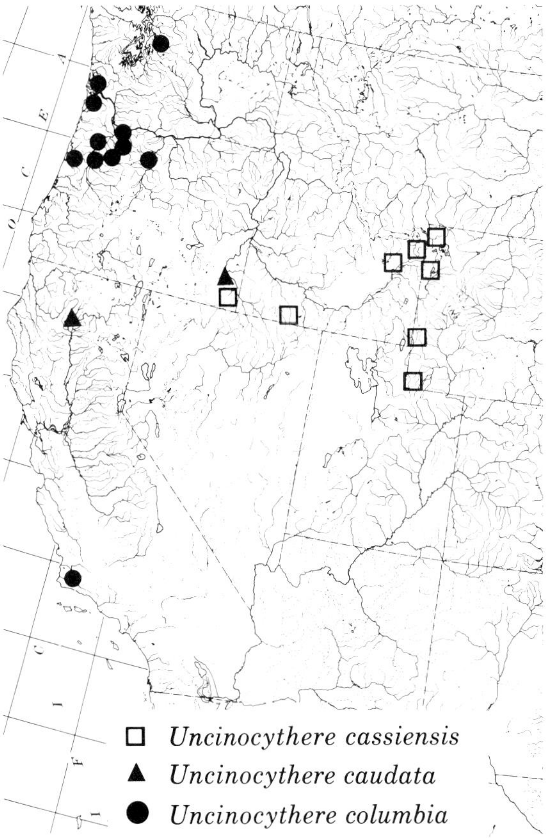

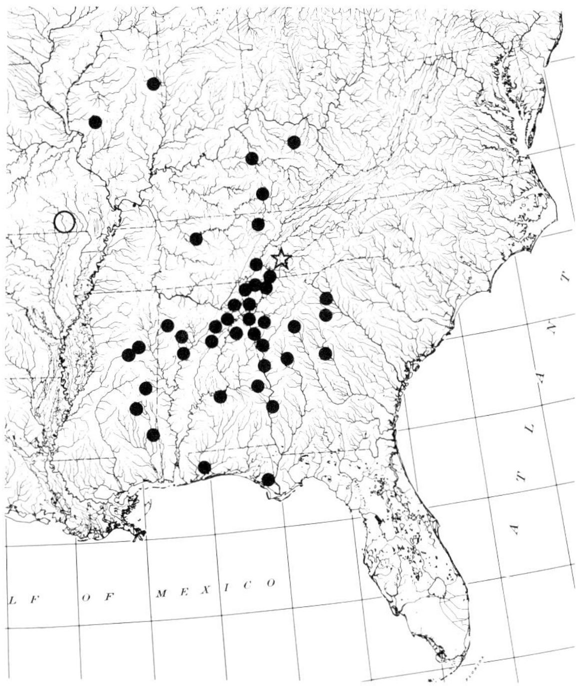

PLATE LV

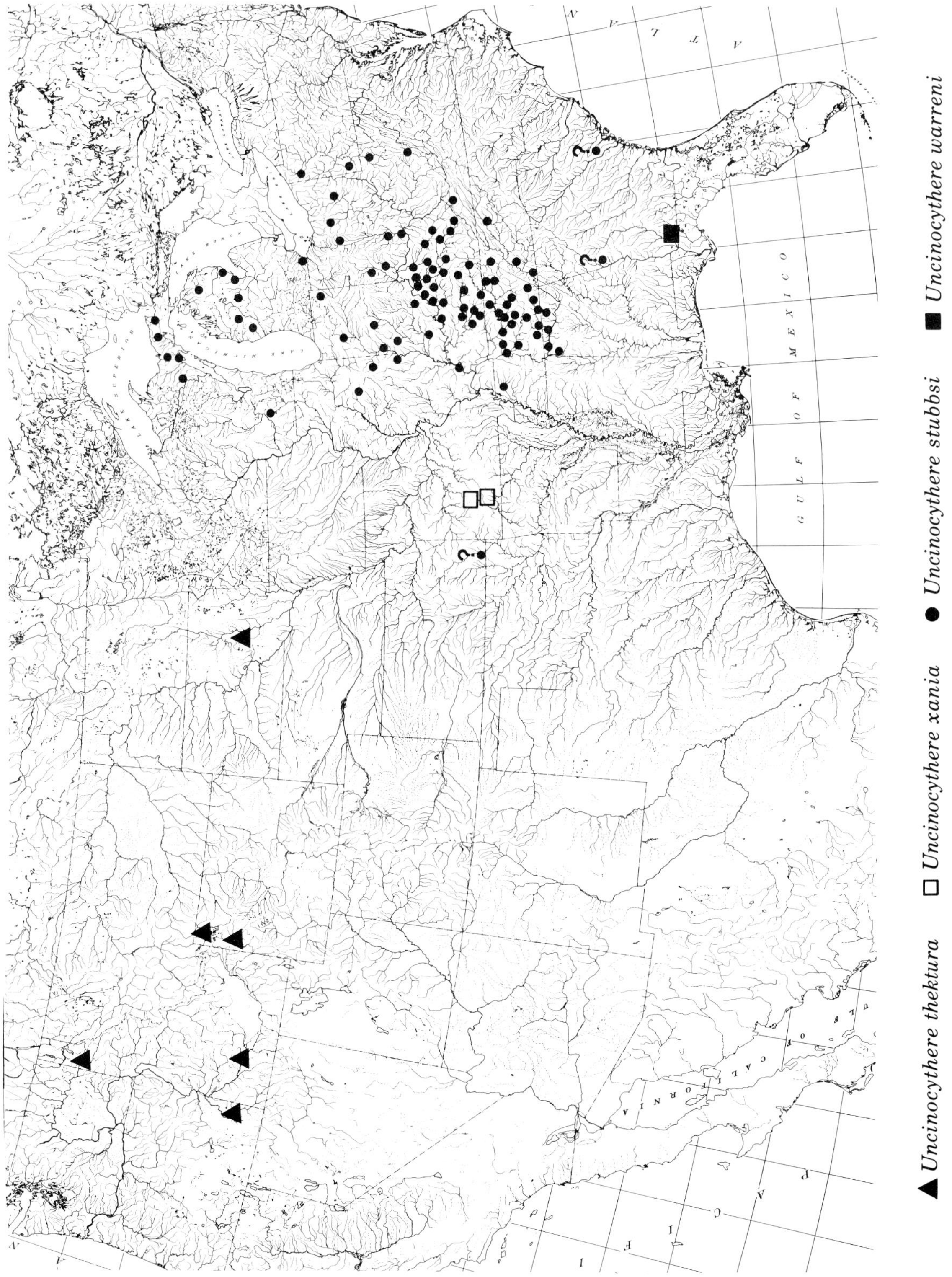
GULF OF MEXICO
▲ Uncinocythere thektura
□ Uncinocythere xania
● Uncinocythere stubbsi
■ Uncinocythere warreni

PLATE LVI

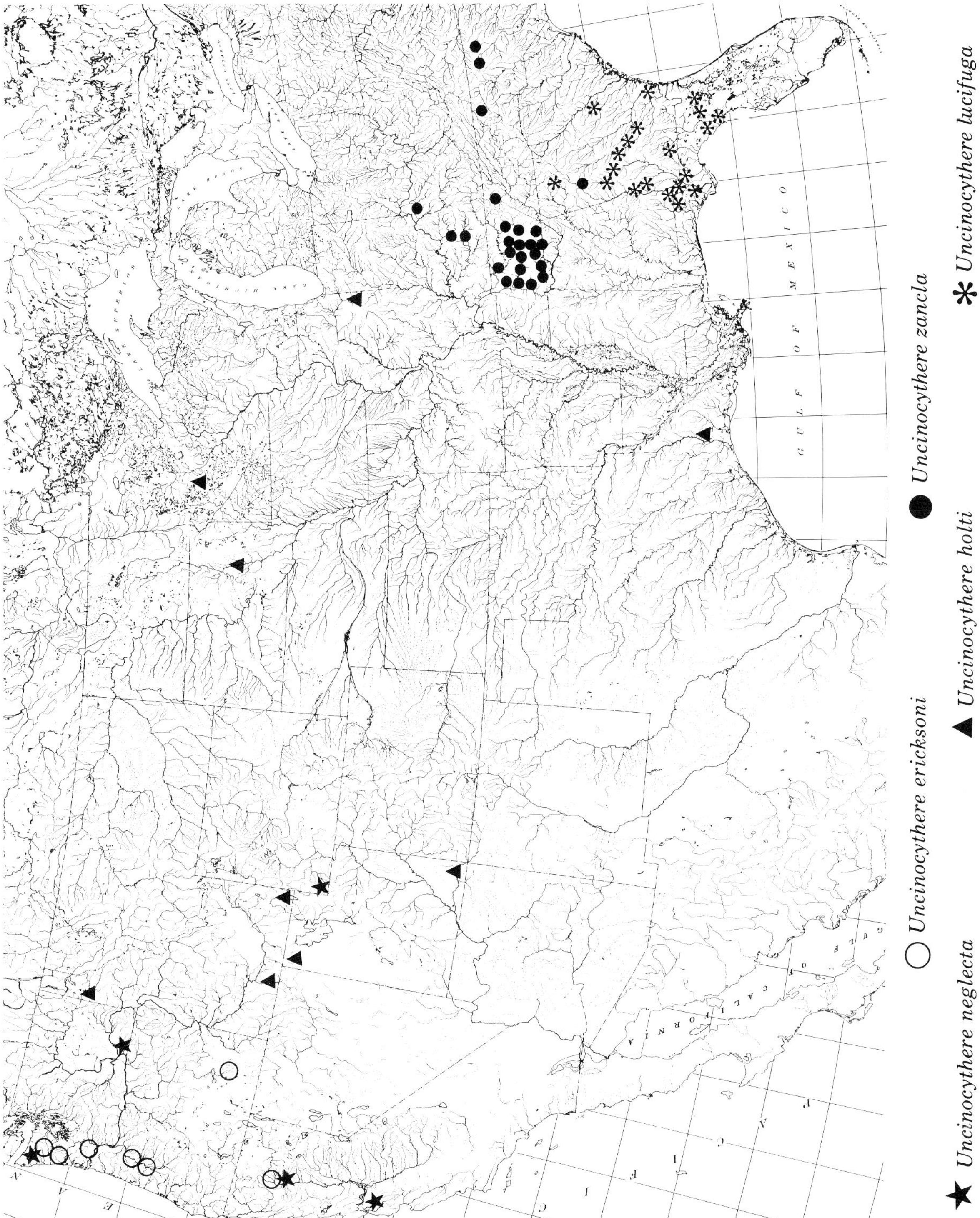
Uncinocythere neglecta
Uncinocythere ericksoni
Uncinocythere holti
Uncinocythere zancla
Uncinocythere lucifuga
GULF OF MEXICO
GULF OF CALIFORNIA
PACIFIC OCEAN

PLATE LVII

□ *Uncinocythere cuadricuspide*

● *Uncinocythere zaruri*

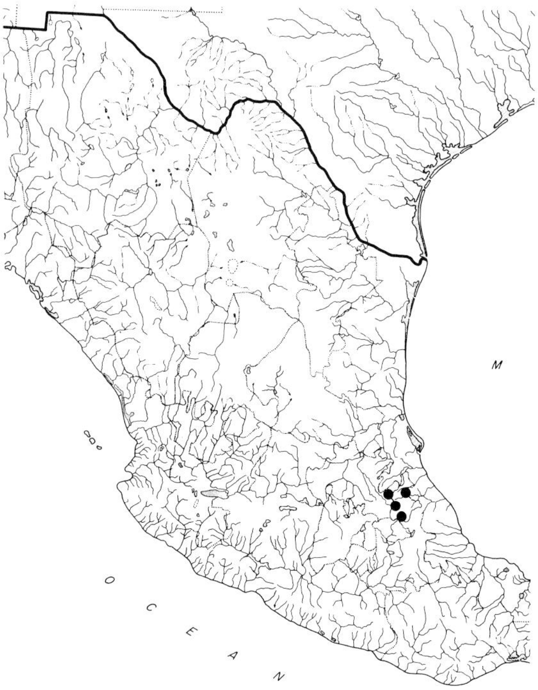

Uncinocythere bicuspide

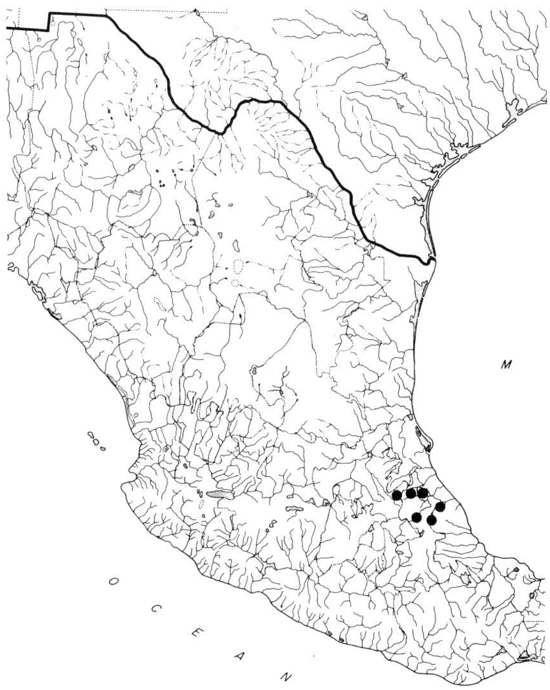

Uncinocythere dobbinae

PLATE LVIII

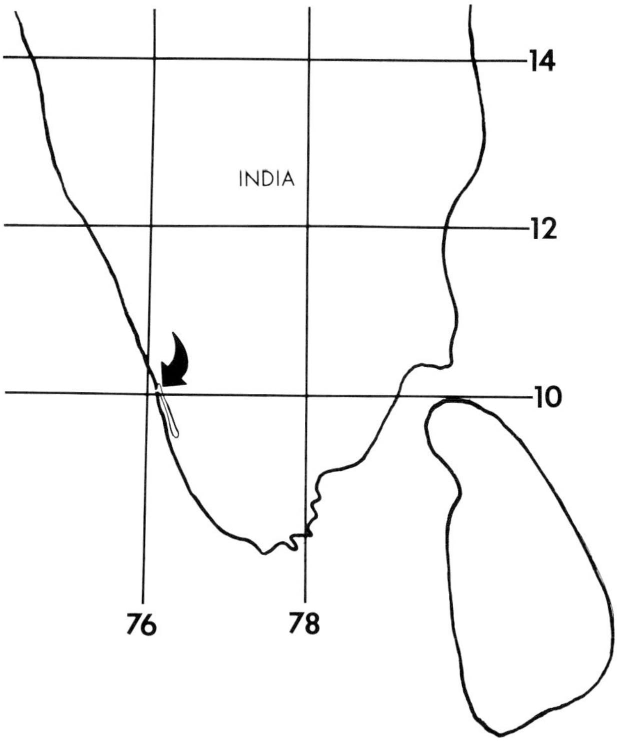

Microsyssitria indica

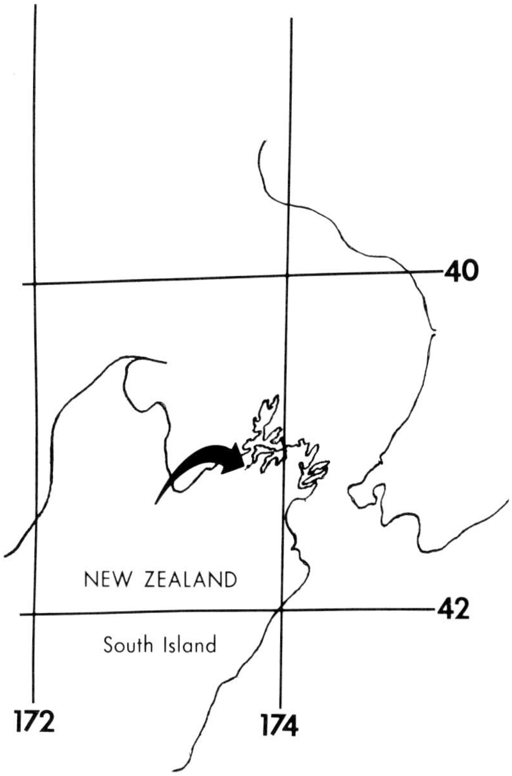

Laccocythere aotearoa

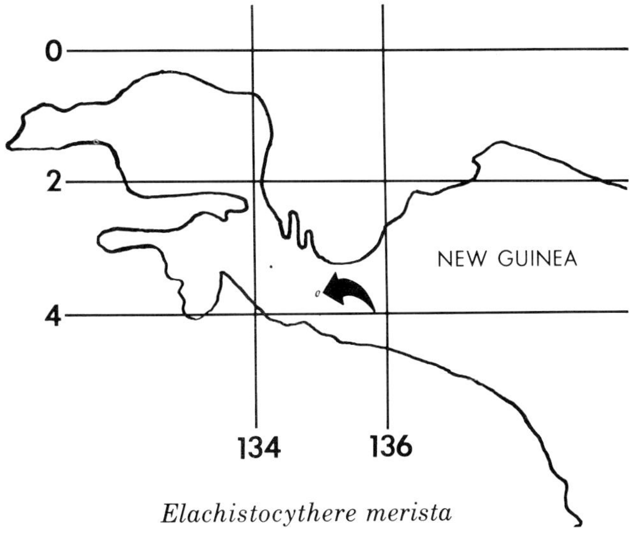

Elachistocythere merista

PLATE LIX

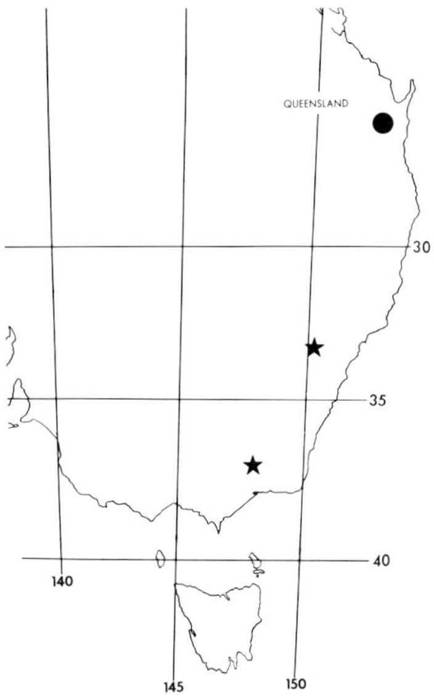

★ *Herpetocythere acanthoides*
● *Herpetocythere australensis*

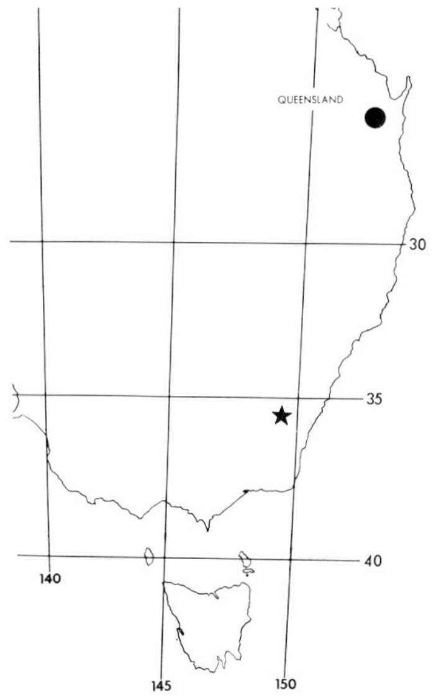

★ *Herpetocythere bendora*
● *Herpetocythere mackenziei*

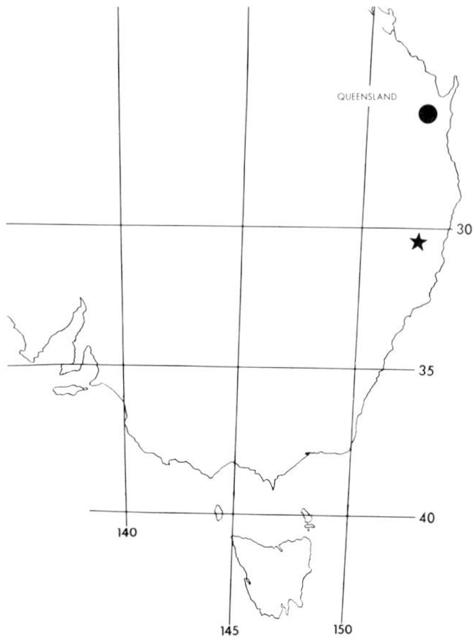

★ *Herpetocythere gnoma*
● *Herpetocythere labidioides*

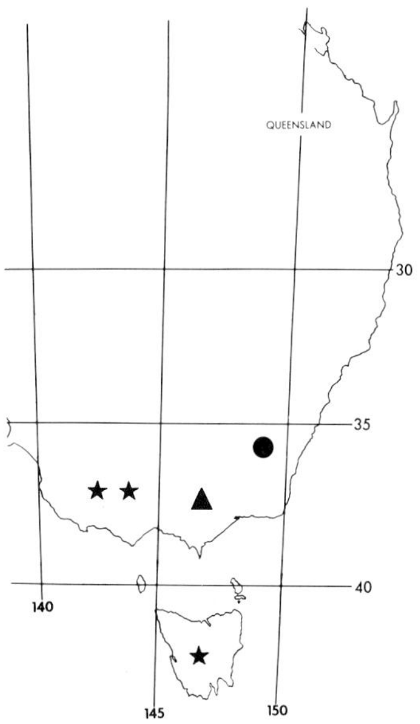

▲ *Notocythere erica*
★ *Notocythere mirranatwa*
● *Notocythere rieki*

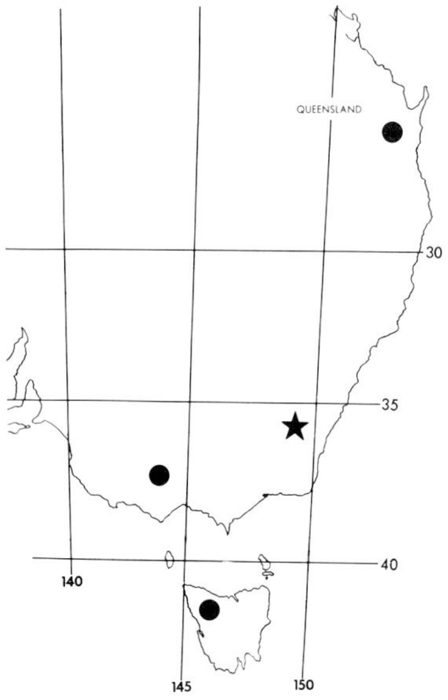

★ *Notocythere blundelli*
● *Notocythere syssitos*

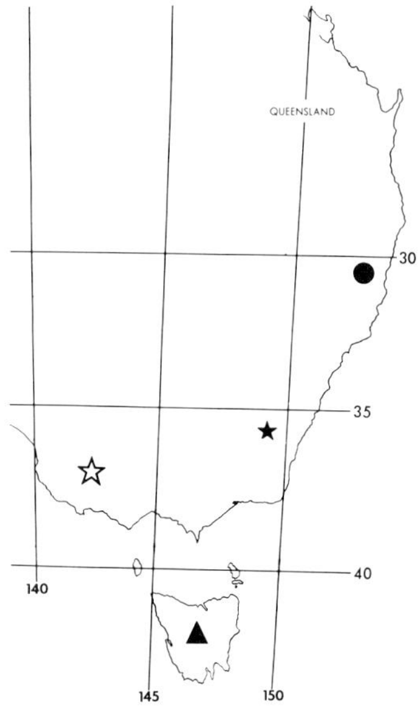

● *Notocythere antichthon*
☆ *Notocythere grampians*
★ *Notocythere synomodites*
▲ *Notocythere tasmanica*

PLATE LX

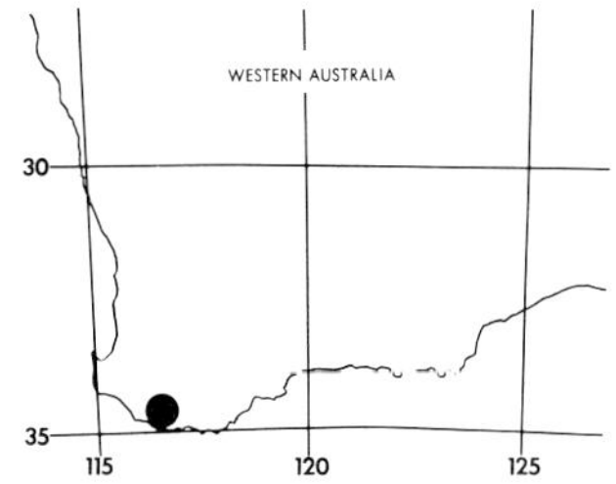

Chelocythere kalganensis

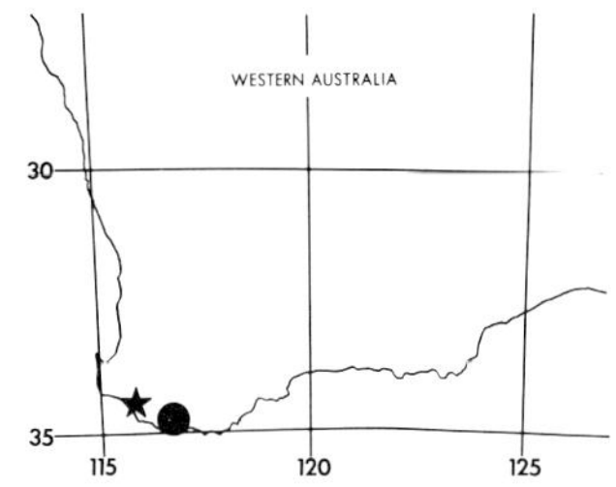

★ *Riekocythere cherax*
● *Riekocythere xenika*

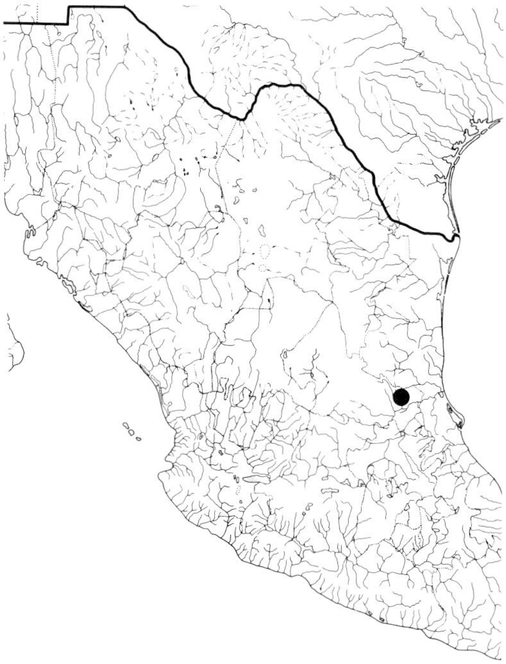

Sphaeromicola coahuiltecae

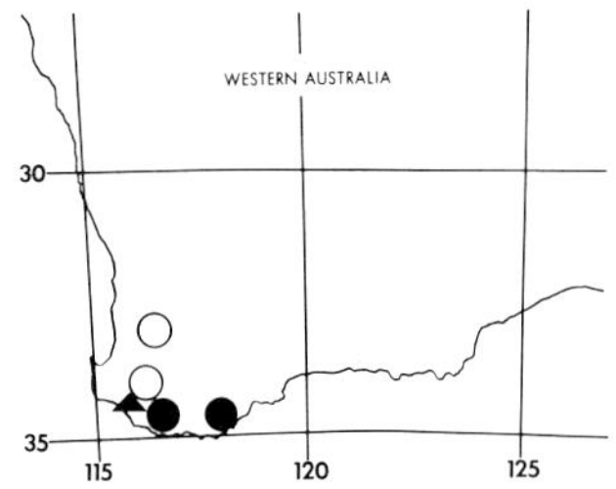

▲ *Hesperocythere klasteroides*
○ *Hesperocythere tallanalla*
● *Hesperocythere xiphoides*

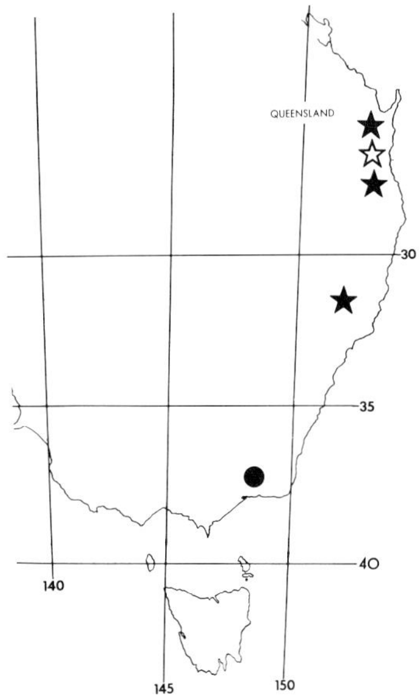

☆ *Lichnocythere synethes*
★ *Lichnocythere tubrabucca*
● *Lichnocythere victoria*

Sphaeromicola cirolanae

PLATE LXI

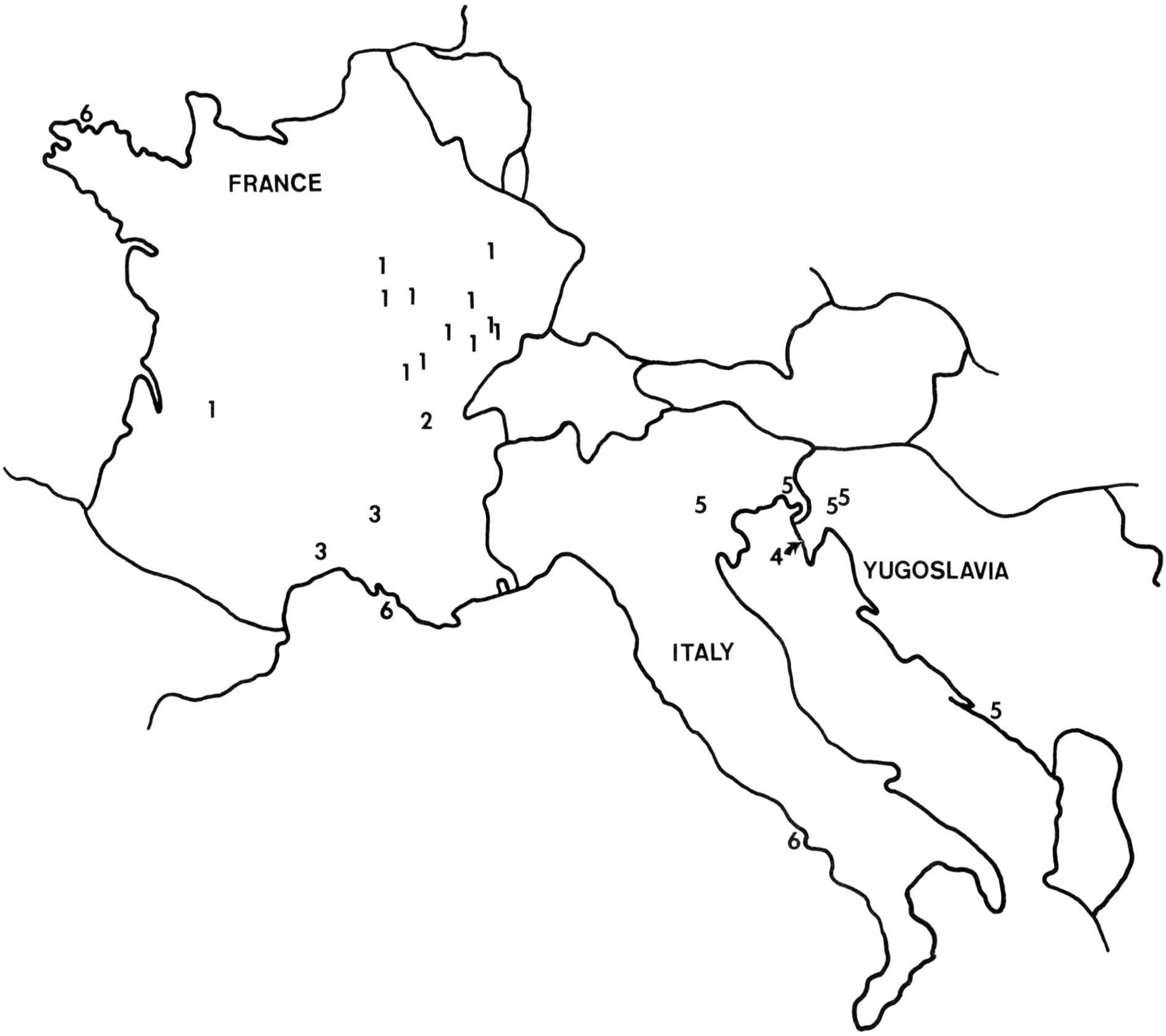

6 *Hartiella dudichi*
3 *Sphaeromicola cebennica*
2 *Sphaeromicola hamigera*
4 *Sphaeromicola sphaeromidicola*
5 *Sphaeromicola stammeri*
1 *Sphaeromicola topsenti*

PLATE LXII

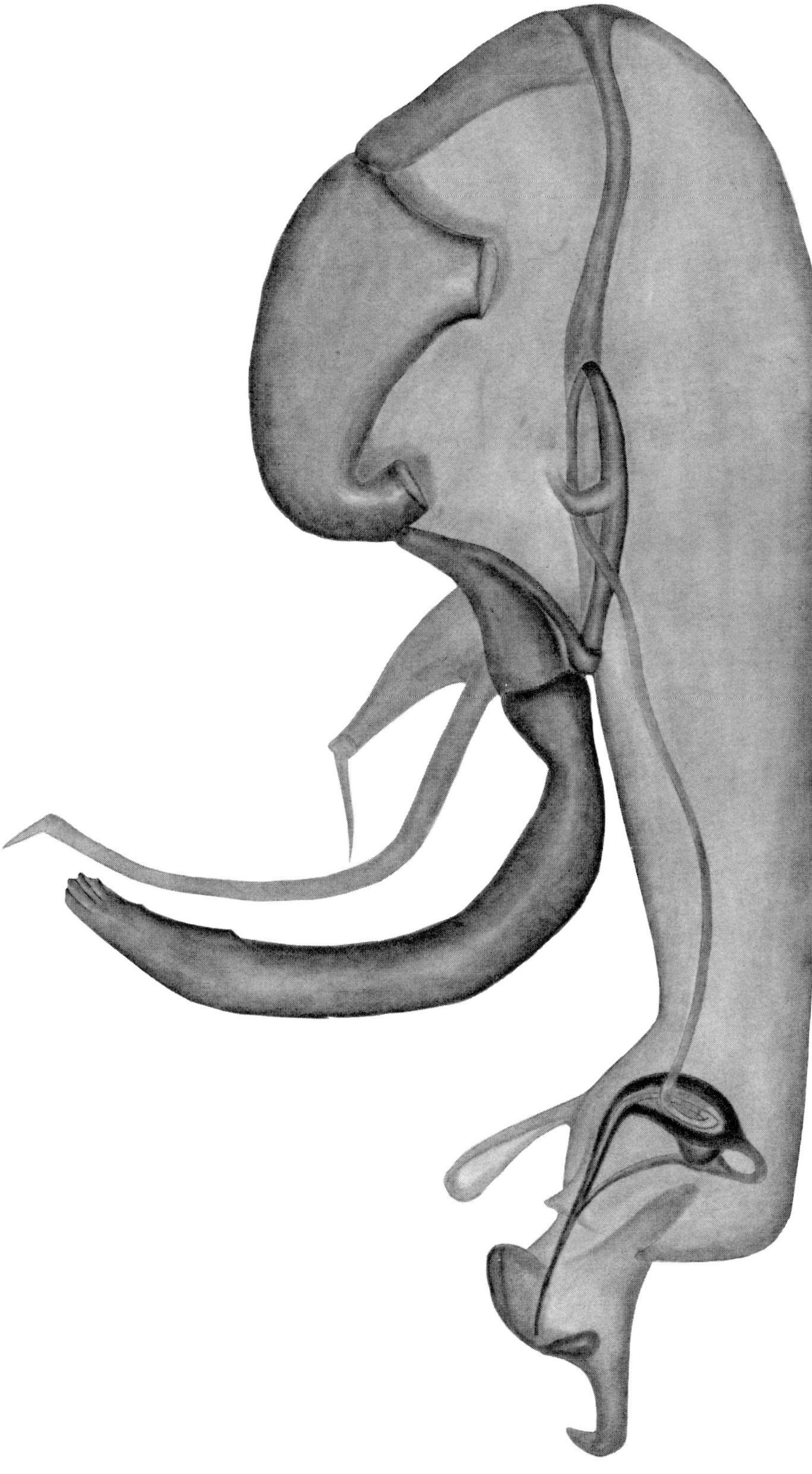

Copulatory apparatus of *Ascetocythere asceta*